Environmental Conflict
and Democracy in Canada

Edited by Laurie E. Adkin

Environmental Conflict
and Democracy in Canada

UBCPress · Vancouver · Toronto

20 19 18 17 16 15 14 13 12 11 10 09   5 4 3 2 1

Printed in Canada with vegetable-based inks on FSC-certified ancient-forest-free paper (100% post-consumer recycled) that is processed chlorine- and acid-free.

Library and Archives Canada Cataloguing in Publication

Environmental conflict and democracy in Canada / Laurie E. Adkin, ed.

Includes bibliographical references and index.
ISBN 978-0-7748-1602-1 (bound); 978-0-7748-1603-8 (pbk); 978-0-7748-1604-5 (e-book)

1. Environmental protection – Canada--Citizen participation. 2. Environmental policy – Canada – Citizen participation. 3. Human ecology – Canada. 4. Environmental policy – Canada. I. Adkin, Laurie Elizabeth, 1958-

GE199.C3 E58 2009   333.720971   C2009-900921-8

Canadä

UBC Press gratefully acknowledges the financial support for our publishing program of the Government of Canada through the Book Publishing Industry Development Program (BPIDP), and of the Canada Council for the Arts, and the British Columbia Arts Council.

This book has been published with the help of a grant from the Canadian Federation for the Humanities and Social Sciences, through the Aid to Scholarly Publications Programme, using funds provided by the Social Sciences and Humanities Research Council of Canada.

UBC Press
The University of British Columbia
2029 West Mall
Vancouver, BC V6T 1Z2
604-822-5959 / Fax: 604-822-6083
www.ubcpress.ca

Dedication

to Dennis Adkin (1919-2005),
who never stopped fighting for a better world,

and

to Olivier, and to all children,
who so much deserve one.

Contents

Illustrations

Preface

Laurie E. Adkin

The essays in this book bring together political economy and actor-centred analyses to identify the roots and to explain the outcomes of a range of environmental conflicts. The contributors try to understand the failures and successes of environmentalists and other actors in contesting the interpretation of a conflict, bringing about procedural or regulatory reforms, or building alliances. They ask what we may learn from these conflicts regarding the potential for ecological democratization in Canada. This is a new direction for Canadian environmental studies and for citizenship theory. On the one hand, with regard to environmental conflicts, citizenship theory has been underutilized as an interpretive lens. On the other hand, ecological citizenship theory has seldom grounded its arguments in analyses of the discourses and practices of social and political *actors*. This collection seeks to provide such linkages by examining a variety of environmental conflicts *as* democratic struggles with multiple social dimensions.

In Chapter 1, I introduce concepts and arguments in green democratic theory and identify what I take to be some of the contributions and shortcomings of this literature, as well as my reasons for framing environmental issues as "conflicts." Here, I focus on several arguments, to which I return in Chapter 18, where I draw out the lessons of the case studies for the conceptualization of ecological citizenship. Most importantly, I argue for the need to connect both normative conceptualizations of environmental citizenship and green (deliberative) democratic theory to an understanding of politics that foregrounds power relationships and hegemonic struggle. To begin this task, I pay close attention to patterns in the practices, understandings, and discourses of actors involved in environmental conflicts and ask how these struggles may be (or are being) interrelated and transformed by their articulation to questions about democratic governance and citizenship. Essentially, in case after case – as demonstrated by the studies in this book – we see how environmental conflicts may be (re)interpreted discursively as struggles about the quality of, and conditions for, the meaningful participation by differently situated groups in societal decision making via political processes and institutions. The environmental struggles of citizens' groups, environmental nongovernmental organizations (ENGOs), First Nations, and many other actors are necessarily also struggles about governance and citizenship; ecological goals cannot be achieved if existing liberal-democratic norms and institutions are not transformed. Ecological sustainability demands justice. In lieu of the pacifying discourse of "sustainable development," counter-hegemonic actors need a theory of ecological change that recognizes the existence of conflict rooted in social injustice and power relationships, and that identifies conditions and resources for overcoming obstacles to change. It is time to set aside the master's tools of "sustainable development" and turn our efforts toward the realization of ecological democracy.

The interpretation of environmental conflicts as having at stake *the meaning of citizenship* – or the democratization of our political, economic, and legal institutions – requires much greater attention than it has received so far. Research needs to address the following questions: How do state-organized consultation processes, monitoring bodies, or opportunities for judicial interventions shape and constrain citizen participation? How do they privilege some interests over others? How do citizens' groups, ENGOs, First Nations, or other actors challenge hegemonic assumptions about societal priorities or about citizenship? What reforms of policy-making processes have these actors proposed? ("Policy" is used in a broad sense here, to refer not only to government legislation or regulations, but also to the absence thereof, to governmental non-decisions or non-interventions, status quo practices or entitlements of the private sector, and developments that alter or secure the boundary between private and public spheres of governance.) What conceptions of citizenship emerge from these struggles?

The authors in this volume employ an eclectic mix of approaches reflecting, among other factors, their disciplinary backgrounds, their ontological perspectives, and the kinds of issues raised by their case studies. Three levels of analysis, however, are considered throughout: the political economy of these conflicts, their institutional contexts, and the discourses of the actors engaged in them (including media and governmental framings of the issues). In addition, Chapters 2, 7, 11, and 13 consider the roles of scientific and other forms of knowledge in environmental conflicts. Overall, the studies in this book present environmental policy as a terrain of complex, ongoing struggles among conflicting interests interacting with structural and institutional frameworks of opportunity and constraint. Choices with regard to discursive strategies – including the formulation of identities and the framing of conflicts – play critical roles in these struggles.

The environmental policy literature in Canada has maintained, for the most part, a separation from work on social movements. Few published studies have conceived of environmental organizations as social movement actors or have examined the relationships among the actors involved in efforts to determine policy outcomes. This book aims to help fill this void by examining experiences in which linkages between struggles have been attempted or by asking what has prevented such linkages from being developed. The chapters examine different "interfaces" between "environmentalists" and other actors or identities. Two chapters investigate the difficult relationships between environmentalists and farmers in the contexts of the regulation of genetically modified crops (Chapter 2) and intensive livestock production (Chapter 3). Chapter 13 highlights the leadership of women in opposing a high-risk landfill, while noting their relative isolation (the divisions within their community, their treatment as "non-experts" and as "emotional" women, their limited financial means to pursue litigation, the apparent absence of involvement by ENGOs). Chapter 10 describes a case in which a divided environmental movement and local "anti-protectionist" sentiments weakened the resistance to the privatization of many of Newfoundland's parks. Chapter 17, on the other hand, reports on a case in which ENGOs successfully linked larger ecological themes and alternatives to exurban citizens' concerns about urban sprawl. Chapter 12 discusses the forces preventing universities from assuming "institutional citizenship" and leadership in relation to the realization of sustainable communities. Chapters 6, 8, 9, and 15 focus on relations between environmentalists and First Nations. Chapter 5 examines the developing relationships between non-Native fishing communities and First Nations in southwest Nova Scotia. These contributions constitute a rich new source of empirical and theoretical work on the understudied relationships between environmental and First Nations' struggles.

Many of the chapters in this book identify the forms taken by struggles for environmental justice in Canadian contexts – an area of research that is only beginning to be developed in this country. In addition to Chapter 13's study of Crane Mountain Landfill and Chapter 16's discussion of brownfield "instant gentrification" in Toronto, seven chapters provide analyses of cases in which "development" or government policies have imposed disproportionate health risks on First Nations communities and failed to meaningfully involve First Nations in decisions that affect their cultures and livelihoods – indeed, their survival.

Although an effort was made in the selection of chapters (given the research available) to include the broadest possible range of conflicts and to represent varying parts of the country, many environmental issues and geographical areas remain unrepresented in this volume. However, I hope that, despite its gaps and shortcomings, this initial collection will help to forge a new direction in Canadian environmental studies and politics – one that embraces a multidisciplinary, multi-dimensional approach to the understanding of environmental conflicts and seeks to uncover their implications for democratic struggle.

Acknowledgments

The patience, persistence, and hard work of many individuals have made the completion of this project possible. Having already devoted months or years of research to their topics, the twenty-five authors went through repeated rounds of selection, editing, review, and revision of their work, beginning with a call for proposals in November 2004. During this time, children were born, parents and other loved ones died, jobs were changed, degrees were completed, and many developments occurred in their areas of research. I am grateful for the authors' stamina and commitment to this collection.

In bringing this book to print, we have been greatly assisted by our editor at UBC Press, Randy Schmidt, a remarkable shepherd. Anna Friedlander handled production matters with cheerful efficiency, and Deborah Kerr did a painstakingly thorough job of the copy-editing. Thank you also to Khyati Nagar of the Arts Resource Centre at the University of Alberta for her expert assistance with the figures. The book's reviewers provided very helpful suggestions that substantially improved the manuscript. We are also grateful for the support of the Aid to Scholarly Publications Programme (ASPP), which is funded by the Social Sciences and Humanities Research Council of Canada (SSHRC), and of the Support for the Advancement of Scholarship Research Fund of the Faculty of Arts at the University of Alberta.

Closer to home, I thank Mustafa Kaya for his loving support throughout, and Olivier Baran Adkin-Kaya for quite properly refusing to forego any bedtime stories on account of Mommy's work. My rock, my teacher, to you both I owe so much.

Environmental Conflict and Democracy in Canada

1
Ecology, Citizenship, Democracy
Laurie E. Adkin

A green democratic theory literature has grown up alongside one that differentiates between social-democratic and neo-liberal models of citizenship, and that examines the gendered nature of citizenship in differing welfare regimes.[1] With rare exceptions (Christoff 1996; MacGregor 2006), the two do not intersect. In my view, it is time to integrate the ecological, egalitarian, feminist, and social needs dimensions of critical citizenship theory. Such an integration is required if we aim to elaborate a more coherent and comprehensive program of reforms that is consistently aware (or at least attempts to be) of the interrelations among ecological imperatives and social needs, gender and racial equality, international solidarity, and the interests and rights of future generations, as well as those of non-human species. Although I do not claim to accomplish this large task in this chapter, I do hope to give a push to the project.

One common thread among struggles for social justice and ecological reform is the call for the democratization of political institutions and processes of decision making in light of the ways in which these privilege elite interests. This link is recognized by Peter Christoff (1996, 161), who argues that ecological citizenship is an "extension" of social citizenship in the form of demands for environmental welfare. Ecological conceptions of citizenship "extend social welfare discourse to recognize 'universal' principles relating to environmental rights and centrally incorporate these in law, culture and politics."

Notably, green democracy goes further than existing conceptions of social citizenship in seeking to include *human "non-citizens"* (non-citizen residents of a territory, residents of other territories, future generations) in decision making, recognizing the supranational nature of environmental concerns and questions of global justice. Green theorists also extend the idea of democratic representation in another important way, arguing that the interests of *non-human species* deserve moral consideration in political decision making (Eckersley 1992, 2004a; Dryzek 1995a).

The environmental justice movements already embody some of the linkages as well as principles of social and ecological citizenship. These campaigns make clear how the poor (often racialized groups) disproportionately suffer environmental harms. But this movement does not demand only, and negatively, that environmental harms be "equally distributed" among classes and racial groups. It opens the door to the consideration of alternative modes of development that avoid and prevent environmental and social risks. Ecological democracy demands that the model of development eliminate or minimize harms and that alternatives be considered (in the realms of science, technology, investment, policy, and institutional design). It demands, most of all, meaningful citizen participation in societal decision making.

In neo-liberal regimes, the supposed needs of the market – most importantly, "freedom" – have primacy over the "social," and the divide between public and private is re-entrenched (as we see with the familialization of welfare). Thus, the opportunities for citizens to intervene politically in decisions that constitute *societal* choices are strictly circumscribed. We may join political parties and vote, but the shrinkage of the state's regulatory role vis-à-vis market forces and the ideological convergence of the traditional parties mean that many questions are excluded from the realm of the political. The conditions for active citizenship are further eroded by the concentration of media ownership (Edge 2007; Winseck 2006) and the underfunding of public broadcasting (see Leys 2001).[2] These developments empower hegemonic interests to determine the framing of political debates (as we see with regard to the debate over family policy in Canada, where the options are constructed, principally, as the universal breadwinner model or the male breadwinner model but *not* the gender parity model, or, in the climate change debate, where the options are presented as "Kyoto" or "no Kyoto" but not "*beyond* Kyoto").[3]

Why, as Robyn Eckersley (2004b, 241) asks, have liberal democratic states proven incapable of resolving ecological problems, choosing instead to "merely displace them over space and time"? First, rarely are there any constitutional obligations to represent the interests of non-citizens (territorially, legally), future generations, or non-human species. Second, "partisan competition places groups and organisations that are well resourced, well organised, and strategically located at a distinct advantage to poorly resourced, poorly organised, and dispersed groups, such as community environmental groups" (Eckersley 1996, 215).

But even more importantly, political and economic institutions in liberal democracies are the product of the articulation of political liberalism (individual freedoms, civil rights, pluralism) to economic liberalism (private-property rights, minimal regulation of markets, atomistic view of human nature). The primacy of private-property rights, the supposed impartiality and efficiency of market forces, the dependence of governments on private investment for job creation and revenue, the values embedded in the system of national accounting, not to mention the global institutions of neo-liberal governance – all this amounts to the equation of rates of capital investment and profit, or rates of GDP or GNP, with the health of "the economy," the reified category upon whose "stability" and growth all our fortunes are said to depend. All social and environmental demands are measured against its needs: How might greater spending on health care affect the economy? How will a decline in consumer spending affect the economy? How will stricter environmental regulation affect the economy? What will be the economic costs of reducing greenhouse gases? So long as capitalist economic institutions and neo-liberal state-society relations are off limits to substantial reform, certain zero-sum trade-offs are unavoidable (for example, the argument that the implementation of "polluter pays" rules will result in a contraction in investment and increased unemployment). As Eckersley (ibid., 215) puts it, "The upshot is that the longer-term public interest in environmental protection is systematically traded-off against the more immediate demands of capital and (sometimes) labour. Indeed, it is precisely this process (and expectation) of trade-off that has been inscribed into the state agencies and decision rules which govern environmental decision-making."[4]

These rules are understood by environmentalists, as well as by other actors. Thus, the current efforts of many international, governmental, or non-governmental organizations charged with promoting sustainable development are aimed at getting business "onside" with targets to reduce various emissions or the consumption of raw materials and energy, to recycle, or to meet new product standards. In regard to greenhouse gases (GHGs), the large body of research and proposals aimed at demonstrating the profitability or minimal costs to business of agreeing to targets reflects the strategic perception that sustainable development must be made acceptable to business interests. This strategy is clearly expressed in the title of one of Stéphane Dion's (2005b) speeches as environment minister, which was addressed to a business conference: "Cutting Megatonnes of GHGs; Making Megaprofits." Dion (2005a) repeatedly represented Prime Minister Martin's Project Green as a vision for "a sustainable, competitive economy, a prosperous Canada." Ottawa's April 2005 plan for reducing GHGs – *Moving Forward on Climate Change: A Plan for Honouring Our Kyoto Commitment* – though criticized by the Canadian Council of Chief Executives (CCCE), gave the business coalition virtually everything it wanted, short of renunciation of the protocol targets (CCCE 2002; CAPP 2002).[5] In this game, as I have argued in connection with Canadian environmental politics in the 1980s (Adkin 1992), large corporate interests ultimately decide the limits of sustainability.

On the other hand, we see resistance everywhere to the numerous negative "externalities" (social and environmental) produced by "free markets." We also see challenges to neo-liberalism's view of the good citizen (the self-reliant individual), which equates individual autonomy with independence from the state and fails to recognize relational identities or social solidarities. The involvement of civil society organizations – from small citizens' groups to large NGOs – in environmental conflicts wedges open discursive space for a reconceptualization of citizenship as participatory, expansive, solidaristic, and ecological. Citizen activists understand that ecological health is a public good and hence a democratic right that can only be enjoyed collectively. As Robyn Eckersley (1996, 227) puts it, social and ecological rights "belong to individuals not only as individuals but also as members of social and ecological communities." Moreover, ecological health cannot be achieved in isolation from the outcomes of other struggles, as their objectives and conditions for success are intertwined. Gender equity, social justice, anti-racism, anti-colonialism, eco-centrism – all intersect with ecological choices. These interdependencies are what ecologists and other social activists need to recognize and to keep in play in the formulation of programs of reform.

Green Democratic Theory and Hegemonic Politics
In the burgeoning literature on environmental democracy and environmental or ecological citizenship, attention has focused either on *normative principles* or the *procedural prerequisites* for deliberative democracy. The former assesses what should be entailed in the practice of ecological citizenship – in particular, the duties or obligations of citizens – and what values should guide ecological citizenship. The latter attempts to determine which rules would maximize inclusive, uncoerced, and informed citizen participation in decision-making processes. In the following sections, I wish to identify some of the contributions and limitations of these two approaches to the conceptualization of ecological citizenship.[6] I will start by outlining some of the current uses by political theorists of such terms as "environmental citizenship," "green citizenship," and "ecological citizenship."

Environmental/Ecological Citizenship: The Concepts
Environment Canada's Freshwater Website (Environment Canada 2008) currently defines environmental citizenship as "a personal commitment to learning more about the environment and to taking responsible environmental action. Environmental citizenship encourages individuals, communities and organizations to think about the environmental rights and responsibilities we all have as residents of planet Earth. Environmental citizenship means caring for the Earth and caring for Canada." In this definition, citizenship is both a personal commitment and a moral injunction that

prods individuals and organizations to act responsibly. This definition also introduces both rights and responsibilities as aspects of citizenship, although neither these nor their agents are specified.

Dave Horton (2006) identifies two main uses of the term "environmental citizenship": First, it is an "individualized project," whose discourse is primarily disciplinary (people need to be made more aware and more responsible for their behaviour); this task may be assisted by state agencies, the education system, or civil society organizations. Second, it refers to the existence of environmental *rights,* primarily through a discourse about social justice. Andrew Dobson (2003) places these uses into different categories: "environmental" and "ecological" citizenship, the first referring to the extension of *rights* to the environmental domain but within a liberal framework, and the second to a "post-cosmopolitan" (non-territorial) focus on *virtues and obligations* (see also Sáiz 2005).

A good deal of green democratic theory focuses on postulating the virtues of ecological citizenship, on ecological citizenship as an ideal set of practices (obligations, duties) or orientations (Barry 1996; Dobson 1996, 2006; Marzall 2005; Mark Smith 2005). John Barry (1999, 231-32) defined "green citizenship" as the practice of *ecological stewardship* and "a responsible mode of acting." Tim Hayward (2006a) privileges the virtue of "resourcefulness." For Dobson (2006), the central "virtue" of ecological citizenship is justice, with obligations and entitlements taking the concrete form of equal ecological footprints for the global human population. James Connelly (2006, 66) conceptualizes ecological citizenship as a set of duties and the virtues (such as frugality, care, patience) appropriate to their fulfillment. Like other theorists of ecological citizenship, he emphasizes the importance of rational democratic deliberation to determining what the duties of citizens should be. And, like Dobson, he proposes that not all citizens will bear the same responsibilities, because they are starting from different positions with regard to their ecological footprints.

Tim Hayward (2006a, 2006b) and Alex Latta (2007) point to a significant problem with the construction of ecological citizenship as the practice of reducing one's ecological footprint in the interest of global justice. Essentially, this creates two categories: the *active* citizens, or agents, of ecological sustainability or justice and the *passive* objects of injustice (those in the global South who have not been able to secure a fair share of the earth's carrying capacity – Dobson uses the term "ecological space" – due to over-consumption in the global North). Dobson's (2006) defence that unequal obligations should instead be understood to produce "citizens with obligations" and citizens who are "recipients of justice" has not mollified his critics, for whom the term "demanders of justice" might have been more acceptable. One might very well understand the discourses and practices of environmental justice movements around the world as exemplary of active ecological citizenship.

Like Environment Canada's definition of environmental citizenship, the idea of "stewardship" for the earth bypasses questions of "equity, exclusion, and justice" (Agyeman and Evans 2006, 200). Yet, whereas the stewardship discourse fails to address the crisis of social justice (among groups of humans), the environmental justice discourse focuses exclusively on human-human relationships, leaving out the problem of human relationships with non-human nature.

John Barry (2006, 23-24) has, more recently, differentiated between "environmental citizenship," which is akin to weak environmental reforms aimed at reducing pollution or increasing resource efficiencies and that are acceptable to corporations and imposed by the state, and "sustainability citizenship," which he defines as "a more ambitious, multifaceted, and challenging mode of green citizenship, which focuses on the underlying structural causes of environmental degradation and other infringements of sustainable development such as human rights abuses or social injustice." Here we are moving toward a more expansive view of ecological citizenship, one linked to counter-hegemonic politics. The role of the state and the problem of how to "green" states also come into play in Barry's discussion of state-centred versus civil-society-based approaches to reform. Ultimately, Barry argues, green states will be produced only by informed, radicalized, "green" citizens and their oppositional practices (resistance) – including civil disobedience. Robyn Eckersley (2004b) also underlines the importance of vibrant, active green publics in bringing about future "green states" and (following Hannah Arendt) Mick Smith (2005, 60) argues that active participatory politics that take place in the public sphere – including non-violent direct action – are "absolutely central to any conception of an environmental citizen."

If the existence of vibrant, informed, and mobilized green publics is essential to transforming citizenship rights and to creating green states, the next question is how can the formation and sustenance of such *green publics* be facilitated? (In other words, how do we escape the chicken-and-egg problem of how green citizenship can be enabled in the absence of a green state and how a green state can be created in the absence of green citizens?) Dave Horton (2006, 128) argues that environmental citizenship develops "within the cultural and political spaces of contemporary environmentalism." Interestingly, by studying the conditions that sustain green activist communities, he identifies three conditions needed to promote green citizenship more broadly. These include sites/places for community organizing, meeting, and socializing; time; and "materialities" such as alternatives to cars, meat, and imported foods. "Citizenship," Horton (ibid., 147) suggests, "is less a quality of individuals than of the architecture that produces and reproduces that citizenship." Such conditions may be created by civil society actors and/or promoted and supported by states.[7] But if we rely exclusively upon the latter, we are back to our chicken-and-egg problem – how, initially, to create green states.

To Horton's list of conditions for ecological citizenship, Nicholas Nash and Alan Lewis (2006, 176-77) add more participatory democracy and the decentralizing of decision making away from elites. Other theorists (such as Dryzek 2000; Mason 1999; Schlosberg, Shulman, and Zavestoski 2006) see deliberative democratic principles and institutions as the means of obliging various actors to justify their desires or preferences in terms of the general good. For David Schlosberg, Stuart Shulman, and Stephen Zavestoski (2006), "critical environmental citizenship" entails institutional and technological *opportunities* for deliberative democracy (which focuses debates on the public good).

The focus on "virtues" and "obligations" speaks more to a liberal-individualist conception of a society made up of undifferentiated citizens (except with regard to their ecological footprints, in Dobson's case), whereas a focus on "rights" assumes the existence of collective subjects and conflicting *societal* interests. Sherilyn MacGregor (2006) observes the tendency of the "ecological citizenship" theorists to treat the citizen as an undifferentiated subject. She shows, for example, how some of the duties and values (such as care) that (male) environmental citizenship theorists wish to promote mean quite different things for the lives of men and women. In other words, citizens are *gendered* individuals. Of course, they are also differentiated by their class, race, and other positions, and any theorization of citizenship rights and duties needs to be examined along these axes as well.

As noted above, Peter Christoff (1996) gives a central place to *environmental rights* in his definition of environmental citizenship, which he views as an *extension* of social and political rights. Just as social rights underpin social welfare states, he argues, environmental rights will underpin green welfare states. Environmental rights are not viewed, however, as being "neutral" in relation to the conditions of capitalist accumulation – any more than social rights are. Christoff (ibid., 161) believes that demands for environmental welfare "at minimum reshape, and often work against, the requirements of capital reproduction and accumulation." This view is borne out, I argue, by an examination of environmental struggles.

Moving Ecological Citizenship to the Terrain of Counter-Hegemonic Politics

The environmental/ecological citizenship literature has sought to envisage what an ecologically sustainable society would look like, from the point of view of values, behaviours, and responsibilities of individual citizens. To some extent, it has crossed over into the terrain of theorization of the green state, which asks what the role of *the state* would be, in relation to civil society, in an ecologically sustainable and socially just society, and what kinds of constitutional rules or rights would guide state decision making. However, much of the ecological citizenship literature has had a "top-down" character, with (as MacGregor observes) predominantly white male political theorists pronouncing on "primary virtues" and the like, and paying scant attention to *collective* subjects and their relationships to one another and to ecological reform. As Alex Latta (2007, 378) puts it, "current debates have tended to focus on the challenge of cultivating 'green' attitudes and behaviours in individual citizens, leaving questions of democracy and collective action on the sidelines." Thus, normative political theory provides various visions of ecological citizenship (especially the desirable value orientations or practices of citizens) but not enough in the way of *strategies of political change*. In some cases, these theorists' conceptions of change seem to amount to persuasion or appeals to individuals to change their "values" – a kind of civilizational view – rather than to a strategy to radicalize and connect real, existing conflicts. As Jason Found and Michael M'Gonigle argue in Chapter 12 on the struggle to steer universities in environmentally sustainable directions, quoting Michel Foucault (1980, 133), "the problem is not 'changing people's consciousness' ... but the political, economic, [and] institutional regime for the production of truth."

Debates in environmental citizenship theory have too often been conducted in what Ruth Lister (2007) refers to as an "empirical void." What we need are empirically grounded analyses of the political, economic, and institutional *obstacles to* change, as well as of the *resources for* change. Like Horton, we should be asking what conditions might nurture and sustain vibrant green public spheres on various spatial scales in the context of global capitalism. But we should also be asking how relationships of domination/subordination or privilege/exclusion *among* subordinate social groups come into play in environmental conflicts and how these might be transformed into relationships of solidarity. These questions are important to a project of social and ecological change that understands politics in terms of hegemonic struggles. They are taken up in more detail in the case studies in this volume and in its concluding chapter.

Deliberative Democracy and Hegemonic Politics

The deliberative democracy theorists provide us with an excellent set of criteria by which to evaluate existing policy-making processes (including land-use consultations, environmental impact assessments, and public hearings regarding proposed developments). But their tendencies to assume that the ideal conditions for deliberative decision making will produce consensus outcomes, and to bracket crucial problems of power relationships, provoke skepticism.

Much of the green democratic theory literature follows Jürgen Habermas in focusing on the procedural requirements for deliberative democracy (see Chapters 7 and 11, this volume). In this approach, as summarized by Michael Mason (1999, 8), "Discourse refers to modes of communication between people in which understanding rests upon, or presumes the possibility of, agreement motivated by convincing reasons rather than by any form of coercion or deception." For John Dryzek (2005, 233), "discursive democracy" refers to procedures and institutions that "involve collective decision-making through authentic democratic discussion, open to all interests, under which political power, money, and strategizing do not determine outcomes." The discourse principle holds that "only those action norms are valid in which all possibly affected persons could agree as participants in rational discourses" (Habermas 1996, 459, quoted in Mason 1999, 8-9). Thus, the procedures of discursive democracy maximize inclusion of those with interests in the outcomes, neutralize strategic power in favour of communicative reason, and seek consensus decisions (identification of common interests or goods). It is through involvement in such contexts that participants may identify general ecological interests, and their preferences may be changed by "ethical reflection on the good life" (Mason 1999, 50). Naturally, discourse theorists of democracy tend to expect that ecological sustainability would be identified, under these conditions, as a common, generalizable interest – one that shapes moral judgment and thinking about the future.

What I find problematic about the deliberative democracy literature is its tendency to assume that rational consensus (without exclusion) on questions of justice is achievable or that the consensus achievable will be acceptable to all parties. (That is, the points that can be agreed upon may not be sufficient to resolve the conflict to everyone's satisfaction.) In regard to environmental conflicts, once certain ecological parameters are established as "non-negotiable," conflict between capitalist accumulation, on the one hand, and societal and ecological interests, on the other, becomes inevitable.[8] For example, in his account of the experience of British Columbia's Commission on Resources and Environment (CORE) – a significant deliberative democracy experiment of the mid-1990s – Michael Mason (ibid., 125-26, emphasis added) reported that

> From the perspective of environmental preservation groups, the commission itself failed to realize that the competing interests evident in provincial wilderness disputes reflect incommensurable values. In such land-use conflicts, consensus may not necessarily be benign, according to wilderness activists, if it is a requirement blocking the political protection of an ecologically vital public interest – the preservation of natural areas beyond the levels of the protected areas strategy (Western Canada Wilderness Committee 1996; Greater Ecosystem Alliance 1994). For the environmental sector,

legislative enactment of a sustainability act would properly recognize certain ecological parameters as nonnegotiable, countering the economic power of extractive resource-based corporations.

We see that the actors do not agree on the value (necessity) of protecting a particular natural area, and no compromise is possible. For the social movement actors, at stake is the defence of "a physical and/or moral territory" (Offe 1984,189). This is an example of the non-negotiable nature of many social movement demands (remove the missiles, provide reproductive choice for women, stop the arms race, ban forest clear-cutting, or toxic chemical pollution, or land mines).

Certain ethical stances do not lend themselves to "rational" universal agreement but nevertheless call for political decisions. When the deliberative process cannot yield a non-exclusive consensus, the decision is deferred to the state. In the case of CORE, the state rejected the commission's recommendation for a sustainability act in which legal requirements for certain "ecological parameters" would be respected, ceding, instead, to pressures from the forestry industry.[9] Mason's (1999, 125) summary of the outcome is instructive:

> Those politicians, fearing the precedent that would have been set by CORE's recommendations on ongoing public involvement, and swayed by the lobbying of civil servants from the dominant land-use agency – the forest ministry – soon moved to weaken the participatory proposals of the commission. This left unchallenged the development biases within the provincial resource bureaucracy and has so far blocked the advance of environmental capacity-building envisaged by CORE ... The major challenge to the communicative resolution of land-use conflicts in British Columbia remains the absence of a comprehensive sustainability act, including an explicitly legislated public participation process and fundamental changes in the legal mandates governing Crown land-use planning.

Another example of conflict between non-negotiable ethical stances and hegemonic interests is provided by the challenges posed to anthropocentric sustainable development discourse by deep ecology and eco-centrism. At stake are how we, as humans, should live with nature and what moral consideration is due to the needs of other species. There is no purely "rational" way of deciding whether it is just for humans to exploit other species, or to destroy their habitats (even to the point of causing their extinction), or to what extent such harms might be justifiable. Nor can there be only one, universal approach to such questions. Mason (ibid., 39) observes that "the procedural neutrality of liberal democratic institutions leaves it to the citizens to decide which notion of the good life to subscribe to." In this case, it is difficult to imagine a non-exclusionary consensus regarding human entitlements – all the more so because parties (species) with obvious interests in the outcomes of deliberation are typically not represented at the table.

Principles of democratic governance – of democracy – cannot really be separated from beliefs regarding the conditions for a good life. Claims regarding the latter give rise to demands or expectations regarding the former. The democratic radicalization of liberal norms in both economic and political spheres is a *condition* for the possibility of achieving ecological and social *needs,* and democratization is understood by citizen activists in these terms (Adkin 1998a).[10] Although there are multiple views of what ecological and social needs are, or of how they should be weighted, I suspect that a broad popular consensus is possible regarding not only their legitimacy, but also the factors that obstruct their realization. The formation of such a consensus, however, implies *exclusion* – the construction of political friends, enemies, and partial allies. Rational argumentation (including in favour of procedural democracy) is, of course, part of this discursive, counter-hegemonic politics, but it cannot be expected to produce consensus in the sense of the dissolution or absence of fundamental conflict.[11] In a Gramscian sense, *consensus has a hegemonic nature* involving moral and intellectual leadership, the active construction of consent, and coercion.

Chantal Mouffe argues that power can never be eliminated from rational argumentation, and legitimacy can never be grounded on pure rationality. This is because conflict and power are constitutive of the political. We must relinquish, Mouffe (2000, 100, 101) says, "the ideal of a democratic society as the realization of a perfect harmony or transparency." In her definition, "'politics' ... indicates the ensemble of practices, discourses, and institutions which seek to establish a certain order and organize human coexistence in conditions that are always potentially conflictual." This conflictual view of social interests, and of politics, stems from a post-Marxist ontology rather than a liberal one.[12] As Mouffe clearly spells out in her later book *On the Political* (2005, 53), the radicalization of democracy "requires the transformation of the existing power structures and the construction of a new hegemony."

Her "agonistic" model of democracy differs sharply from that of Habermas or Rawls, as she (2000, 101-3) explains in this passage from *The Democratic Paradox:*

> The central question for democratic politics ... pace the rationalists, is not how to arrive at a consensus without exclusion, since this would imply the eradication of the political. Politics aims at the creation of unity in a context of conflict and diversity; it is always concerned with the creation of an 'us' by the determination of a 'them.' The novelty of democratic politics is not the overcoming of this us/them opposition – which is an impossibility – but

the different way in which it is established. The crucial issue is to establish this us/them discrimination in a way that is compatible with pluralist democracy ... An adversary is an enemy, but a legitimate enemy, one with whom we have some common ground because we have a shared adhesion to the ethico-political principles of liberal democracy: liberty and equality. But we disagree concerning the meaning and implementation of those principles, and such a disagreement is not one that could be resolved through deliberation and rational discussion ... I agree with those who affirm that a pluralist democracy demands a certain amount of consensus and that it requires allegiances to the values which constitute its 'ethico-political principles.' But since those ethico-political principles can only exist through many different and conflicting interpretations, such a consensus is bound to be a 'conflictual consensus.'

More recently, drawing on psychoanalytical theory, Mouffe (2005, 28) writes, "The theorists who want to eliminate passions from politics and argue that democratic politics should be understood only in terms of reason, moderation and consensus are showing their lack of understanding of the dynamics of the political. They do not see that democratic politics needs to have a real purchase on people's desires and fantasies and that, instead of opposing interests to sentiments and reason to passions, it should offer forms of identification conducive to democratic practices."

I think we might say, simply, that political discourse needs to have roots in people's lived experiences – it must "speak to" the conflicts that they experience in their daily lives but in a way that shows how these conflicts are more than individual experiences and how they call for collective action. Elsewhere (Adkin 1998a), I have made an argument for the necessity of articulating ecological and social struggles by means not only of democratic discourse, but also by recognizing "real experiences of deprivation and alienation" and identifying the conditions necessary for happiness. We should also remember that people are moved to engage in "politics" (to inform themselves, to participate in discussions or debates, to protest, to organize various kinds of events, to join organizations, and so on) when they feel that something important is at stake and that something can be done, collectively, about it. When there is no sense of societal importance, of urgency (for a group, for nature, for society), or of the possibility of effective action, there is little motivation for involvement.

Thus, it is fair to ask of green democratic theory – as Mouffe does of Habermasian theory more generally – whether its understanding of politics is adequate to the challenges that confront radical ecologists. When, with whom, and on what terms should we seek "consensus"? When is it necessary to differentiate between friends and enemies? What is negotiable? What isn't? If we agree that the construction of an "us" and a "them" is critical to counter-hegemonic politics, what, then, is the role of deliberative democracy?

One might argue that the normative and deliberative approaches to ecological democracy are *embedded in* counter-hegemonic politics, or, in Italian Marxist Antonio Gramsci's (1971) terms, embedded in a *war of position*. Gramsci viewed class struggle as a "war of position" waged in every sphere of society (economic, political, cultural) and at every level of intervention (given that the outcomes of local conflicts are always determined, in some measure, by regional, national, or international developments and alignments of forces). Gramsci sometimes depicted war of position as "trench warfare," in the sense that every inch of ground is hard-fought, and the struggle is both offensive and defensive. Although strategy and leadership are important, the "war" can be won only by the efforts of countless individuals who sacrifice, plan, and work for every advance. The key terrain of this trench warfare is that of ideas and beliefs. The main tasks of the counter-hegemonic forces are to call into question the legitimacy and naturalness of the ruling order and to replace these with a vision and a program of their own. Although we need not follow Gramsci in his reduction of counter-hegemonic struggle to a binary class opposition, we may certainly appreciate the relevance of his trench warfare metaphor for understanding the nature of social movement struggles today.

Gramsci's concept of a "historical bloc" may also be reconceptualized in the contemporary context to refer not only to the efforts of the working class to constitute an ideological and organizational movement capable of shifting a society's ruling norms and path of development. A historical bloc may also be understood in terms of a social movement for ecological democracy that is capable of articulating diverse struggles for equality, freedom, and recognition without erasing their specific origins or meanings. Indeed, the attempts, through war of position, to construct or to hold together an existing historical bloc are the essence of "politics" at any level. Insofar as democratic counter-hegemonic movements can constitute historical blocs, possibilities are opened for democratizing reforms and the strengthening of communicative rationality.

The demands emerging from environmental and social justice struggles create common stakes in a project of democratization. They offer a *connecting thread* among diverse struggles, reinterpreting these as democratic struggles. That is, procedural criticisms and demands for the transformation of liberal democratic institutions *arise from* a plethora of real conflicts around ecological risks and social needs, as these are experienced by differently situated individuals and groups. Procedural democratic forms are also *ends in themselves* because civic self-determination and autonomy may be viewed as integral to a good life. Looked at in this light, these struggles reveal the potential for the formation of a historical bloc with an ecological-democratic identity.

Conceptualizing Ecological Citizenship

I use the term "ecological citizenship" to encompass the democratic, social, and environmental *rights* defended or demanded by social actors, along with the *responsibilities* for the well-being of other humans and other species that have emerged from political-ecological discourse. I further use the term to refer to the *practices* of citizens as they seek to secure or enhance these rights and to act responsibly. These practices, which occur in both the private and the public spheres, are both personal and political. In this sense, I concur with Liette Gilbert and Catherine Phillips' (2003, 314) assessment that "citizenship should be critically understood not only as rights granted by a government, but also as practices through which the limits of established rights are (re)defined and (re)affirmed ... Citizenship is therefore not only a set of formal rights, but rather is a continual process of construction and constitution."[12] "Ecological citizenship" is, in effect, a shorthand term for a more complex articulation of ecological concerns to social justice and participatory democracy. I prefer this term to "environmental citizenship," which, I think, has taken on connotations of a narrower sphere of concerns, as well as a mere moral claim or injunction. Like the term "sustainable development," "environmental citizenship" is rapidly being co-opted by corporations, business associations, and governments in the ways suggested by John Barry (2006) and Agyeman and Evans (2006).

In my view, the case studies in this book entirely corroborate Gilbert and Phillips' (2003, 319) view that "the fight for participation in control and/or decision-making ... is clear in environmental rights struggles." Environmental struggles seek to protect or regain citizenship rights that "have been vulnerable to state and market domination: rights to information, to expression, to culture, to identity and difference (and equality), to self-management, to the city, to nature and to services" (ibid.). We can go further than this: environmental struggles identify ways in which democracy must be *extended* or radicalized. In Chapter 18 of this book, I examine in greater detail some of the "deficits" in citizenship that have been highlighted by environmental struggles in Canada, as well as the democratizing reforms that have been called for by the diverse actors engaged in these conflicts. As I argue, these struggles – countless, varied in their intersections, and waged at multiple spatial scales – may be viewed as the trench warfare of democracy.

Notes

1 This is a large literature, but see Janine Brodie (2002), Gøsta Esping-Andersen (1999), Julia S. O'Connor (1996), Diane Sainsbury (1994), and Janet Siltanen (2002).

2 To some extent, these developments may be countered by the growing use of the Internet.

3 These ideal-type categories – widely used in feminist theorization of welfare states – were introduced by Carole Pateman (1988) and Nancy Fraser (1994). The male breadwinner model refers to social insurance and welfare policies, as well as a wider range of social

policies and employment legislation, that assume that the typical family has a full-time male breadwinner and a full-time female homemaker. The universal breadwinner model attempts to facilitate women's employment in the wage sphere by, for example, providing state subsidies for child care and is for this reason viewed as promoting gender equity. The gender parity model (which doesn't exist anywhere) aims to divide both paid work (public sphere) and unremunerated care and domestic work (private sphere) equally between men and women.

4 John Dryzek (1995a, 16), likewise, argues that liberal democratic institutions are not well designed to resolve environmental problems: "Interests may be placated in proportion to their material political influence, and compromises may be achieved across them, but wholesale ecological destruction can still result. Resilience in liberal democracy is inhibited by short time horizons ... and a general addiction to the 'political solvent' of economic growth."

5 To oppose ratification of the Kyoto Protocol by the Chrétien government, the CCCE formed a coalition – called the Canadian Coalition for Responsible Environmental Solutions (CCRES) – with the Canadian Chamber of Commerce (CCC) and the Canadian Manufacturers and Exporters Association (CME). It appears that the CCCE, CCC, and CME began to organize this coalition in 2001, in response to the federal government's 2000 Action Plan on climate change.

6 This is necessarily an outline, rather than a fully developed and comprehensive critique.

7 Connelly (2006), among others, argues that states should create the structural conditions for citizens to act ecologically without being coerced to do so.

8 The discourse theorists themselves recognize the existence of some rather fundamental obstacles to the realization of the conditions for deliberative democracy. For example, Michael Mason (1999, 99) asks, "How will economic sovereignty be wrested from those powerful vested interests who gain so much from the current asymmetries of social power?" John Dryzek (1995b, 112) argues that "all liberal democracies currently operate in the context of a capitalist market system. Any state operating in the context of such a system is greatly constrained in terms of the kinds of policies it can pursue. Policies that damage business profitability – or are even perceived as likely to damage that profitability – are automatically punished by the recoil of the market."

9 The Harcourt NDP government, which initiated CORE, was followed by the NDP government of Glen Clark, which was anti-environmentalist. For a full account of the CORE experience, see the excellent case study in Mason (1999, ch. 3), as well as Chapter 7, this volume.

10 By "democratic radicalization of liberal norms," I mean such reforms as the entrenchment of social and environmental rights and their precedence over private property rights, the institutionalization of the precautionary principle, the adoption (where it does not exist) of proportional representation in electoral systems, and the implementation of measures to facilitate meaningful citizen participation in societal decision making.

11 Andrew Dobson (1996, 141) argues, in fact, that discursive democracy may be viewed as "a tool for criticizing non-green decisions in actually existing democracies."

12 As Mouffe (2005, 54-55) puts it, referring to her work with Ernesto Laclau, "Despite its shortcomings, we see the Marxist tradition as having made an important contribution to our understanding of the dynamics of the capitalist system and its consequences over the ensemble of social relations. This is why, contrary to Beck and Giddens, we acknowledge the crucial role played by economic power in the structuring of an hegemonic order."

13 Alex Latta (2007, 388) further elaborates the concept of citizenship as "fundamentally dynamic, a set of institutions and practices that are contested and reshaped by citizen activism."

2
Unsatisfactory Democracy: Conflict over Monsanto's Genetically Engineered Wheat
Peter Andrée and Lucy Sharratt

In May 2004, biotechnology company Monsanto announced that it would halt its efforts in Canada and the US to commercialize Roundup Ready (RR) wheat, a variety of hard red spring wheat genetically engineered to be resistant to its herbicide Roundup (Monsanto 2004b).[1] Monsanto's decision was proclaimed a victory by the civil society organizations (CSOs) that had fought RR wheat since 2001, including many organizations and communities that had been campaigning against the introduction of genetically engineered organisms (GEOs) to Canadian agriculture since the 1980s. The decision was seen as a particular victory for farm organizations that had come together in unprecedented unity against this GE product. Stewart Wells, president of the National Farmers Union (NFU), stated, "Clearly, Monsanto is backing off because the opposition to genetically modified wheat is overwhelming" (quoted in NFU 2004).

This decision to suspend commercial pursuit of RR wheat was actually the second major Canadian defeat for Monsanto, the world's largest biotechnology company and now the world's largest seed company. The first occurred in 1999 when the Canadian government rejected Monsanto's recombinant bovine growth hormone (BGH), a genetically engineered veterinary drug designed to increase milk production (see Sharratt 2001). However, these two cases in which GE products were stopped in their tracks stand in stark contrast to the general experience of the Canadian regulatory system: since 1995, over fifty varieties of twelve different GE crops and foods have been introduced to the Canadian agri-food system.

In this chapter, we address two questions: First, how did Monsanto's striking decision to remove its RR wheat from regulatory consideration come about, especially in light of the success of the biotechnology industry as a whole in bringing other GE products to market? Second, what does this case tell us about the power of citizen resistance to GE products in Canada? Answers to these questions provide insights into the potential for the kind of ecological democracy envisaged by Robyn Eckersley, John Dryzek, and other green political theorists, as discussed by Laurie Adkin in the first chapter of this volume.

According to Eckersley (2004b, 111), ecological democracy in the context of risk issues rests on the principle that "all those potentially affected by a risk should have some meaningful opportunity to participate or otherwise be represented in the making of the policies or decisions that generate the risk." Dryzek (1999, 266) agrees, arguing that democratic equality should "extend across the perpetrators and the victims of environmental risks." These authors conclude that the ideal of ecological democracy is thus best realized through a deliberative model of democratic decision making. Furthermore, because non-human beings cannot participate in democratic discourse and would probably have to be spoken for by proxy, Eckersley (2004b) proposes that the outcomes of *ecological* democracy are likely to be inherently precautionary. That is, in an effort to ensure that all species affected by risk decisions are considered as if they were themselves part of the deliberation, ecologically democratic decisions would err on the side of not imposing new risks on nature when significant uncertainties exist about the potential impacts of those decisions.

Dryzek (1999) encourages us to see ecological democracy not simply as a future ideal-type. Rather, we can already identify elements of ecological democratization in the way that a number of environmental conflicts are currently being addressed in the public sphere. The example of RR wheat may provide a case in point. On the surface, a kind of ecological democracy appears to have been at work here. Despite efforts by both industry and government to manage this issue and keep a lid on controversy, the outcome was actually the result of widespread deliberation in the public sphere. This deliberation involved intense debate about a broad range of potential benefits and risks of GE crops and food for farmers and the food industry, the general public, and many species of plants and animals. Furthermore, scientific uncertainties regarding the impacts of GEOs did play an important role in the debate, and the weight of precautionary considerations helped create a situation in which Monsanto had little choice but to back down.

In this chapter, we argue that there were indeed elements of ecological democratization in the process that led to Monsanto's decision, and these are to be lauded as a sign of what is possible when citizens join forces to oppose the private interests of a large corporation. However, we emphasize that there is significant divergence between the outcome in this case and the way in which Canadian policy and regulation treat GEOs generally. Indeed, this case highlights resistance on the part of government and industry to the institutionalization of ecological democracy.

Our analysis reveals a distinct absence of democracy in the formal regulation of the products of agricultural biotechnology, despite the strong public demand for a democratization of the process. Expressly constructed to facilitate commercialization, the Canadian regulatory system is actually incapable of dealing with GE products that raise widespread concern among both consumers and farmers. As currently constituted, this system is also far from precautionary. In the case of RR wheat, we argue that political and economic considerations were ultimately more important than risk concerns in achieving a seemingly "precautionary" outcome. In sum, though the RR wheat outcome suggests a semblance of ecological democracy in the public sphere, it remains highly unsatisfactory (hence our title) because it took place as a reaction *outside* and *against* the state, rather than being facilitated *by* the state.

This chapter begins with an overview of the Canadian regulatory system for GE crops and foods. We then turn to the struggle over RR wheat, pointing to some of the key dynamics that led to Monsanto's decision to remove its product from regulatory consideration. In the concluding section, we compare this story with the ideal of ecological democracy. We argue that a more formally democratic and precautionary approach to GE regulation needs to replace the protracted conflicts witnessed to date and the valuable, but ultimately unsatisfactory, democracy achieved through public protest.

Background

The conflict over the potential commercialization of RR wheat in the early 2000s must be understood in the context of earlier debates regarding the regulation of agricultural GEOs in Canada and elsewhere. In these debates, which began with the advent of genetic engineering techniques in the 1970s, CSOs raised a wide variety of concerns. These ranged from questions about the potential of GEOs to introduce new environmental contaminants and health threats to concerns about the way that genetic engineering, along with new interpretations of intellectual property rights laws that allowed for patents on GEOs, would reinforce the trend toward corporate concentration in the global food system. Underlying all of these issues were ethical concerns about engineering organisms outside of the boundaries of their own reproductive capacities.

The biotechnology industry's response to such concerns was to argue that they were best addressed through government regulation. As a good example of a hegemonic strategy designed to establish a new consensus, the industry then worked with governments and academics closely aligned with its interests – first in the US and then Canada – to establish a regulatory approach that would give its products an official stamp of approval while being minimally onerous (see Andrée 2005). In Canada, this minimalism was evident in four of the key features of the regulatory system established in 1993 (Canada 1993). First, this system had a narrow focus on "science-based" safety considerations, thereby excluding socio-economic or ethical issues.[2] Second, the safety data upon which decisions were made were classified as "confidential business information" and unavailable for scrutiny whether by outside experts or the public. Third, the regulatory system looked at the *products* of "modern biotechnology" on a case-by-case basis, rather than treating GE as a distinct technology and GEOs as a unique class of organisms requiring technology-specific risk assessment. In the case of seeds, this was achieved by regulating GEOs inside the new category of "plants with novel traits." This category was constructed to include GE seeds but also those with distinctive traits developed through conventional breeding and other technologies such as mutagenesis. Finally, approvals were granted for GE products if they could be deemed "substantially equivalent" to their closest non-engineered counterparts based on a limited set of criteria (for more detail, see Andrée 2002).

As the first GE crops and foods came to market in Canada in 1995, a variety of consumer, farm, and environmental CSOs campaigned to oppose them, arguing that regulatory judgments based on "substantial equivalence" reflected sloppy science while ignoring key socio-economic and ethical issues. These groups also asserted that, at a minimum, Canadian consumers should be given information and choice in the marketplace via product labelling. However, these activists faced an uphill battle. Without labelling, targeted boycotts of GE foods were difficult, and the Canadian government had fully accepted the view that transgenic organisms – as a class – did not pose any unique risks, so was vehemently opposed to mandatory labelling.

In the early 1990s, the European Community (EC) adopted a decidedly different approach to the regulation of GEOs. Instead of employing the euphemistic category of "plants with novel traits," the EC singled out "genetically modified organisms" (GMOs) as a distinct regulatory category. A GMO was defined as "an organism in which the genetic material has been altered in a way that does not occur naturally by mating and/or natural recombination" (EEC 1990). Another difference in the EC approach was the absence of "substantial equivalence" as the basis for assessment of environmental impacts. Instead, the directive established flexible procedures for determining what would constitute "evidence for safety" or "environmental harm" in practice. A final critical difference was the space that the EC approach left for democratic intervention, which existed at two levels. At the national level, risk assessment bodies in many EC member states were composed of interested parties, including environmentalists as well as scientists. Then, at the supranational level, EC-wide decisions on whether to commercialize specific GMOs were subject to a majority vote when one or more countries raised objections to their adoption (Dunlop 2000). This process became particularly important when a widespread public backlash emerged against GM foods in the wake of the UK's bovine spongiform encephalopathy (BSE) health and farming crisis of 1996.

That spring, after some patients showed signs of the human variant of spongiform encephalopathy, the British government was forced to retract ten years of official statements that eating beef from cows afflicted with BSE posed no risk to consumers. The public also learned that the Agricultural Ministry had suppressed evidence of risk and silenced its own scientists (Levidow 1999). The BSE controversy, with its vindication of the consumer voice and scientific caution over misguided regulatory entrenchment, enabled two major victories for critics of GE. In 1997 the European Parliament decided to require labelling of all GE food ingredients. Then, in 1999, a *de facto* moratorium came into being when five European member countries stated that they would not authorize any further GEOs until stronger regulations were in place (European Council and Commission 1999).

Conflict over Roundup Ready Wheat
Roundup Ready wheat, like the RR canola, soy, cotton, and corn that preceded it, is genetically engineered to be resistant to glyphosate. Glyphosate is a broad spectrum herbicide, meaning that it kills almost all green leafy plants within days after contact. In 2001 Roundup was the best-selling herbicide in the world (Monsanto 2001). This situation was brought about, in large part, by the popularity of the first generation of RR crops released in the 1990s. In Canada, Monsanto was selling RR soy and canola as well as insect resistant (Bt) corn. Wheat would have added another major (staple) crop to what remains a limited product line.

For North American farmers and grain exporters, the greatest concern raised by the prospective introduction of RR wheat was the potential loss of export markets. Wheat is Canada's most important crop in terms of area seeded, total production, and export revenue. In Manitoba, for example, it is grown on 43 percent of the total farm acreage, or about two of every five years of crop rotation (Thomas, Leeson, and Van Acker 1999). Canadian farmers grow approximately 27 million tonnes of wheat each year, valued at $3.8 billion (Canada 2005). Approximately 70 percent of this wheat is exported, through the Canadian Wheat Board (CWB), to Asia and Europe (Stokstad 2004; Burroughs 2005). These were the very same markets that had made clear, in the late 1990s, that they were not interested in GE foods. This position was confirmed in a series of customer surveys undertaken by the CWB, beginning in 1998, which found that a large majority of the millers and other food processing companies that bought their product did not want GE wheat (Rosher 2005). Moreover, many of these buyers did not want to buy from regions or countries that had GE wheat in production (even if only a small percentage of production), because of contamination risks.

The fears of major market shutdowns were joined by at least three significant agronomic and environmental concerns raised by the potential introduction of RR wheat. Initially articulated by farm groups and environmentalists, they were later substantiated through public-sector research. The first issue was that, because RR wheat was resistant to glyphosate, farms that had planted it in a growing season would find it difficult to control in the following season should it sprout as an unwanted volunteer (a plant that grows from seeds of a previous crop). The pre-seeding chemical "burn-off" of volunteers and other weeds using glyphosate has become a normal part of reduced-tillage farming. Many prairie farmers grow both wheat and RR canola. In order to continue using reduced-tillage practices, or in order to rotate RR wheat with RR canola or other RR crops, farmers would have to apply other herbicides to kill off the glyphosate-resistant volunteers in subsequent crops (Van Acker and Entz 2001). A University of Saskatchewan cost-benefit study, released in 2002, confirmed that increased management costs related to volunteer control would very probably offset potential gains through yield increases (Furtan, Gray, and Holzman 2002). The question of volunteer control was also an environmental issue because the additional herbicides required would probably be older and more environmentally damaging than glyphosate.

A second agronomic/environmental concern was that RR wheat, or the RR genes, could be expected to spread to neighbouring farms, causing new herbicide tolerant weed problems. In other words, weed management changes, and the associated costs, would even be required of those farmers who chose not to grow RR wheat, because of the likelihood that the RR trait would spread to other wheat varieties either through pollen flow, the mixing of seeds that inevitably occurs in grain handling and storage, or other seed contamination (Grenier 2002; Furtan, Gray, and Holzman 2002). Evidence that RR wheat would probably become a weed in the fields of these farmers came to light through the Canadian experience, in the late 1990s, with varieties of herbicide tolerant GE canola (see Downey and Beckie 2002). Canola plants with tolerance to three different herbicides can now be found growing in fields, along roadsides, and even in towns across the prairies, requiring the use of yet other herbicides to eradicate them. One consequence of the spread of GE canola was the inability to grow certified organic canola due to losses from contamination, and the resulting loss of this market and rotation crop for many organic grain farmers.

The third key agronomic/environmental concern around the potential adoption of RR wheat, first raised by a group of reduced-tillage farmers in Brandon, Manitoba, in the fall of 2001, relates to the first two: The spread of RR wheat volunteers/weeds across the fields of both adopters and non-adopters could threaten the viability of reduced-tillage farming in Canada (Van Acker 2005). Reduced-tillage farming is practised on about half of prairie farms (CFIA 2003). This management practice has environmental benefits including retention of soil moisture and organic matter conservation, reduced soil erosion, and carbon sequestration. It also leads to economic benefits for farmers because of reduced management costs. Reduced tillage could be threatened in two ways by the introduction of RR wheat: First, through an increased presence of weeds (including wheat volunteers with glyphosate resistance); second, through a growing population of other weed species that developed glyphosate resistance under the selective pressure of heavy Roundup use. In either (or both) of these cases, research suggested that the increased costs associated with controlling the glyphosate-resistant weed/volunteer populations with additional herbicides could lead farmers to return to tilling (Grenier 2002; Van Acker, Brulé-Babel, and Friesen 2003).

These specific concerns about RR wheat formed the backbone of the economic and environmental case that critics developed against the product in the early 2000s in Canada. However, it was clear that a wider set of issues was inspiring farmer and consumer animosity toward RR wheat. These concerns were connected to the power of Monsanto itself and to the experience of communities dealing with this company. Strikingly, farmers across Canada (and the United States) were finding themselves in direct conflict with the corporation. For example, the contamination losses in organic canola led organic grain farmers in Saskatchewan, represented by the Saskatchewan Organic Directorate, to pursue a class action lawsuit, a process begun in 2002, against Monsanto and Bayer for compensation of lost sales.[3] The case of *Monsanto v. Percy Schmeiser* was also gaining worldwide attention.

Schmeiser, a Saskatchewan farmer, had been sued by Monsanto for growing patented RR canola without paying royalties to the corporation. Schmeiser admitted that the canola on his farm did contain Roundup Ready genes but asserted that these genes were unwanted contamination caused by seeds falling from passing trucks or from pollen drifting in from neighbouring fields (Anonymous 2004a). His lawyer argued that, because Schmeiser did not actively use the proprietary trait (by not spraying with Roundup), Monsanto's right to enforce its patent protection should not extend to this particular farmer. Schmeiser lost his first case in a Saskatchewan Federal Court in 2001. This decision sent shock waves around the world as farmers realized that contamination by proprietary GE plants could cost them the right to save their own seed for replanting. Schmeiser appealed the lower-court decision, and the case was heard by the Supreme Court of Canada, receiving prominent attention as a David versus Goliath story until its conclusion shortly after Monsanto's 2004 decision on RR wheat.[4] With the potential commercialization of RR wheat, wheat farmers also faced the possibility of losing the right to save their own wheat seed, and this would have significant repercussions. In Manitoba, half of all wheat seed is saved rather than purchased (Van Acker and Entz 2001).

Concerns about Monsanto's influence over government decision making were long-standing, dating back to the protracted fight and ultimate government rejection of Monsanto's bovine growth hormone (Sharratt 2001). These concerns were raised by RR wheat critics when it was discovered that the product had been developed in partnership with Agriculture and Agri-Food Canada (AAFC) and that the department was poised to receive royalties from commercialization. Why did the Canadian government devote more than $500,000 of public resources (Bueckert 2004) to developing a biotech "lemon," as RR wheat was termed in one academic report of 2002 (Furtan, Gray, and Holzman 2002)? And more importantly, how could the regulatory body conduct a full and impartial environmental assessment when, ultimately, the responsible minister had also been responsible for partially funding the project? These social and political issues were at least as important as many of the environmental concerns raised around RR wheat, especially for pro-ducers. One survey of farmers undertaken by researchers from the University of Manitoba in 2003 found that "corporate control of the food supply" ranked second after market issues among respondents concerned about RR wheat (Mauro and McLachlan 2003).

Monsanto's attempts to build acceptance for RR wheat in the face of these concerns reflected what it had learned from the GE battles of the 1990s. In 2000 Monsanto began to work proactively through formal dialogue with the main stakeholders in the North American wheat chain. This dialogue, which took the form of a Wheat Industry Advisory Committee, reaffirmed that the industry had major reservations regarding the release of GE wheat. In response, Monsanto developed a pledge specifically focused on RR wheat. Introduced in January 2002, this pledge stated that the company would not commercialize RR wheat until six conditions were met. These included regulatory approvals in the United States, Canada, and Japan; the identifica-tion of buyers; putting marketing arrangements in place in major export markets; and establishing "appropriate" grain-handling protocols to "provide a meaningful choice for customers" between biotech and conventional grain (Monsanto 2004a). The fact that Monsanto was willing to consider protocols that would allow for choice between GE and non-GE wheat – an idea it had resisted in the context of earlier products because segregation could open the door to rejection in the marketplace – shows just how much pressure the company was under regarding this particular product.

In 2002, when Monsanto submitted applications to Health Canada and the Canadian Food Inspection Agency (CFIA), for food safety and environ-mental assessment respectively, it was not worried about market concerns hampering its application, because the regulatory process did not include a socio-economic assessment of the potential impacts of a new GE crop. In terms of the agronomic concerns around volunteer management raised by some critics, Monsanto believed that these issues did not differ from the problems posed by any other herbicide tolerant (HT) crop, all of which could be treated with herbicide mixes. Furthermore, the CFIA had never turned down an application for an HT crop, including a conventionally bred HT wheat variety developed by competitor Syngenta. On the subject of volunteers potentially becoming an environmental problem for non-adopters and reduced-tillage farmers, Monsanto representatives argued that, because wheat is self-pollinating, the risks of gene flow to related cultivars would be considerably less than with GE canola (Van Acker 2005). In fact, Monsanto asserted that RR wheat would lead to environmental benefits through *increased* adoption of reduced tillage, since the product would allow wheat farmers to seed directly into unplowed fields treated with Roundup (Grenier 2002).

The first voices to be raised against Monsanto's plans for RR wheat came from CSOs that had been campaigning against GE foods and crops since the mid-nineties. Groups such as Greenpeace rejected GE wheat and saw the market concerns of the wheat industry as a strategic opportunity to create more allies in the agricultural sector, a long-term goal in the anti-GE movement that had proved elusive. In 2001 Greenpeace and the NFU organized a diverse temporary alliance of over 210 organizations that included traditionally conservative groups such as the Canadian Federation of Agriculture (CFA) and the CWB. This alliance released a joint letter to Prime Minister Jean Chrétien registering collective opposition to RR wheat (NFU et al. 2001). The letter noted that the signatories had a diverse set of reasons for their opposition – from agronomic, health, environmental, and market concerns to reservations regarding the adequacy of the Canadian regulatory system for GEOs. Significantly, this letter was launched a year before Monsanto even applied for regulatory authorization, providing another indicator of the heightened level of civil society concern regarding this particular product.

The unlikely partnership of mainstream farm associations with environmental groups was made possible through the work of Greenpeace campaigners who understood the necessity of building alliances inside the agricultural community and who recognized the timing of a political opportunity. Some of the industry and farm groups, however, faced resistance from within their own membership to this public partnership with environmentalists (Rosher 2005). As a result, the coalition of 2001 did not last into 2002, splitting to effectively engage in what became a two-pronged attack on Monsanto and the Canadian government. One alliance was spearheaded by the CWB, whereas the second was led by Greenpeace, the NFU, and the Council of Canadians.

While conducting their publicity campaign against RR wheat, the CWB and industry associations with similar interests countered the Monsanto-sponsored Wheat Industry Advisory Committee with an industry group of their own, which developed a more stringent list of conditions that they felt must be met prior to the commercialization of any GE wheat. In February of 2003, the CWB-initiated Grain Industry Working Group on Genetically

Engineered Wheat published its conditions, including market acceptance, fully functional segregation systems, agronomic information, and a cost-benefit analysis. Most importantly, it wanted Monsanto to engage in a consultation process before any decision to commercialize RR wheat was taken; the process would "solicit agreement by farmers, grain companies, customers, technology developers, and marketers that all elements are in place, that costs are appropriately shared, that liability is clearly outlined and acceptable to all parties, and that the benefits of commercial production of GM wheat outweigh the risks and costs" (GIWGGMW 2003, 13). Given that these conditions were not met (and that the segregation condition was not yet even technically possible), the CWB and its partners focused their efforts on getting Monsanto to formally withdraw its application to the CFIA. They also called on the Canadian government to establish a "fourth pillar" for GEO regulation (alongside food, feed, and environmental safety) that would take the form of a comprehensive cost-benefit analysis.

For their part, Greenpeace and their anti-GE allies worked on many levels in the campaign against RR wheat. They bought advertising and published reports, fact sheets, and newspaper editorials presenting their arguments (MacRae, Penfound, and Margulis 2002; NFU 2003; Wells and Penfound 2003). In the winter of 2003, the Council of Canadians and partners organized a series of town-hall meetings in twelve cities across the Canadian prairies where they raised their concerns and heard from farmers. CWB representatives also spoke at these meetings. By summertime, RR wheat had generated so much negative publicity that 61 percent of Canadians polled did not want genetically engineered wheat (Greenpeace 2003). Even more striking, 87 percent of farmers were opposed to growing RR wheat (Mauro and McLachlan 2003). In June, Greenpeace activists "quarantined" an RR wheat field testing facility in Manitoba. The following spring, the Council of Canadians, along with Friends of the Earth Québec, organized a campaign that asked constituents to mail slices of bread to their member of parliament as a form of protest against the possible introduction of GE wheat to Canada.

Canadians were joined by strong consumer response in countries that imported North American wheat. For example, a consortium of Japanese consumer groups, representing over a million consumers, came to Canada of its own accord to meet with politicians to express its opposition to RR wheat. Meanwhile, Greenpeace International continued its strategy of supporting the closure of global markets to GEOs and solicited statements from millers and other companies in Europe, Japan, and elsewhere confirming that they would not buy Canadian wheat if RR wheat was commercialized.

A third prong in the RR wheat campaign emerged over the course of the conflict from within the public research sector. Some of this research has already been referred to above. Other data that proved important in developing the case against RR wheat came from research on canola. R.K. Downey and H. Beckie's (2002) work at the Saskatoon Research Centre of Agriculture and Agri-Food Canada (AAFC) on outcrossing rates from canola, followed by a University of Manitoba study (Friesen, Nelson, and Van Acker 2003), confirmed that certified non-GE canola seed in Western Canada was already contaminated with GE herbicide-resistant traits. These findings dispelled the myth that the purchase of certified seed would allow a farmer to avoid having a crop contaminated with unwanted transgenes. A.L. Brulé-Babel, R.C. Van Acker, and L.F. Friesen (2003, 11) modelled potential outcrossing rates for GE wheat and found that "it will likely not be possible to maintain low levels of [GE] trait contamination in non-[GE] crops." These studies effectively dispelled Monsanto's claim that, because wheat is a self-pollinator, transgene spread would not be a problem, either agronomically (in terms of creating new weeds) or economically (by contaminating crops that could no longer be sold as GE-free).

Evidence that Monsanto's self-pollination argument was losing ground in the scientific community can be found in a March 2004 decision of a sub-committee of the Prairie Registration Recommending Committee for Grain, the group responsible for establishing the protocols Monsanto used for testing its RR wheat lines. In its decision, the subcommittee expanded the buffer zone requirements around RR wheat trials from three to thirty metres because of the growing realization that transgene spread was not as rare as first assumed (Rance 2004). Grain traders such as Cargill (a close business partner of Monsanto) were also very interested in the emerging research and contacted at least one of the scientists involved, Rene Van Acker, at his university office to get the details. Cargill was concerned because the emerging research suggested that, should containers at distant ports be rejected due to unanticipated GE contamination, grain shippers could face the possibility of having to pay for their return (Van Acker 2005).

The Government Response
In the face of the mounting clamour over Monsanto's RR wheat, how did the Canadian government respond? Parliament became engaged with the growing controversy through House of Commons Committee on Agriculture and Agri-Food hearings that began in April 2003. These hearings convened three times, providing a national podium at which the key actors in the debate could articulate their arguments.

The obvious conflict between Monsanto's demands and those of Canadian farmers brought tremendous pressure to bear on the government and on the minister of agriculture, Lyle VanClief, in particular. Prior to the parliamentary hearings, the media repeatedly attempted to get VanClief to speak to the market concerns of western wheat producers and to commit to a cost-benefit analysis of RR wheat (as proposed by the CWB and partners) before it would be approved in Canada. VanClief was extremely reluctant to make such a public promise because it would have fundamentally challenged the science-based regulatory approvals process that the Canadian government had trumpeted since 1992. Such a commitment could also cause problems in the context of Canada's trade relationships. Under the North American Free Trade Agreement and the World Trade Organization agreements, Canadian firms can expect to face only science-based environmental and human health regulatory measures outside of the country. In exchange, Canada promises to uphold this same standard. Preventing a Canadian product from being commercialized in Canada on *economic* grounds – though itself not a transgression of trade rules – would leave the Canadian government in a difficult position when seeking fair (that is, science-based) treatment for its GE products in other countries.

Despite these kinds of concerns, the pressure finally had its intended effect. In March 2003, VanClief admitted that Ottawa was considering mechanisms for ensuring that some kind of market impact assessment would take place before RR wheat was commercialized: "I'm not about to see our wheat market spoiled. The exact process (for reviewing market factors) has not yet been determined but the recognition is clearly there that that concern has to be addressed" (quoted in Jeffs 2003, H1).

VanClief was referring to the work of an interdepartmental committee that had been examining mechanisms for placing limits on some types of genetic innovations in plants and animals in order to protect Canada's commercial interests (AAFC 2005). According to a senior representative from Agriculture and Agri-Food Canada (AAFC), this committee was working with the Canadian biotech industry to develop a voluntary policy framework for "commercialization of novel agricultural products ... [that] makes sure the concerns of consumers and buyers of novel products are met" before the product is released (quoted in Anonymous 2004b). Although the biotechnology industry was actively working on developing such a voluntary framework, there were still no concrete proposals on the table in early 2004. AAFC stated that, if industry failed to come up with a voluntary system, the federal government would be prepared to "lay down the rules" (Anonymous 2004c). However, it is clear that the AAFC committee was uncomfortable with this and even unsure about how to do it in a way that would meet domestic interests while factoring in international considerations: "We were working within the department on what, if push comes to shove ... we can do. And ... it wasn't pretty, what we would have had to do, as a result of the various pressures that we were under. It wasn't going to be very elegant ... so we resisted it as much as possible ... It would have had to be some kind of political solution, which everybody agreed would not have been very good" (AAFC 2005).

Roundup Ready wheat also appeared to be entering new terrain for environmental safety assessments in the early 2000s, despite the fact that all previous HT crops had been approved for release in Canada. Reports of multiple herbicide resistances in some varieties of canola were leading the CFIA to look more carefully at herbicide tolerant traits in general (Phil MacDonald 2005). In 2002 the CFIA had begun a technical consultation process, drawing largely on academic, government, and industry input, in order to determine how best to handle what appeared to be a growing problem of herbicide-resistant weeds (whether volunteers or wild relatives of cultivars). The solution proposed by the CFIA, and accepted by industry only because it did not want to see a stronger regulatory approach, was once again that of voluntarism. Companies seeking approvals of new herbicide-resistant crops would be required to submit detailed herbicide tolerance management (HTM) stewardship plans. These plans would have to demonstrate that any risks posed by new crops could be realistically managed by farmers without compromising the sustainability of conventional agricultural practices.

Aware of this requirement, Monsanto had been working to develop an HTM stewardship plan for future RR wheat farmers, but it was facing a growing number of questions from agronomists, weed scientists, and eventually the CFIA, when they all gathered for a CFIA-sponsored meeting on HT crops in September 2003. With the emerging research on wheat outcrossing, seed purity, and the potential impacts of glyphosate-tolerant wheat volunteers on non-adopters, including farms using reduced-tillage systems, it was not at all clear whether a practical stewardship plan could be developed for RR wheat. As Phil MacDonald (2005) of the CFIA notes, a key sticking point was the question of suitable management options for non-adopters of the technology who would be left having to control glyphosate-tolerant weeds that they had previously controlled with glyphosate. "Would those [options] have been apparent if the [RR wheat] application had been in long enough?" MacDonald (ibid.) asks. "Maybe ... I am not sure how they were going to address the stewardship issues. I am not saying they couldn't have, but I think they had enough concerns of their own that they would not be able to satisfy what we were looking for." The CFIA sent a "deficiency letter" to Monsanto in September 2003, containing a list of technical questions that would require complex field experiments to answer (PBO 2003). Monsanto never responded to this letter (Phil MacDonald 2005).

Why did Monsanto decide to withdraw its application for commercial approval of RR wheat in May 2004? Our research suggests a confluence of factors. The strong, united voice of the farmers and the CWB, the sustained and growing public protest in Canada, and the market shutdowns abroad were all the wrong conditions for Monsanto and, more importantly, were all forcing Ottawa to hint that it was preparing to step into the fray in some fashion in order to protect Canada's commercial interests. The CFIA deficiency letter, along with the signals that Monsanto was receiving from the big grain trader Archer Daniel Midland (Rampton 2004) and Cargill's worries about the potential loss of non-GE wheat due to contamination, probably sealed the product's fate. Monsanto (2004b) claims that "anti-biotech groups" did not influence its decision. This is clearly not the case, given the importance of both consumer and farmer resistance in this story and the role that civil society groups played in nurturing the resistance to GEOs, and then to RR wheat in particular, from the mid-1990s to 2004.

Conclusion

Returning to the criteria of ecological democracy presented at the beginning of this chapter, we can draw several conclusions from this case. First, we can indeed identify some elements of ecological democratization. Engagement by a wide range of actors, including farmers, environmentalists, consumers, and scientists, did occur. These individuals and communities participated in this public debate, both as stakeholders representing specific interests (material or otherwise) and as citizens concerned about developing Canadian agriculture in a way that reflects the common interest of Canada as a whole, ecologically and economically. The result was highly charged deliberation within the public sphere on a wide range of pros and cons of GE products and the genetic engineering project in general. Scientific uncertainty also figured centrally in this debate, in the form of questions that had been asked of Monsanto and never fully answered. And in the end, Monsanto's decision appears to have been to retreat before the considered opinion of the vast majority of participants in this public deliberation became institutionalized through the state. At the level of the public sphere, this outcome does look like a kind of ecological democracy.

One central problem, however, is that, though it is important to recognize these elements of ecological democratization in the public sphere, this case shows us how far Canada is from having an institutional form of ecological democracy in GEO regulation. The outcome in the case of RR wheat was not produced via formal channels that fully and democratically considered the risks, benefits, and uncertainties of the GEO in question. Instead, it evolved in the context of a regulatory crisis: a majority of Canadians, along with many politicians, wanted to see the product held back for a variety of important reasons, but many of these were simply irrelevant to the regulators. One source for this critical disconnect is that the regulatory system established in 1993 addresses only a limited range of science-based concerns.

The notion that environmental and health regulation must be science-based has become a central tenet of neo-liberal governance. However, it is problematic on several levels – and not simply because examples have shown that science-based decisions are influenced by specific economic interests (see Sharratt 2001). Science-based decision making is often structured to recognize particular issues as worthy of consideration while ignoring others, whether intentionally or through the tacit judgment of certain groups of "experts" (Wynne 1994). In the context of GEO regulation, for example, a system focused on "plants with novel traits" downplays the uncertainties and risks associated with the process of genetic engineering itself by focusing on the traits expressed. As a result, environmental regulators study the likelihood that the new trait exhibited by a crop plant would enable it to become an invasive weed, but important gaps could remain in their knowledge of the GE product even after it had been approved as safe.

A science-based process also validates only some expert knowledge as legitimate. Consequently, the risk evaluations of farmers, consumers, and other lay people are often discredited, even though these groups have valuable experience and insights to offer the process, and they will be affected by decisions made. For example, during the RR wheat debate, farmers argued that the growing agricultural use of glyphosate appeared to be related to the incidence of fusarium head blight in wheat (a fungal disease that can eventually have negative health effects on wheat consumers). The CFIA repeatedly rejected NFU calls to sponsor research into this question, and the issue was never a formal part of the regulatory review of RR wheat (S. Wells 2005). Peer-reviewed research published since Monsanto's May 2004 decision supports the claims of these farmers (Fernandez et al. 2005). These examples suggest that science-based regulation can be easily biased and that the bias of Canadian GE regulators does not generally err on the side of caution.

The inability to ensure that science-based decision making is truly disinterested and comprehensive suggests that we need a very different decision-making model for dealing with GEOs. For a number of reasons, we ourselves call for a more ecologically democratic model for GEO decision making. This model would actually be more robustly objective because it would require thorough deliberation on both the scientific and non-scientific issues that are raised by GEO applications. Widespread inclusion in deliberation would make available the knowledge and lived experience of communities and their risk evaluations, and it would help to avoid the unfair displacement of risk onto vulnerable communities (whether human or non-human). A more inclusive deliberative process that entails consideration of the needs and interests of the non-human world would also be more risk-averse, as Eckersley (2004b) notes, and this is clearly desirable, given the irreversible nature of large-scale GEO introductions to the environment. Notably, this conclusion is consonant with the ideas of other authors in this volume who look at the science/citizen interface, such as Darren Bardati's call for "civic science" (Chapter 7) and John Parkins' notion of "extended peer communities" (Chapter 11).

The ideal of ecological democracy as a model for GEO regulatory decision making may seem far-fetched in today's context, but it is important to recall that, in the case of RR wheat, we have already seen a more deliberative process operate in the public sphere. Furthermore, some countries have experimented with elements of this model in their GEO regulatory systems. In the Netherlands, for instance, multiple stakeholders participate in GEO reviews (Bergman 2002). In India, a new GEO must demonstrate its economic viability in addition to its environmental safety (India 1998). However, though an economic cost-benefit analysis may be an important element of a more deliberative approach, we wish to emphasize that this kind of evaluation does not, in itself, satisfy the need for democratic debate regarding the ethical, social, and economic issues raised by individual GEOs. Thus, the proposal made by the CWB during the RR wheat debate – that Ottawa should establish cost-benefit analysis as the fourth pillar of GEO assessment – is only the beginning of the criteria that need to be incorporated.

The Canadian government has made no formal moves to adopt a more deliberative decision-making process (even to include simple cost-benefit analysis). Even a recent attempt to increase transparency – listing GE products submitted for approval and inviting comments regarding them – relies on corporations to volunteer this information. This continued lack of action may be due to the fear that an institutionalized requirement for public participation could jeopardize the entire GE project, something that Ottawa is unwilling to do, given its close ties with the biotech industry and its commitment to the existing science-based process.

In the absence of formal democratic processes in GEO decision making, we have what Laurie Adkin refers to in Chapter 18 as "trench warfare"; each side in these debates works through the public sphere to generate consensus for their point of view. These protracted struggles may show hints of democratic deliberation, but they are ultimately unsatisfactory. In this particular case, the victory of CSOs critical of GE depended on the exploitation of powerful economic arguments to influence government and corporate decision making. For civil society activists, harnessing economic arguments to achieve environmental ends is a precarious strategy at best. Where such arguments against a new technology do not exist, the prospects for environmental outcomes are even less likely. In addition, farmers were at the forefront and proved a considerable political force in both the BGH and the RR wheat conflicts. However, Canadian farm associations (with the possible exception of the National Farmers Union and some organic farm organizations) should not be relied upon to speak on behalf of nature in agricultural disputes in general and GE conflicts in particular.

Despite the RR wheat victory, for the engaged citizen who has been witness to the various conflicts over the commercialization of GEOs since 1995, the dynamic that pits CSOs against governments and industry in Canada has reinforced cynicism toward government in the context of broader anti-globalization critiques. Unfortunately, this cynicism may help to eat away at the power and legitimacy of the state as a political actor at the very moment when a democratically accountable state is most needed to protect citizens and the environment from private interests.

Notes

1 In this chapter, we use the term "genetic engineering" to describe recombinant DNA technology, referred to in European Union regulations as "genetic modification" (EEC 1990). Although these terms can be used interchangeably, the biotechnology industry in Canada has made a concerted attempt to confuse the public regarding their meaning. The industry frames recombinant DNA technology as "modern biotechnology" in a continuum of technologies, such as traditional plant breeding, that "modify," "manipulate," and "alter" nature. This has enabled misleading statements like "we have been modifying crop plants for centuries" in debates over the risks of genetic engineering (Derek Burke 1998, 1845).

2 In environmental, health, agriculture, and other areas of policy, the term "science-based" began to be used in the 1990s in Canada. As we explain in this chapter, this policy criterion has social functions that need to be carefully dissected. For another explanation of why this criterion for policy decisions has been problematic, see Schrecker (2001).

3 In December 2007, the Supreme Court of Canada ruled that, under Saskatchewan law, the farmers' claim could not be certified as a class action and so the case has not been heard.

4 The Supreme Court of Canada decision of 2004 upheld the Monsanto patent on Roundup Ready canola but did not award Monsanto a royalty of fifteen dollars per acre, because Schmeiser "earned no profit" from having the RR canola genes in his field (*Percy Schmeiser and Schmeiser Enterprises Ltd. v. Monsanto Canada Inc. and Monsanto Company*).

3
Regulating Farm Pollution in Quebec: Environmentalists and the Union des producteurs agricoles Contest the Meaning of Sustainable Development

Nathalie Berny, Raymond Hudon, and Maxime Ouellet

In relation to the conflicts surrounding agricultural pollution in Quebec since the 1990s and, in particular, the salient issue of intensive hog production, we examine the strategy and resources of the Union des producteurs agricoles (UPA) and its member organization, the Fédération des producteurs de porcs du Québec (FPPQ), as well as those of other actors engaged in efforts to determine the outcome of successive rounds of regulation.[1] Despite the UPA's privileged access to the state, the victories it achieved in the 1990s have been challenged in recent years, leading us to ask what factors explain the apparently greater concerns about the environment in Quebec government policies and how durable this change will be. Our analysis focuses on the modes of both structuration and representation of interests in order to distinguish the various resources available to the actors involved in these struggles. Growing concerns about the ecological sustainability of the system of food production (both globally and locally) are raising questions about the adequacy of existing participative procedures to fuel citizen mobilization and place new issues on the governmental agenda.

In the 1960s, the concept "environment" found its way beyond scientific circles to refer to interactions between human activities and nature. People became more aware of the increasingly harmful, albeit diffuse, polluting effects of modern agriculture, which had been industrialized following the Second World War (Stonehouse 1997, 106; Weersink et al. 1998, 311). Since that time, the competitive pressures of globalized trade have further intensified the environmental impacts of chemical- and capital-intensive agriculture. Free-market-oriented farm policies have become widespread and have resulted in a concentration of land ownership: in 2004, the Quebec minister of agriculture noted that 70 percent of total revenues in agriculture came from 20 percent of farms (Turcotte 2004a). Moreover, the livelihood of small producers has become less secure with the termination in 1993 of the national tripartite stabilization plans (Braga 1996, 369).[2]

Land-use problems are linked to global dynamics, but the debates about competing visions of agriculture are experienced locally. As illustrated in our study, most local farm pollution problems fell under the jurisdiction of the provincial government; accordingly, the actors who were organized and represented at the provincial level were most effective in influencing agricultural pollution policies and regulations. However, since 2005, the privileged access of commercial agricultural interests to provincial policy makers has been significantly challenged by environmentalists, non-agricultural rural residents, and citizens increasingly concerned with environmental sustainability and human health. The UPA, as an occupational interest group, can no longer claim to be the only stakeholder in matters of environmental regulation of agriculture. The corporatist arrangements that prevailed in the decision-making process have not succeeded in limiting the scope of the issues under debate. Other groups, including environmentalists, have voiced their visions of a sustainable future for the sector, in particular through the Bureau d'audiences publiques sur l'environnement (BAPE). Established in 1978 within the framework of the Loi modifiant la Loi de la qualité de l'environnement, this unique, independent agency is interesting as an experiment in deliberative democracy, albeit one limited to a consultative role.[3] The part played by the BAPE in the policy process will be examined in the sixth and concluding sections of this chapter, where we discuss the shift in political culture toward increased demand for public participation in decision-making.

The Institutional Framework of Agricultural and Environmental Regulation

In Quebec, agricultural production has been geographically concentrated in the fertile St. Lawrence Lowlands, where urban expansion has also been very intense (see Figure 3.1).[4] Exurban development has radically lessened the importance of agriculture in traditionally rural areas and has produced a troubled cohabitation between farmers and the so-called neo-rurals (people who have more or less recently moved from urban settings to the countryside) (Deslauriers 1995, 174). Yet, agriculture and food processing remain very important to the Quebec economy: "All in all, Québec's agri-food sector (including tobacco) represents 432,941 jobs, i.e., 12 percent of total employment, and a 6.5 percent contribution to [the] Québec economy's gross domestic product (GDP) ... Since 1994, farm and agri-food products trade has steadily increased, going from $1.3 to $3.5 billion in 2004" (UPA 2007).

The problematic image of agriculture is closely linked to hog farming, although intensive agriculture in Quebec is not restricted to this sector: single-crop farming of corn, for example, increased by 52 percent between 1991 and 2001 (EQ 2003, 86). However, the geographical concentration of

Figure 3.1 A view of the farms of St-André de Kamouraska bordering the
St. Lawrence River, May 2006. Photo Marjorie Ouellet.

hog production has fuelled many local conflicts. The influence of hog
producers stems from their contribution to total agricultural revenue, which
accounts for 16.6 percent. In 2005 the value of hog exports was $1.06 billion
(FPPQ 2007).

The odour produced by hog farming has been the most salient problem
for the surrounding populations. However, the effects of agriculture on the
environment and human health are much wider. Soil deterioration, freshwater contamination (surface and groundwater), and loss of biodiversity are
caused by the large quantities of organic waste produced by farms and by
the fertilizers and pesticides used in intensive agriculture. Hog manure production increased ninety times from 1951 to 2001, outstripping manure
production from other types of farming and amounting to 30 percent of the
25 million cubic metres of organic matter produced by Quebec's farms (EQ
2003, 6). In particular, fertilizer/manure runoff has entered the rivers of the
St. Lawrence Lowlands; agriculture has become responsible for one-third to
two-thirds of the pollution in these rivers (ibid., 92).

Until the mid-1980s, the protection of agricultural lands and the phenomenon of diffuse pollution were addressed as distinct regulatory concerns.
The key legislative instrument governing the former is the Act respecting

the preservation of agricultural land (LPTA), passed in 1978 (see Table 3.1).[5] The main purpose of the LPTA was to curb the unbridled urbanization of agricultural spaces. In 1996 Bill 23 was passed, amending the LPTA, which became the Law on the protection of the territory and of agricultural activities (LPTAA).[6] The LPTAA has been subject to various amendments reflecting the ongoing conflicts between the interests of farmers and the pressures of urbanization (or the problems of "cohabitation").[7]

With regard to environmental protection, regulations and guidelines were applied, beginning in 1981, to livestock farming and manure storage structures. These regulations neglected, however, to set standards for the spreading of manure or chemical fertilizers. As the limitations of the "end-of-the-pipe approach" to pollution control became better understood, the idea of environmentally sustainable agricultural practices penetrated policy thinking, to be summed up in the term "agro-environmental" policy.

By 1988 Environnement Québec (EQ) was prepared to adopt a more preventative, or "sustainable," approach to the protection of water, air, and soil.[8] However, stakeholder consultations were unable to reach agreement on new regulations. EQ set up a new mechanism for dialogue *(table de concertation)* in 1992 to re-initiate the process of revising the existing (1981) regulation on farm pollution. Meanwhile, Agriculture Quebec (AQ) created task forces – namely, the Brière Committee in 1989 and the Ouimet Committee in 1992 – which were also charged with bringing together the views of stakeholders and experts.

By 1995 the ongoing negotiations over reforms had reached a deadlock: implementation of a "water-air-soil" regulation published by EQ in the August 1994 *Gazette officielle* was postponed to allow further consultations. Meanwhile, the LPTA was being further reworked. The Parti québécois (PQ), winner of the 1994 election, promised to carry out the agreement by which, on 24 May 1995, EQ and AQ committed themselves to treating the creation of the EQ regulation and the reworking of the LPTA as parts of a coherent framework. In June 1996 the government finalized the content of Bill 23 (which sought to amend the LPTA). An EQ regulation on pollution originating from agricultural sources (the RPOA) came into effect one month later (see Table 3.1).

Negotiations surrounding Bill 23 and the RPOA are revealing with regard to how environmental issues have come to shape agricultural policy in Quebec and the changing strategies of the UPA. Our observations are based on documents produced by various actors and on press releases and reviews (Berny 1999). This empirical material has been complemented by interviews we conducted in 1999 and 2000 with individuals from AQ, EQ, the UPA, and the Union québécoise pour la conservation de la nature (UQCN, called Nature Québec-UQCN since 2005).

Table 3.1

Legislative and regulatory interventions concerning farm pollution in Quebec

Abridged designation	Title	Year	Summary
LQE	Loi sur la qualité de l'environnement	1972	The LQE asserted a general ban on polluting activities and the right to a healthy environment. The EQ was authorized to prepare regulations to apply these principles.
LPTA	Loi sur la protection du territoire agricole	1978	The LPTA (Bill 90) was enacted to protect land suitable for agriculture. It zoned agricultural land and defined the conditions of dispensations.
–	Règlement sur la prévention de la pollution des eaux par les établissements de production animale	1981	This regulation mainly targeted livestock farming and manure storage structures.
Bill 23	Loi modifiant la Loi sur la protection du territoire agricole et d'autres dispositions législatives afin de favoriser la protection des activités agricoles	1996	This act aimed to harmonize the two laws that govern rural land use. It added new dispositions to the LPTA and to the law that endorses the regulating power of municipalities (Loi sur l'aménagement et l'urbanisme).
RPOA under LQE	Règlement sur la réduction de la pollution d'origine agricole	1997	The purpose of the RPOA was to protect water and soil from agricultural pollution, livestock farming, and storage structures. In particular, it regulated the practices of spreading manure and chemical fertilizers.
REA under LQE	Règlement sur les exploitations agricoles	2002, 2005	Adopted to replace the RPOA, the REA planned to balance the spreading of fertilizers/manure and the needs of the soil. It privileged a dynamic approach based on results and adapting regulations. It has been modified several times.

**Corporatism in Quebec Agro-environmental Policy:
Revising the LPTA and Creating the RPOA**

The strategies of the UPA have played an important role in the revision of regulations governing agricultural production. This influence reflects a long-term "corporatist" arrangement between AQ and the UPA (Skogstad 1990). Accordingly, the UPA benefited from privileged access to AQ, a status that had been strengthened by the legal right to monopoly granted in 1972. An analysis of bureaucratic politics, developed by Calliope Spanou (1991), offers a useful framework to explain this corporatist mode of policy making.

Starting from a common hypothesis in the literature on interest groups – that administrations seek support from their constituencies in order to gain an advantage in their relations with other administrations in the competition for revenue and authority – Spanou (ibid.) makes a distinction between the *intervention circle* (the public targeted by the regulations of a policy field) of a given administration and its *support circle* (the public whose interests benefit from their implementation). In our case study, EQ regulates various producing activities, whereas AQ provides programs to support farmers, whose income is markedly subject to uncertainty. Hence, AQ and EQ have differing opportunities for mobilizing support. In the case of AQ, the support circle *is* the intervention circle. As a result, AQ and the UPA tend to have similar views about the role of agriculture and its regulation. Comparatively, EQ appears weaker than AQ, as its intervention circle – potential polluters under the 1972 LQE (Loi sur la qualité de l'environnement) – clearly differs from its support circle, which is composed of people and groups highly aware of the environmental impact of agriculture and committed to environmental protection. Their access to decision makers appears much less developed than that of the UPA, and they undoubtedly experience greater difficulty in mobilizing resources. EQ and AQ are, finally, bound to defend diverging or conflicting interests. From the beginning, the conflicts between these opposing views and interests were not resolved through compromises or accommodation. The UPA retained its monopoly on framing agricultural questions, despite the contestation of corporatist arrangements by other sectors of civil society.

Framing Sustainable Development: The Right to Produce
versus the Control of Farm Pollution

The UPA anticipated the implications for its members of the growing public awareness of environmental issues. In the early 1990s, it appropriated the theme of sustainable development, aiming to frame the problems and the solutions under discussion – in particular, it focused on identifying the relevant and legitimate participants in both the decision-making process and the implementation stage.

The UPA used a definition of "sustainable development" that did not necessarily accord priority to environmental protection; indeed, the emphasis was on the sustainability of the economy, as in "sustainable agriculture." Its publications on environmental matters made it clear that sustainable development did not entail "revolutionizing the model of production, but rather combining sensibly short-term profitability, the productive potential of agriculture, the demands of consumers and the population as a whole, and the protection of the environment" (UPA 1993, 18; UPA 1997, 3). Desiring not to threaten profitability, the UPA unceasingly and firmly expressed its preference for making training and information concerning good agricultural practices attractive to farmers, as opposed to taking a "command-and-control" approach. Progressively, it linked the "right to produce" with "sustainable agriculture" to stress the necessity of taking the livelihoods of farmers into consideration.

Moreover, it presented agricultural production as not merely a particular interest (farmers' livelihoods), but as a *public good* (supplying the population with food and generating export revenue). This background is key to understanding a UPA strategy that otherwise appears quite puzzling – alternating between rejecting coercive rules and supporting EQ regulations when doing so seemed necessary to limit the municipal power to create regulations for rural land use.

To begin with, revisions of the law concerning protection of agricultural land (the LPTA), and of the EQ agricultural pollution regulation, were on the agenda as these measures had proven inadequate.[9] AQ, in charge of the LPTA, was actually sharing its jurisdiction with local authorities. The UPA argued that norms concerning dust, noise, and odour pollution should be defined by the provincial government. An opposite stance was promoted by the Union des municipalités régionales de comté du Québec (UMRCQ) and the Union des municipalités du Québec (UMQ), which pushed for legally defined powers to regulate agricultural pollution (UMRCQ 1995; UMQ 1996).[10] Somewhat paradoxically, the UPA asked EQ to get involved with the revision of the LPTA, demanding that environmental law be enforced on the grounds that provincial environmental quality control law took precedence over municipal regulations. By acknowledging the competence of EQ, the UPA tried to frame the rules at the provincial level. At the same time, it competed with EQ to impose the legitimate vision of sustainable agriculture.

In fact, the UPA urged that the the LPTA reform and the creation of the RPOA should be merged into one coherent process, a step that it characterized as protecting the "right to produce." Since the late 1980s, the UPA had been developing a discourse about the threats posed by urbanization to agriculture (UPA 1995). It asked not only for protection of land *for agriculture*, but also for better protection of *agricultural activities*, and it used this latter notion to influence the scope of the ongoing environmental reforms. It succeeded in suspending the adoption, first, of Bill 123, presented by AQ in November 1993 to modify the LPTA, and, second, of the regulation plan published by EQ in August 1994. In the UPA's view, Bill 123 did not guarantee protection of agricultural activities, and the projected EQ regulation imposed too many limitations on farmers.

Collaboration between EQ and AQ, agreed on in the spring of 1995, modified the negotiation process by broadening the range of stakeholders and hence the scope of the issues. Nonetheless, the ability of these newly included partners to influence the outcome was limited by a significant discursive shift in the framing of the debate. On 24 May 1995, following consultations with the UPA, municipal unions, their supervisory ministry, the Ministère des Affaires municipales (MAM), and the Ministère des Ressources naturelles (MRN), AQ and EQ agreed to implement the "right to produce." Henceforth, noise, dust, and odour *pollution* were to be conceived of as *neighbourhood nuisances*. In the future Bill 23, these problems were to be treated as nuisances, to be dealt with at the local level. The "water-air-soil" regulation that EQ had been preparing for years was finally converted into a water-soil approach (as noise, dust, and odour were forms of air pollution). Moreover, EQ was charged with enforcing the right to produce. The agreement detailed the principle of "immunity" – that is, if farmers obeyed EQ regulations, municipalities could not restrict their activities. In the end, the May 1995 agreement reinforced the corporatist management of environmental issues.

Notably, the UPA metamorphosed the "right to produce" into the concept of sustainable development or, more appropriately, "sustainable agriculture." This became obvious in 1996 when it challenged EQ's objectives in the name of the viability of agricultural activity. Significantly, after the agreement of 24 May 1995, AQ's draft bill was entitled "The Protection *and the Sustainable Development* of Agricultural Activities" (emphasis added). The municipal unions and other opponents of the UPA, on the other hand, continued to distinguish "the right to produce" from "sustainable agriculture" (UMRCQ 1995, 4). They feared that the first article of AQ's bill, which mentioned the importance of "favouring sustainable development," would give the Loi sur la protection du territoire agricole (LPTA) legal priority over municipal regulations.

Ultimately, the title of AQ's Bill 23 did not mention sustainable development: instead, it referred to "the protection of agricultural land [and] of agricultural activities." In reality, conflicting positions were not reconciled with the 24 May 1995 agreement. On the contrary, the agreement resuscitated the rivalry between AQ and EQ. The prioritization of the right to produce demonstrated the relative strength of actors in the conflict and the superior weight of AQ within the Cabinet as well.

The Predominance of Interest Group Politics: October 1996 to July 1997
During this period, the UPA attempted to impose its conception of sustainable development and to contain pollution-reduction measures. It played its game by combining three key components of political decision making: timing, institutionalized access to decision makers, and expert advice. Each component also points to EQ's structural weaknesses and to the marginal role played by environmental groups in the phase when the issues were being defined.

First, following the adoption of Bill 23 in June 1996, timing greatly contributed to the UPA's winning strategies. The contents and effects of this legislation remained somewhat indefinite since the regulating power of municipalities over agricultural activities still depended on "guidelines" to be adopted by the provincial government. AQ and EQ differed regarding what these guidelines should entail. For example, on the issue of distances between farms and residential areas, AQ and EQ presented two sets of norms, that of the former being more lenient. The implementation of the EQ's regulation on pollution from agricultural sources (RPOA) was due in October: at this date, the parliamentary committee on agriculture, food, and fishing had to decide between the two regulatory blueprints. The UPA launched an offensive against EQ and its minister, promoting the standards elaborated by AQ. To cope with the situation, an interdepartmental committee was created in October 1996 to reconcile the respective views of AQ and EQ; it included members from both AQ and EQ as well as officials from the MAM and the Ministère de la Santé et des Services sociaux (MSSS). The UPA was the only interest group invited to present its views to the committee. This unique relationship was reinforced by the intervention of Premier Lucien Bouchard, who invited the environment minister to find a solution agreeable to the UPA.[11]

Second, unlike the UPA, which had institutionalized access to decision makers, environmental groups were practically ignored by AQ in the discussions on Bill 23. Also, the UPA had been consistently involved in negotiations on the new EQ regulation. The Ouimet Committee worked exclusively with municipal unions and the UPA. Moreover, environmental groups were excluded from the hearings of the parliamentary committee on Bill 23. In the spring of 1995, the UQCN and À Court d'Eau (Water Shortage) – a local association committed to the protection of the Assomption River – were formally allowed to participate in the consultation undertaken by EQ to discuss the RPOA. However, other environmental groups organized a debate outside the political and administration arenas. Among them were many associations that emerged in the 1990s from local-level mobilizations against the expansion of hog farming. For example, in February 1996, the Mouvement vert Mauricie asked for a moratorium on the burgeoning hog-farming industry to stop several local projects. The Union des citoyens du monde rural (UCMR), which united two local groups, was created in June 1996 and also demanded a moratorium on behalf of the rural population (Marc Saint-Pierre 1996).[12] These groups also sought to delay the adoption of Bill 23, which they considered unacceptable.

These organizations fuelled a wave of contestation that grew as it incorporated environmental demands and concerns about human health. The conflicts between farmers and neighbouring residents were widely reported in the media.[13] Alliances were formed, linking both organizations and issues such as pollution and human health. Seven groups published a manifesto pushing for a provincial moratorium on hog farming and for public hearings regarding sustainable agriculture, thereby linking these two demands (UQCN 1996b). Other groups lodged a complaint pursuant to NAFTA rules against the Quebec government for its "chronic failure to enforce its laws, regulations and policies when it comes to agricultural matters" (Francoeur 1997).

Organizing a debate about hog farming lay within EQ's competence; in accordance with the requirements for environmental impact assessment, the EQ minister may request the BAPE to call for public input on projects submitted by developers. By holding public hearings, he could possibly have countered the UPA's strong role. However, the minister was incapable of acceding to the demands of environmentalist groups, who proved to be very critical of him. Instead, the interdepartmental committee set up in October 1996 was entrusted with the organization of the debate. In retrospect, EQ officials and the minister appeared to be relegated to the fringes of the political process each time the UPA achieved direct access to the office of Premier Bouchard – each time the negotiations between EQ and AQ appeared to be in a deadlock. In these conditions, the UPA had the opportunity to negotiate directly, in September 1996 and probably thereafter, with EQ on disputed points of the RPOA (Gagné 1996b; Milette 1997). The UMRCQ (1996) denounced the bilateral contacts between the UPA and Cabinet during the discussions of the interdepartmental committee.

As a third political resource, the UPA eventually used expert opinions to challenge EQ authority. The EQ's approach to the question of distances separating farms and residential areas was stigmatized by the UPA president, who termed it "environmental terrorism" (Pellerin 1996). The UPA charged EQ with being deaf to the reality of agriculture in Quebec. Pushing for (economically) sustainable agriculture, it also disputed other aspects of the regulations such as the phosphorus level in fertilizers and the time period during which manure spreading is permitted. In reality, the UPA strengthened the idea that only the producers had legitimate knowledge about agricultural matters, and it used the script of sustainable agriculture to convince the public that its cause was just and, thereby, to preserve electoral support for the PQ, as a party that had demonstrated a pro-producer bias.

The final result was stamped with the UPA's view, even with regard to the larger problem of agricultural pollution. For example, in July 1997, the government determined the guidelines defining the rules to be applied by municipalities to distances between farms and residential areas, which were in accordance with the norms proposed by AQ. Noise regulation was maintained in its existing form. The EQ's RPOA left many contentious points unresolved. In a search for a decent trade-off, EQ submitted them to expert committees prior to resuming public consultation.[14]

At the end of the 1990s, the outcome of the farm pollution debate and the influence of EQ and AQ appeared to be largely determined by factors related to corporatist practices and interest group politics. However, later developments call into question the durability of the UPA's predominance beyond the political and administrative fields. The environmental-agricultural conflicts were not so much resolved by the regulations arrived at in the 1990s as postponed and displaced to other fields of action.

Since then, environmentalists and neo-rurals have enjoyed some success in getting their concerns on the table and in forcing more open decision-making processes. The BAPE has become a significant resource for actors and issues excluded from the decision-making process. Over time, a greater proportion of the projects submitted to the BAPE by the minister of the environment have reached the stage of public hearings, increasing from only 20 percent in the 1980s to 40 percent in the 1990s and 75 percent in 2004-05 (Baril 2007, 41). The BAPE must make room for a variety of views and representations. This achieves an essential requirement of deliberative democracy: exchanges of arguments allow the expression of conflicts as well as the emergence of compromise. Although the audiences led by the BAPE are mainly consultative in nature, they also allow the environmental and neo-rural mobilizations to enlarge the scope of the issues under debate within the bureaucratic sphere. Research has demonstrated elsewhere – in France, for example – that opponents of spatial management projects usually associate their claims to the public interest, as opposed to particular interests, in order to gain credibility (Lolive 1997). This strategy may help to compensate for the structural weaknesses of environmental interests identified by Spanou (1991).

Nonetheless, the question remains: how radical is this change? Gilles Côté (2004) draws a qualified balance sheet in his study of citizen participation in the environmental and social impacts assessment in the case of Alcan. Citizen participation must not be taken for granted in Quebec, and setbacks are possible.

Corporatist Intermediation of Interests versus a New Political Culture? Assessing Current Trends

Though its final version was less rigorous than initially planned, the 1997 RPOA made it apparent that environmental issues now ranked higher in public policy priorities and that regulations that "intrude directly into farmers' practices" had become inevitable (Montpetit and Coleman 1999, 698). Recent legislative developments suggest shifts toward a more comprehensive agro-environmental policy.

The aftermath of the adoption of Bill 23 indicated that the UPA had succeeded only in delaying the public debate and regulatory change. The Règlement sur les exploitations agricoles (REA), which revised the RPOA in June 2002, required farmers to comply with new rules. As a direct consequence of the BAPE's public hearings on the management of freshwater resources, it included a greater reduction of phosphorus use. Moreover, a 2001 moratorium practically prohibited further hog-farming development in Quebec for two years in legally defined zones. Predictably, the president of hog farmers' association, the FPPQ, found the new REA "out of proportion" with the problem, and the UPA president complained of public and governmental ignorance regarding agriculture (Dufresne 2002). In contrast, environmentalists approved of the government's stand (*Le Quotidien* 2002). In December 2003, the moratorium was renewed for an additional year and partially lifted one year later. Again, the UPA leadership stressed the efforts made by farmers to address environmental and nuisance concerns (Turcotte 2004b) but could no longer control the flow of events. Following the 2002-03 BAPE hearings on hog production, the government tabled a plan to ensure its "sustainable development," to be enforced by frequent inspections of production units (EQ 2005, 47). In return, the moratorium was lifted on 15 December 2005.

If the hog sector's expansion was under the spotlight, agriculture as a whole had finally been integrated into a coherent policy that resolved previous political controversies. First, a new Règlement sur les exploitations agricoles (REA), which focused on the intensive cultivation of corn, was published in December 2004 to assess the impact of fertilization practices on river basins. EQ concentrated its efforts on enforcement. Finally, Bill 54 (Loi modifiant diverses dispositions législatives concernant le domaine municipal), assented to in November 2004, significantly altered the balance of forces in the determination of agricultural practices. It created a system of public consultations regarding the authorization of hog-farm activities, and it empowered municipalities to regulate this matter, under some conditions (UMQ 2005, 16).

To remind the public that farmers were working to fill their most basic need, the UPA (1999) published a special forty-page section, "Nourrir le monde," in Quebec newspapers. However, the concept that agriculture always served the interests of all in society had become less self-evident. With the "proliferation of factory-style animal agriculture," pollution concerns had intersected with human health risks (Horrigan, Lawrence, and Walker 2002, 445). The public had learned more about agricultural practices and knew which ones were most environmentally sustainable. A new political culture had emerged, which focused on the participation of citizens and more transparency in public decision making. Given this changing context, the UPA has been less successful at what Danny Trom (1999, 44) calls "the politicization of the singular," that is, making a particular interest synonymous with the public good. A poll commissioned by the Fédération des producteurs de porcs du Québec (FPPQ) showed a huge gap between the perceptions of the public and the efforts undertaken by producers (FPPQ 2001).

Since 1999 the evolution of the context has forced the UPA and its constituent parts to adopt a more collaborative attitude and to demonstrate their commitment to environmental matters. The UPA promoted its efforts on the environmental front in two recent press releases: "L'Union des producteurs agricoles partenaires au Salon national de l'environnement 2006" (31 May 2006) and "Succès de l'activité-bénéfice de la Fondation de la faune du Québec sous le thème: pour la biodiversité des cours d'eau" (1 June 2006). In the context of the newly announced Quebec policy on sustainable development, the FPPQ (2005) presented a paper to the minister of environment that asserted a commitment to sustainable hog production. In 2006 the FPPQ reported its partnership in twenty-seven research projects related to environmental protection (FPPQ 2007). Though the UPA's use of purely corporatist tactics has declined, its director general still calls for a "coherent" approach to sustainable agriculture in which environmental goals are balanced by measures aimed at sustaining the economic viability of farms (Larose 2007).

The Ongoing Sway of the Economic Argument

The crisis recently undergone by the hog sector affords us some insight into what has really changed in terms of how agro-environmental policy is made and what interests it reflects. Environmental and rural non-farming lobbying played a large role in bringing about stricter environmental regulation of hog production in Quebec. However, the intensification of international competition for shares of the global pork market, combined with the relatively strong Canadian dollar, has recently generated a crisis of profitability within the pork industry. The meat producers characterize "environmental constraints" as "aggravating factors" in their declining profit margins (Duhamel 2007). They complain that they cannot compete with the economies of scale obtained in the United States or Brazil, where production facilities are even larger (Barcelo 2007, 41).

This crisis was illustrated by the case of Olymel, the second-largest producer of fresh meat in Canada and the eighth largest in the world. It closed a number of Quebec plants in 2005 and 2006, and a huge debate raged in early 2007 concerning the fate of its Vallée-Jonction plant, which received 20 percent of the hogs in Quebec in 2006. The government intervened in the conflict between the company and the eleven hundred unionized plant workers. The organization of an industry-wide consultation called for by the Confédération des syndicats nationaux (CSN 2007) was refused by Premier Charest. Instead, Olymel hired former premier Lucien Bouchard to act as its mediator with the government and the union. Government and corporate shareholders framed the alternatives as either adaptation (a "flexible model of production") to "the imperatives of globalization" or closure (Turcotte 2007; Annie Saint-Pierre 2007). The Olymel workers accepted several concessions related to their wages, their pension, and insurance plans (Laprade 2007). The fact that alternatives to forcing down the wages and benefits of workers were not seriously discussed (for example, the proposals from Maurice Doyon, quoted in Barcelo 2007, 44), while environmental standards were also blamed for making Quebec pork products less competitive in international markets, indicate the challenges facing farmers, food industry workers, and environmentalists with regard to the restructuring of the agrifood sector.

Parallel to the crisis in the hog-production sector, the comprehensiveness of environmental regulation of agriculture was diminished several times.[15] By the end of May 2006, the minister of municipal affairs had decided not to create protected strips of land along three rivers in the Nicolet-Yamaska municipalité régionale de comté (MRC), thereby overturning a ruling adopted in March by the municipal authorities. Apparently, the minister had made that decision under combined UPA and AQ pressure, and in the face of opposition from EQ (Francoeur 2006a). About the same time, Mario Dumais, an economist at the Université du Québec à Rimouski, published an opinion piece in *Le Devoir* (Dumais 2006) in which he concluded that intensive agriculture was necessary to meet the explosion of needs in food.[16] Moreover, one of his university colleagues reported from a survey conducted in 393 Quebec municipalities that the protection of agricultural land had detrimental effects for the vast majority of regions (Francoeur 2006b).

These last developments may be interpreted as a counter-offensive by private agricultural interests and their "captured" ministry. Business associations in Quebec, too, were developing strategies to counter the opposition to proposed "development" projects in various sectors. A study was commissioned by the Fédération des chambres de commerce du Québec (FCCQ) to evaluate the obstacles to "wealth and job creating projects." Selecting two projects defeated by public opposition, the author, a business professor, surveyed media coverage of the projects and concluded that investors and developers should implement more effective information and media campaigns (Rabeau 2006). Offering a litmus test to identify legitimate participants in public consultation processes, he advised developers to question the representativeness of so-called citizens' movements and to demand "more transparency" regarding the governance of groups that received public funding (ibid., 83). Moreover, he suggested that developers distinguish groups that were "really" affected by the projects from those who "systematically" and "ideologically" opposed development; the former might eventually be made "partners." Finally, he advised business associations to ask the government to "review the role and the decision-making criteria" of agencies in their assessment of projects; the BAPE, for instance, should be modernized in light of the "new economic realities" (ibid., 82, 84).

Louis Bernard (2007, 5), formerly a top Quebec civil servant and a candidate for the PQ leadership, has expressed similar views. He specifically urged the government to take a clear stance with regard to projects that it supports; the BAPE would then be requested to examine *how* the project would be realized, not *whether* it would be realized.[17] Beyond the questioning of the BAPE's functions, nothing less than its termination has been advocated by André Caillé, former president of Hydro-Québec (Boisvert 2006).

The government's decision not to protect the Nicolet-Yamaska river areas (mentioned above) may confirm the belief that public office holders are prone "to alter final rules to suit the expressed desires of business commentators, but do not appear [ready] to alter rules to match the expressed preferences of other kinds of interests" (Webb Yackee and Webb Yackee 2006, 135). It may also fuel the thesis that "institutions" like the UPA are more effective in lobbying than are voluntary groups advocating for public interests – in particular when these institutions benefit from the indirect support of other producers (Lowery, Gray, and Fellowes 2005).

The founding of the Union paysanne (UP 2001) appeared for a while to challenge the monopoly position of the UPA. The UP disputed the corporate model of interest representation, as it allowed individual citizens to become members, in accordance with the assumption that sustainable agriculture "hinges on appropriate relationships 'between people *and people* and nature'" (Gertler 1999, 121, quoting Patricia Allen, emphasis in original). However, this vocal organization did not, in the end, jeopardize the UPA's special relationship with AQ. Internal dissensions resulting from the inclusion of citizens as full members and the diversity of interests embodied by the UP made it difficult to reach decisions. Though interests within the UPA do not correspond to a monolithic whole, the material and occupational interests of "captive" members are more easily shared, particularly when threatened by external forces.

Notwithstanding the influence of the UPA, public opposition to intensive livestock operations (ILOs) remains strong, and much depends on how serious the environmental and health problems will become, on how effectively the opposition to ILOs will be articulated to a sustainable agriculture discourse, and on the ability of various actors to unite around this alternative vision. Much is at stake in the current juncture. In 2006 the Charest government appointed a commission on the future of agriculture and the agri-food sector (Pronovost Commission). In its report, released on 12 February 2008, the commission (Pronovost Commission 2008, 243) "diagnose[d] a sector of the economy in serious disarray and unsure of its own way" and, significantly, recommended the opening up of the system of laws, regulations, structures, and modes of operation "to dialog within the agriculture and agrifood sector and with civil society."

Conclusion: Environment and Democracy

Our account of the struggles to shape agricultural regulation in Quebec has shown the importance of interest-group politics in public decision making. However, in the aftermath of Bill 23, citizens' interests imposed new conditions on the conduct of interest-group politics and on the corporatist arrangement between the UPA and AQ, requiring them to accommodate broader social concerns about the environment and health, and to respond to proposals for a more environmentally sustainable model of agricultural production. Environmental organizations and neo-rural citizens' groups have occasionally proved to be highly effective in focusing public attention on the environmental problems associated with industrial agriculture and in defining the stakes of the conflict in ways that subvert the "right to produce" discourse. The UPA has been obliged to develop its own discourse of "sustainable development," to counter the more radical versions offered by the Green Party and others.

EQ has recently been renamed the Ministère du Développement durable, de l'Environnement et des Parcs (MDDEP) and was given additional means to implement its expanded mandate. Since the Sustainable Development Act entered into force in April 2006, it is "assisted" by the sustainable development commissioner (the founding president of UQCN), who has the mandate to prepare "comments concerning the principles, procedures and other methods used in the area of sustainable development by the Administration [and to make] recommendations respecting the carrying out of the Act" (Auditor General of Quebec 2007, 1). The Sustainable Development Act also mandates the tracking of progress in the achievement of standards and targets. This monitoring of results – one of the conditions of ecological democracy identified by Christoff (1996) – could fuel public debate as the ministry also makes available to the public data about agricultural practices.

More transparent reporting can be expected to diminish the gap between experts and citizens, and thereby undermine one advantage of the producers' organizations – their claim to hold a monopoly over "expert knowledge" of farming practices and needs.

Information and knowledge are crucial in the mobilization of citizens. However, more is required to empower citizens to decide. As exemplified by our case study, the institutional settings remain of utmost importance in the discussion and negotiation of collective norms. In this regard, the BAPE as a model of deliberative democracy is still questionable. Does the BAPE equalize the expression of conflicting interests in order to encourage mutual persuasion and consensual collective solutions? The BAPE's recommendations may be contested, even by forces opposing development projects, as observed in the Rabaska case.[18] The president of the BAPE himself acknowledges that the process must go beyond consultations to include participation in decision making, with elected officials arbitrating between interests (Cosgrove 2005). However, because of the increasing exposure of Quebec producers to market competition, there are growing pressures to limit the BAPE's functions. Environmental and producer/owner organizations seek to shape policy outcomes on a very uneven playing field. State agencies, mainstream political parties, and producer/owner organizations are wedded to the same model of economic growth; the state is not neutral. The status quo is further reinforced by the web of international trade and financial relationships in which the province is implicated. In this context, democratization of the decision making at national and international levels also matters. At the provincial level, the creation of environmentally sustainable food production will greatly depend upon the outcome of struggles to extend deliberative democracy through vehicles such as the BAPE.

Notes

1 The UPA is the provincial association that, since 1972, has been legally recognized as the representative of Quebec agricultural producers organized within territorially or sectorally based federations.

2 The national tripartite price stabilization plans were established by federal-provincial agreements for a number of commodities. In the case of live cattle, for example, the program guaranteed a certain percentage of a long-term average of quarterly feeding margins. Producers, the federal government, and the provincial government each provided a third of the costs of the voluntary program (Braga 1996, 374).

3 The BAPE's mandates are described on its website: http://www.bape.gouv.qc.ca/sections/english/.

4 By 1991 only one per every four farms that had existed in 1951 still survived, and 49.5 percent of the lands formerly allocated to agriculture served new purposes (Rouffignat 1995).

5 The LPTA was supplemented in 1978-79 by the Act respecting land use planning and development (LAU).

6 The text of Bill 23 (which came into effect in 1997) is available at http://www.mrclotbiniere.org/pdf/11g_pl23_art78_79.pdf. The full text of the Act respecting the Preservation of agricultural land and agricultural activities (LPTAA), R.S.Q., c. P-41.1, with all amendments

consolidated and amended up to August 2008, is available at http://www.canlii.org/qc/laws/sta/p-41.1/20080818/whole.html.

7   For example, in December 2000 the minister of agriculture, fisheries, and food tabled Bill 184 to amend Law 23 respecting the protection of agricultural land and activities. In particular, Bill 184 sought to prevent municipalités régionales de comté (MRCs) – Quebec's municipalities are grouped into eighty-eight counties, or regional county municipalities – from adopting bylaws that contradict the general aims of the preservation of agricultural land and the protection of farming activities.

8   To avoid confusion, we refer throughout this chapter to Agriculture Quebec (AQ), Environnement Québec (EQ), and Ministère des Affaires municipales (MAM), although the titles of these ministries changed during the period examined.

9   For instance, many storage structures did not comply with EQ regulations. See Louis-Gilles Francoeur (1995).

10 The UMRCQ is now known as the Fédération québécoise des municipalités (FQM).

11 This was mentioned by the press, including the UPA's weekly paper. See Jean-Charles Gagné (1996a).

12 In early October 1996, the UMRCQ also asked for public hearings on hog-farming expansion.

13 Conflicts also occurred between small farmers and large hog operations. See Konrad Yakabuski (2002).

14 In fact, following a new analysis in June 1998 by the consultative committee, the RPOA was twice modified: first, in September 1998, to determine the authorized dates for the spreading of liquid and solid manures, and, second, in April 1999, to more precisely define the so-called phosphorus loads in manures. In the latter case, the ruling was less restrictive than was originally desired by EQ.

15 For instance, the relative absence of controversy concerning the implementation of a relaxed version of the REA in the last months of 2005 could be explained by the contacts between UPA people and many members of the Quebec National Assembly. See Francoeur (2005).

16 Mario Dumais had been an economist at the Coopérative fédérée (an agricultural cooperative that, in combination with its affiliated cooperatives, has the fourth-largest sales in Quebec) and a commissioner of the Pronovost Commission (which was created in May 2006 by the Quebec government to examine the future of agriculture in the province).

17 The BAPE's recommendations are made to the minister, who makes the final decision. Bernard's comments suggest that the minister has felt obliged to accept the outcome of BAPE inquiries.

18 The Rabaska case refers to a project to build a liquefied natural gas (LNG) terminal on the south shore of the St. Lawrence River in front of Quebec City – a project approved by the BAPE. A summary account of that case may be found in Raymond Hudon, Christian Poirier, and Stéphanie Yates (2008).

4
Modern Enclosure: Salmon Aquaculture and First Nations Resistance in British Columbia
Donna Harrison

Assertion by coastal First Nations of their constitutional right to consultation in regard to industrial activities in their traditional territories has lent support to environmental organizations opposing Atlantic salmon aquaculture in British Columbia. In response, the aquaculture industry has also sought partnerships with coastal First Nations, signing at least six agreements from 2001 to 2004. This chapter examines the case of the Ahousaht First Nation, which initially opposed salmon farms but later partnered with a subsidiary of the Norwegian company Mainstream. The Ahousaht's decisions are interpreted here in relation to the wider oppressive relations of enclosure fostered by the state, which contain Ahousaht activism by funding aquaculture expansion and by eliminating alternative economic options – namely, the commercial salmon fishery. It is argued that, in utilizing the contractual agreement as a means to influence industry, Ahousaht resistance has been altered but not abandoned. I also ask what First Nations' struggles for cultural autonomy and survival mean for movement toward ecological democratization in a colonial settler state such as Canada.

Impacts of Commercial Salmon Fishery Restructuring
"You've heard of being between a rock and a hard place?" I was asked when I visited Ahousaht in 2005. "Well, we're on a rock, in a hard place." This is the story of that place. Ahousaht, with over eighteen hundred members, is the largest of the fourteen Nuu-chah-nulth tribal nations located on the west coast of Vancouver Island. The Ahousaht reserve, where approximately half of the population lives, lies northwest of Tofino on Flores Island, about a thirty-minute ride by water taxi. The nation has never signed a treaty with the Canadian government.

Like many coastal First Nations peoples, the Ahousaht have been highly dependent on the aquatic environment for income (see Figure 4.1). Historically, coastal First Nations' knowledge and skills were crucial to the development of the commercial salmon fishery; in the highly racialized labour force

Figure 4.1 Fisher in Ahousaht, 2005. Photo Donna Harrison.

that developed, they became a means through which cannery capital kept down the costs of labour (Muszynski 1996). As the industry modernized, the emerging petty commodity form of production, where small-boat owners work for themselves or for a percentage of the catch and not for a wage, accommodated First Nations rooted on coastal reserves and loosely corresponded to traditional forms of goods production. The commercial fishery, which includes shellfish as well as finfish (such as herring, salmon, tuna, cod, halibut, and hake) remains to this day the largest single employer of First Nations people in the province.[1]

That economic base was eroded by a radical federal restructuring initiative launched in 1996, which rationalized the salmon industry, the largest and most lucrative of all the West Coast commercial fisheries. In response to declining numbers of salmon returning to spawn, and to declining world market prices, government policy aimed to protect the resource and to make the fishery more "cost-effective." From 1996 to 2000, the state purchased 54 percent of all commercial salmon licences through three rounds of voluntary "buy back."[2] It also tightened seasonal and area capture restrictions, initiated a series of gear and catch monitoring modifications, and divided the coast into smaller fishing areas that each required a salmon licence. For the first time, more than one salmon area licence could be "stacked" on a boat, so that, with the purchase of all of the relevant area licences, a given vessel could fish the entire coast.[3] The cost of maintaining access to the salmon fishery skyrocketed. At the same time, reduced total allowable catches restricted fishing opportunity, especially in the latter half of the nineties. Only those who could afford additional area licences or who had access to other means of reliable income could wait out the sweeping closures that followed on the heels of fleet rationalization. The removal of alternative fishing privileges (for example, halibut, sablefish, groundfish trawl and sea cucumber and urchin fisheries) from the salmon A licence to additional licences (which began in 1974 but escalated rapidly after 1989), worked to prevent cash-strapped fishers from targeting these alternative species to make up for lost commercial salmon-fishing opportunities. Most of these fisheries are now transformed into individual quotas, which are traded like other commodities in an open market and sell to the highest bidder. Federal fishery restructuring therefore had the effect of enhancing the presence of large capital vis-à-vis small producers in the industry.

Economically vulnerable participants faced elimination. A sort of cannibalism ensued as fishers began to absorb more licences in a frenzy to remain economically viable. At the same time, the market value of commercial salmon licences rose. In 2002, the average gillnet licence doubled its 1994 value, soaring to $110,000; the average troll licence increased 123 percent over 1994 to $120,000 (Ecotrust Canada and Ecotrust 2004, 11-12). Between 1996 and 2005, state and private licence sales reduced active salmon vessels by 60 percent (from thirty-six hundred to fifteen hundred) (Canada. Department of Fisheries and Oceans 2005).

First Nations participation in the commercial fishery fell from 31 percent of the overall commercial fleet in 1996 to 21 percent in 2003 (Gordon S. Gislason and Associates 1998, 2-6; James 2003, 6). When special Native-only licences are excluded from this total, First Nations representation plummets to only 3 percent of all commercial licences (Ecotrust Canada and Ecotrust 2004, 20).[4] Historically high levels of participation in alternative fisheries (in 1951-52 First Nations individuals owned, on average, 23 percent of alternative fishing licences) were also eroded (Allen Wood Consulting 2001, 23). By 2004 total First Nations participation in alternative fisheries amounted to only 10 percent, including all communal and reduced-fee Native-only licences mandated by the constitutional right of First Nations to fish for food (Ecotrust Canada and Ecotrust 2004, 20).

The number of licences registered in Ahousaht fell by more than half, from twenty-eight in 1994 to just thirteen by 2002; not a single seine licence remained. The small community of Kyuquot, a neighbouring Nuu-chah-nulth village, lost twenty of its twenty-five licences during the same period (ibid., 19). Social conditions worked against First Nations participants retaining their licences. Because land on reserves is not owned outright (Aboriginal title involves a *sui generis,* or unique, interest in land, as opposed to a fee-simple interest), it cannot be used as collateral for loans; Ahousaht fishers could not mortgage their homes to borrow funds for another licence. Further, their federal tax-free status afforded them capital gains tax exemption should they sell a licence, so their licences in fact yielded higher value when re- deemed (non-Aboriginal fishers paid up to 40 percent on half the profit made on the licence in capital gains tax). The greater comparative impover- ishment of First Nations fishers added significant incentive to sell up.

Fisheries restructuring contributed to profound economic hardship for coastal communities such as Ahousaht. Fishing opportunity was nearly eradicated for the next three and a half years for the West Coast troll fleet in which Ahousaht commercial fishers were participants. In 1998 some West Coast trollers made only $3,000 (*Vancouver Province* 1998). Employment in Ahousaht was halved – down by 46 percent – in the first year after the re- structuring began (*Vancouver Sun* 1998). Prior to 1996, the commercial salmon industry had provided more than seventeen thousand jobs, on average (Gordon S. Gislason and Associates 1998). In 2001 there were only fifty-four hundred commercial fishers (all species) and thirty-nine hundred plant workers in the processing of captured and farmed fish (British Columbia. BC Stats. Ministry of Management Services 2002, 24). Population losses in coastal communities after 1996 were especially severe in the North Coast region (above Vancouver Island) and on the north and west coasts of Van- couver Island (including Ahousaht), where production jobs declined nearly 20 percent (Coastal Community Network 2002).

First Nations, of course, have litigated to maintain access to the fishery, arguing that such access is based on Aboriginal rights existing before colonial settlement. But litigation remains expensive and time consuming, with outcomes determined by a justice system that holds a "narrower vision of self-governance" than do First Nations (Sullivan 2006, 46). When negotia- tions between eleven of the Nuu-chah-nulth Nations and the federal and provincial governments regarding the nature and extent of Aboriginal fish- ing rights broke down in 2003, the First Nations turned to the courts. After four weeks of testimony beginning in April 2006, the trial was adjourned until July 2007 to allow lawyers representing the Canadian government time to prepare their case. In July, a ruling that only those First Nations without overlapping title claims could proceed to the next phase sent many of the member nations, including the Ahousaht, scrambling to resolve internal territorial disputes by the imposed November 2007 deadline. The trial re- sumed in February of 2008. A verdict is not expected until after 2010 (August 2005a, 16). Several expert Nuu-chah-nulth witnesses have died (Wiwchar 2005b). There is general concern that federal fisheries policy may render meaningless whatever rights are confirmed in the verdict – that sweeping proposals for individual transferable quotas will intensify capital investment in the industry and further marginalize First Nations players, spelling "the end of First Nations participation in the commercial fishing industry" (August 2005b, 4). The unilateral implementation by Fisheries and Oceans Canada of a three-year integrated groundfish management plan in 2006, which included individual quotas, prompted the Nuu-chah-nulth to litigate, arguing that the plan should have triggered their constitutional right to be consulted. But a 29 May 2007 ruling in favour of Fisheries and Oceans acknowledged only "limited" impact on the Nuu-chah-nulth and concluded, therefore, that no legal trigger was necessary (Steel 2007, 6). The tribal council is appealing the decision.

Another vehicle for the recognition of Aboriginal sovereignty, and hence of Aboriginal rights to participate in the commercial fishery, is the treaty process. But relations between the British Columbia Treaty Commission and the Ahousaht are adversarial and without expectation of impending settlement. Thus, formal recourse to the legal arm of the state has so far failed to defend or protect their commercial fishing rights and has failed to stave off the profound economic hardship experienced due to the disappearance of fishing opportunities.

The economic downturn deepened the poverty and social marginalization bequeathed by a legacy of institutional racism. In British Columbia, Aboriginals are, on average, less highly educated, poorer, and sicker than their non-Aboriginal counterparts. In 2007, for example, the overall provincial unemployment rate of 4.1 percent was nearly three times higher for non-Metis First Nations persons living off-reserve (11.4 percent) and is probably higher still for those living on-reserve (British Columbia. BC Stats 2007a). Only 5 percent of First Nations members possess a university degree, compared to 22 percent of non-Aboriginals, resulting in fewer recognized skills and lower earnings (British Columbia. BC Stats 2007b). The North Island statistical region, which includes Ahousaht, has the greatest Aboriginal poverty in the province: here, 30.3 percent of Aboriginal families live on less than $20,000, whereas, across BC, 24.5 percent of First Nations fall into this category (Canada Revenue Agency 2003; British Columbia. BC Stats 2001). Only 10.8 percent of the BC non-Aboriginal population lives on less than that.

Federal fisheries-restructuring policies therefore collectively increased the vulnerability of individual First Nations fishers and their families. Regrettably, an epidemic of attempted suicides swept through the community in the new millennium, prompting the Ahousaht to solicit external help to respond to the despair, grief, and sense of hopelessness engendered there.[5]

Aquaculture
At the same time the salmon fishery was scaled back, salmon aquaculture expanded. Salmon farming – rearing fish in open pens close to shore – is free from the danger, seasonality, and unpredictability of traditional fishing methods in open waters, and it promised sustained employment for coastal communities. First appearing on the West Coast in the late 1970s, it began

to challenge the "wild" commercial salmon fishery only in the mid-1990s. In the last decade of the twentieth century, farmed salmon production ballooned in Canada, surpassing the commercial salmon catch in tonnage in 1998. By 2004 the sixty-two thousand tonnes of farmed salmon produced was more than double the wild product (British Columbia. Ministry of Environment 2005). Today, the province is the world's fourth-largest farmed salmon producer, behind Norway, Chile, and the United Kingdom. The meteoric rise of salmon farm production locally has been mirrored globally, and together with increasing supplies from Pacific capture fisheries in Russia and Alaska, has contributed to sharply depressed salmon prices and threatened the viability of the traditional capture fishery in BC and elsewhere.

Just three corporations dominate the BC industry, which is fiercely competitive and characterized by global overproduction.[6] One of these companies – Marine Harvest Canada – controls half of the total British Columbia output and about 20 percent of world farmed salmon production. Low profit margins provoked mechanization strategies limiting the number of aquaculture employees; so, although industrial production has increased five-fold in the last decade, employment levels have never exceeded twenty-one hundred direct jobs for the entire sector (including shellfish and finfish) (British Columbia. BCStats. Ministry of Management Services 2002). Not one farm site, and only one farmed fish processor, is unionized; many jobs are contracted out, and often wages do not compare to historic commercial fishing incomes.

The heralding of fish farms as "environmentally sustainable" has been called into question by persistent environmental problems. Companies struggle with disease outbreaks, escapes, algae blooms, and aquatic pollution. A plethora of environmental organizations monitor and report on salmon-farming activity, including the David Suzuki Foundation, Ecotrust, Forest Action Network, Friends of Clayoquot Sound, the Georgia Strait Alliance, the Living Oceans Society, the Sierra Legal Defence Fund, and Watershed Watch. Environmental issues centre on three potential dangers emanating from salmon farms: pollution (including the release of nutrient concentrates, antibiotics, and pesticides into the ocean, and the effects of excessive noise and light on the aquatic environment); disease and parasite transfer (to wild salmon and other aquatic organisms); and human health consequences (resulting from ingesting the antibiotics, dyes, heavy metals, and chemical residues found in the flesh of farmed salmon). The killing of predatory mammals, such as seals, which poach farmed salmon and damage nets and equipment, has also been an issue.

In British Columbia, as in Chile, these concerns are heightened because of the potential threats to wild Pacific salmon stocks represented by the introduction of a foreign salmon species. About 93 percent of all salmon farmed in BC are "Atlantics"; chinook and coho, both Pacific species, make up the remaining 7 percent (British Columbia. Ministry of Environment 2007). Additionally, most of the farmed Pacific salmon originate from a singular egg stock; thus, escapees could cross-fertilize with local wild salmon stocks and destroy local and highly diverse gene pools as well as, perhaps, the ability of wild salmon to return to spawning grounds (Zuckerman 2002). Other risks include increased juvenile predation and the colonization of Pacific species' habitats and spawning grounds.

The environmental risks from salmon farming are not distributed evenly among the provincial population, but disproportionately affect those living in the sheltered and often remote areas of the coast and who are most dependent on aquatic resources. This is especially the case for First Nations, in whose territories a majority of BC's salmon farms are located and upon whom a disproportionate burden of the impacts of industrial salmon farming has fallen. Not surprisingly, many First Nations have viewed the multiplying salmon farms as a threat to their livelihood. The Union of British Columbia Indian Chiefs opposes sea-based open-net pens. The BC Aboriginal Fisheries Commission also rejects salmon farms as they currently operate. In May of 2005, a First Nations Summit endorsed a moratorium on salmon farms near the mouth of the Skeena River.

Provincial and federal policy, however, facilitates aquaculture development. In an extensive review of salmon aquaculture, the provincial Environmental Assessment Office, despite acknowledging that First Nations harvesting of aquatic resources was adversely affected, concluded that salmon farming did not threaten human health. The review listed forty-nine recommendations to improve environmental standards and monitoring, and effectively approved salmon farming. After its release, the province allowed for some expansion and relocation of farms, despite a moratorium on salmon farming (in place since 1995). In January 2002, the newly elected provincial Liberal government officially ended this moratorium and weakened some fish farm regulations (such as net inspection requirements, which may affect escape prevention). Internal provincial procedures tend to advance the interests of industry. For example, the one-month deadline for input into tenure applications does not allow enough time for informed and consensus-driven feedback from the affected community, and First Nations feel pressured not to further delay these applications with requests for more time (Darrell Campbell, Ahousaht Fisheries officer, pers. comm. 8 July 2005).

Lack of provincial support for Aboriginal concerns about aquaculture is underscored by federal policy, which has been shaped for years on the premise that farmed salmon represent little to no threat to wild Pacific species. Indeed, Fisheries and Oceans Canada became an unabashed promoter of salmon aquaculture, funnelling millions into industry research and development. It also effectively subsidized industry associations such as the Canadian Aquaculture Industry Alliance and the British Columbia Salmon

Farmers Association (BCSFA), paying compensation for disease outbreaks and spending more than half a million dollars "to develop data to support the registration of aquaculture pesticides – even though the use of these pesticides may contravene Section 36 of the *Fisheries Act*" (Cox 2004, 65).[7] The auditor general criticized Fisheries and Oceans Canada for failing to meet its obligation to respect and enforce the strong Fisheries Act legislation with regard to the protection of wild salmon and their habitat from the effects of fish farming (see Canada. Office of the Auditor General of Canada 2000). It was not until 2003, twenty-eight years after habitat protection provisions were proclaimed as part of the Fisheries Act, that the first full environmental assessment of a fish farm was completed. Flagrant promotion of salmon aquaculture at the federal level thus compounded the obstacles confronting opponents of the industry.

But widespread public support for fish farms (and, specifically, First Nations acceptance) remains elusive. In 2003 the aquaculture industry launched a sophisticated public relations campaign, with the help of the infamous firm Hill and Knowlton (Cox 2004, 83).[8] With the encouragement of the province, corporations have attempted to win over First Nations communities with business agreements. To date, six bands have signed contracts with fish-farming companies.[9] A seventh, the Homalco, revoked an earlier decision to work with Nutreco. Although these agreements tend to silence overt dissent, provide operational security for capital, and grant companies opportunities to seek further access to government resources – such as public funds slated for Aboriginal economic development (see ibid., 50) – they also afford opportunities for the exercise of limited Aboriginal sovereignty.

Salmon Farming in Ahousaht

The first salmon farms began to appear in territory claimed by the Ahousaht in the late 1980s, but their numbers did not increase for a decade. In 1998 there were three farms; by 2005, sixteen farms existed inside the territory, with the entire Clayoquot Sound containing a total of twenty-three. This made Clayoquot Sound, a UNESCO-dedicated biosphere reserve, the second-most dense salmon-farming area in BC, next to the twenty-seven farms located in the Broughton Archipelago.

Seafood is consumed in great variety and quantity in Ahousaht. Fish, clams, oysters, prawns, mussels, and "specialty" items such as sea urchins, sea cucumbers, gooseneck barnacles, herring spawn on kelp, and whale not only provide sustenance and income, but also symbolize wealth and render significant meaning to cultural events and the celebration of political alliances. Seafood is, in fact, one of the most important signifiers of Ahousaht cultural health and identity; its destruction or diminution represents a serious threat to the ongoing viability and vibrancy of the community. Concern for the well-being of wild salmon is matched by awareness of the potential negative impacts of salmon farms on other species. Fish feed, containing pollutants, antibiotics, and other residues, can be eaten by animals that humans, in turn, consume. Gulls mass over farms during feeding times, ingesting pellets that potentially contaminate their eggs, which are also part of the Ahousaht diet. Fecal wastes from the suspended enclosures, dispersed by ocean currents, may suffocate nearby shellfish. In 1999 underwater inspection of an abandoned farm site revealed no life on the surrounding seabed, and a dredge brought up nothing but "slimy excrement" (Danard 1999).

This renders the meaning of, and impetus for, Ahousaht environmental activism significantly different from that of organizations with membership largely drawn from racially dominant or mainstream social strata, whose visceral connection to "nature" as a wellspring of a specific cultural identity has been long severed. It is, therefore, not only the desire to protect the sustainability of the commercial fisheries, but also the potential threat to Ahousaht cultural identity that motivates their activism.

Direct action was first provoked in the mid-1990s, after a private landowner contacted the band about a cloud of dark material emanating from a nearby Mainstream (then Pacific National Aquaculture – PNA) salmon farm. A review of the situation revealed that some PNA documents concerning the site appeared to have been falsified. Angered by the existence of a regulatory double standard, where commercial salmon fishermen faced increasing surveillance, but salmon farms did not, some members of the community began protests at farm sites. They demanded better management of the farms and a switch to closed containment systems, and they spoke of compensation for the pollution of their territory. They were joined in their struggle by the Friends of Clayoquot Sound (hereafter, Friends), a local environmental watchdog, as well as by the David Suzuki Foundation. PNA, a subsidiary of the Norwegian government-owned Cermaq ASA, operated fourteen of the sixteen territorial fish farms and had "a solid track record of negligence and irresponsibility" (Friends of Clayoquot Sound 2003). As the sole non-member of the BCSFA until 2003, PNA was classified even by this organization as a rogue company. Disease outbreaks, escapes, and sudden mortalities plague its operations.[10]

In February 1997, more than two hundred Ahousaht members in a flotilla of vessels boarded the Blue Heron PNA salmon farm site and demanded its closure. They protested that they had not been consulted in regard to farm site location or operations that were harming the harvest of shellfish, herring, and wild salmon in the area. At the company's request, they agreed to provincial mediation, but mediation rapidly soured, and so the Ahousaht initiated provincial court proceedings (which were later dropped). A second flotilla later that year served eviction notice to another farm that had continued to remain in operation although its provincial licence and tenure had lapsed (Lavoie 1997; Keith Fraser 1997). Under the threat of legal action, and pressured by the company for a different tenure site, the provincial Ministry of the Environment ordered the removal and relocation of the Blue Heron farm in August. The new site was also located inside Ahousaht territory and was therefore still objectionable; nevertheless, the band felt hopeful that future consultation would lead to the acceptable relocation of other sites. The head of Ahousaht Fisheries at the time remarked, "The companies all want to talk to us now" (quoted in Curtis 1997, A1).

But the victory was short-lived, as neither level of government appeared to fully support Ahousaht demands for industrial curtailment. Indeed, in July of 1997, federal government scientists visited the Blue Heron site and reported no evidence of damage to other species, thus refuting the basis for the Ahousaht's claims of environmental degradation and loss of fishing and foraging opportunities. That same year, the province quietly offered the band $2 million to proceed with a corporate joint venture to farm chinook salmon – a proposal that failed to win community support.

Campaigns by the band used the media to expose company transgressions to the public and force the province to enforce regulations. For example, when an algal bloom killed over a hundred thousand Atlantics in 2001, film footage taken by the band confirmed that many were left rotting inside the pens for more than a month (Leggatt 2001, 13).[11] After the January 2002 escape of an estimated eight to ten thousand adult Atlantic salmon and an unknown number of juveniles into the waters of Clayoquot Sound, publicity from a 2002 Ahousaht flotilla in which PNA was symbolically evicted guaranteed an investigation by the provincial government. This revealed that the escapes resulted from negligence on the part of PNA (a company boat had sliced through anchor cables that held the pens in place) and was not due to "Mother Nature ... showing us who is the boss once in a while," as a company spokesperson had earlier proclaimed (quoted in Simpson and Wilson 2002, B8). As a consequence of that investigation, PNA pled guilty to eleven of seventeen charges relating to fish escapes and overstocking (one farm was licensed for 250,000 fish but contained nearly 1 million). But the results were disappointing: it was fined a paltry $2,500, hardly an adequate sum to deter further violations. That further violations did occur is confirmed by the facts that, in 2002 and 2003, the province issued a hundred non-compliance notices to PNA (Friends of Clayoquot Sound 2008) and, by December of 2004, PNA had received more fines than any other aquaculture company in BC (although at this time, only three companies had ever been fined in the province). These notices and fines, however, appear to have done little to change the company's day-to-day operations. For these reasons, the Ahousaht felt only minimally supported by the state in their struggle to remove PNA from Clayoquot Sound.

Although the community as a whole did not support salmon farming, this did not mean that individual opinion was consistent or unanimous. Internal debate over the short- and long-term impacts, and possible benefits, of salmon farming influenced how the community responded to the issue. Most commercial fishing opportunities had not been reinstated, and the community was reeling from economic hardship – a misfortune compounded by contraction of employment in the forestry sector. Over time, though some remained adamantly opposed to salmon farming, others looked increasingly toward the employment and influence that might materialize were the band to adopt a more accommodating position regarding the aquaculture industry.

After 2000, the province began to encourage corporations to convince First Nations to allow new and relocated tenures inside claimed territories (Sullivan 2006, 55). PNA used this opportunity to quietly work out, with supportive council members, the basis of a business agreement that included a joint environmental policy, agreeable hiring practices, and investment management designs. When the rest of the elected council failed to endorse the proposed agreement in 2001 (Stackhouse 2001), the British Columbia Salmon Farmers Association (BCSFA) hired a band member as liaison officer with the company to build trust and promote aquaculture to the community. By November of that year, council had reversed its earlier position. It issued a statement endorsing fish farming and began to hammer out what was to become a successful protocol agreement with PNA (which had changed its name to Mainstream), signed on 4 September 2002. The protocol agreement and Ahousaht membership in the BCSFA were considered to be significant public relations accomplishments for Mainstream and were followed by other, similar contractual arrangements with resource companies. That same year, the band negotiated with Interfor for logging rights and jobs in its claimed territories, and, in 2008 – after a six-year battle – it narrowly endorsed mining exploration on the spiritually significant Catface Mountain (Friends of Clayoquot Sound 2008; Lavoie 2008).[12]

The 2002 Protocol Agreement

Although the details are confidential, the protocol agreement with Mainstream is known to secure operations stability for the company in return for recognizing the territorial title of the Ahousaht and compensating them for use of their territory. Stipulations include that the company must work with the band to "raise environmental standards," consult it before expanding operations, contract a local seine boat to be on call in the event of fish escapes, and provide a ten-year guarantee of farm and processing jobs for Ahousaht (Zuckerman 2003). The agreement provides, in principle, for the sharing of research on the environmental effects of the farms and allows for the Ahousaht to demand the closure of any given farm – or all of them – based on research results. "It's not the ideal solution," said one band councillor at the time, "but knowing that the government was going to open things up anyway, one of the best things we could have done was to take control and have a positive influence" (quoted in ibid.). After the protocol agreement was signed, the band council officially terminated the working relationship with Friends on salmon-farming issues, suggesting that another purpose of the protocol agreement may have been to prevent the sort of critical collaboration that had developed between these two groups.

Six years later, Mainstream and the band continue to support the agreement. The company claims that it has improved its communication with, and respect for, the Ahousaht (British Columbia. Legislative Assembly 2006, 160). In 2007 band representatives approved twenty-year tenures for thirteen of Mainstream's local salmon farms (Drews 2008). Media reports suggest that, in combination, the protocol agreement with Mainstream and the Interfor partnership agreement have reduced unemployment in the community to 65 percent, down from 80 percent in 1998 (Lavoie 2008).

Some of what was promised by the protocol agreement has yet to materialize, however. It calls for 50 percent of farm jobs in Ahousaht-claimed territories to belong to Ahousaht people by 2005 and promises numerous industry-training programs (Wiwchar 2005a). But by July 2005, the parties were still negotiating employment targets (Darrell Campbell, Ahousaht Fisheries officer, pers. comm. 7 July 2005). And although employment with the company fluctuates, depending on harvesting times, Ahousaht participation rates have been lower than expected. Sometimes, employment numbers have dipped very low, indeed. In 2004, when the company asserted that 40 percent of employees in the traditional territory were of Ahousaht ancestry, the band wrote to the province to correct this claim, stating that only six Ahousaht, and sometimes only four, were employed on Mainstream fish farm sites (Anne Atleo, elected chief councillor, Ahousaht Administration, letter to David Weir, Land and Water BC, 20 April 2004, obtained through a provincial Freedom of Information and Privacy Act application). In the summer of 2005, thirty band members worked for other fish farm companies, whereas only about twenty worked for Mainstream (Darrell Campbell, Ahousaht Fisheries officer, pers. comm. 7 July 2005). Ahousaht processing workers, laid off after 2003, speculate that they were not called back to work because of a company preference for transient workers, who do not stay long enough to climb the three-tiered pay scale. In 2006 a company representative suggested that, on average, Mainstream employed about forty-five Ahousaht – a figure still short of the protocol target (British Columbia. Legislative Assembly 2006, 159).

The company has been less than forthcoming with research data and is slow to forward results to the band. The Ahousaht, for their part, have continued to conduct a variety of monitoring and research activities on their own behalf and in conjunction with other agencies such as the BC Aboriginal Fisheries Commission and individuals at the University of Victoria. They have found that average local prawn harvests (measured in pounds) have declined sharply (Darrell Campbell, Ahousaht Fisheries officer, pers. comm. 7 July 2005). As prawns are bottom feeders, their abundance levels may signal the general health of the seabed. Inspection dives by community members reveal a loss of living matter on the ocean floor surrounding farms. Disease remains a prominent concern, and as "the question of expansion looms large," the Ahousaht test every spring for the presence of lice in all species of salmon smolts (ibid.). Exploding lice populations from fish farms may explain the recent chinook population crashes in Clayoquot Sound, and some preliminary research results indicate that wild smolt samples near farms are carrying lethal lice loads almost thirty times higher than at control sites (Drews 2008; Hume 2008).

The band continues to oppose industry applications for expansion and renewal. Documents received under the BC Freedom of Information and Privacy Act detail its negative response to company applications to expand six sites in 2004, especially because of the potential impacts of increased waste and excess feed residues on wild salmon stocks and shellfish. In some cases, site expansion was deemed objectionable because of its adjacency to salmon migratory routes, because it might impede navigational access, or because it might create new navigational hazards. The band also warned that the applications breached both its constitutional right to consultation and the right to effect change during the application process, as stipulated in the protocol agreement (Atleo to Weir, 20 April 2004).

Provincial Ministry of Land and Water representatives accompanied the Ahousaht Fish Farm Committee on a farm site inspection and appear to have made adjustments in order to accommodate some of the committee's concerns (and, no doubt, to avoid litigation). In early 2008, given two years of "alarming" salmon returns to the Megin River, the Ahousaht refused to support Mainstream's application for tenure renewal of its Dixon Bay salmon farm, located near the river mouth. Although they were careful to publicly affirm their support of salmon farming, their refusal is motivated by suspicions that lice from the farm are lethally infesting wild juvenile smolts as they migrate out to sea (Drews 2008). Recent Canadian scholarship suggests that salmon farms significantly and negatively impact wild salmon populations (Ford and Myers 2008) and that lice infestations from farms are decimating pink and chum runs in the Broughton Archipelago (Krkosek et al. 2007). Industry and government, however, downplay the impacts of lice transfer; the issue therefore generates much controversy in the press and other public forums. By denying tenure renewal because of probable lice infestation, the Ahousaht disrupt the ideological conformity of discourses supporting the industry. In so doing, they recapture (some) control over the process of salmon farm development.

Capital, Enclosure, and Resistance

The decision by the Ahousaht to sign a protocol agreement and to work with their former adversary cannot be regarded as a surrender, as was suggested by a local headline: "If You Can't Beat Them, Join Them" (Zuckerman 2003). They remain active and honest critics of the farms in their area, and they attempt to hold regulatory bodies and industry accountable. A more accurate analysis of the decision might understand it as a tactical shift in their attempts to influence the terms of exploitation of the ocean environment. Direct action, in the form of flotillas and the like, had achieved limited success; it had attracted public attention and garnered support but accomplished little substantive change. In the words of one elder, "We tried all the things we could without going to war. We fought the government to change the rules; we fought the industry to clean it up; and we weren't getting anywhere. At least now we have a say in what they do" (quoted in ibid.). Given the failure of the provincial and federal governments to meaningfully address concerns over the environmental, social, and health risks of industrialized salmon farming, a contractual agreement with industry appealed as a "timely" route, offering immediate advancement of Ahousaht interests (Darrell Campbell, Ahousaht Fisheries officer, pers. comm. 8 July 2005).

There is considerable confluence between the struggle to control the impacts of industrial salmon farming and the assertion of sovereign rights, including autonomous social and economic development. This confluence is inspired by the injustice of continual subordination to colonial and racist power structures, experienced daily by First Nations as "numerous inscriptions" of "small and grand violences" (Sullivan 2006, 60). Their comparative poverty, the loss of commercial fishing opportunities, the reluctance of the state to recognize Nuu-chah-nulth fishing rights (as evidenced by ongoing litigation), and the lack of state support for their resistance to salmon farming are examples of this sort of personal and social violence. Sovereignty – the reinstatement of full control over resources and habitats – offers the material and economic basis for indigenous social alternatives that subvert these inherited colonial structures. Therefore, the decision to influence Mainstream through shared contractual obligations probably appealed as an opportunity to engage in "transgressive performances of sovereignty" that influence the parameters in which state and industry representatives conduct themselves (ibid.). For example, the protocol agreement, in recognizing Ahousaht title and paying compensation for it, may be used to strengthen their position in treaty negotiations with Ottawa.

Thus, the interpretation of the protocol agreement with Mainstream as a means of exercising rights to self-determination bridged internal divisions within the community. At the same time, the agreement was negotiated under conditions of fundamental inequity in the material, juridical, political, and other resources available to the actors in this conflict. As such, it constrains the Ahousaht's agency by drawing them further into global capitalist economic and social relationships that may ultimately foreclose possibilities for alternative ecologically sustainable livelihoods (Eckersley 2004b). In this globalized economy, profitability is (re)generated by dispossessing others of assets and income as much as through the generation of new products (Harvey 2003). Worldwide, successive waves of privatization – of state, public, and common property – rob millions of people of ecological and material wealth. In other words, capital *encloses*. Salmon farms facilitate enclosure by converting seascapes from multi-use aquatic commons accessible to many, to single-use private-property tenures (Sullivan 2006, 47).

Ongoing opportunities in a vibrant commercial fishery, which also facilitated regular access to the food fishery, enabled Ahousaht labour, to some degree, to remain free from the totalizing impulses of capitalist market relations. But federal fisheries restructuring dispossessed marginal players of rights to aquatic access and livelihood, plunged Ahousaht (and other coastal communities) into economic crisis, and compelled those expelled to find immediate market alternatives. The restructuring restricted access to autonomous forms of labour such as self-employment through fishing. Systemic racism inherent in Canadian society and colonial policies, in discouraging the labour mobility of First Nations, rendered them highly exploitable, further reinforcing the subordination of Ahousaht labour to capital.

As a result, autonomous Ahousaht social life is also threatened. Access to reserve land and coastal waterways as well as the legal affirmation of Aboriginal rights to food fisheries and other resources guarantee that *some* basic needs can be provided outside of the market economy. Free access to some of the means to life works as a buffer against subordination to capital and creates a social barrier against the capitalist transformation of the oceanic commons (De Angelis 2004). Ahousaht extra-economic cultural activity, including ceremonial feasts, historically prevented the marketplace from wholly converting these social forms of exchange into arenas for the accumulation of profit. Social spaces where the Ahousaht and other coastal First Nations enjoy "entitlements and commons disconnected from market logic" help to sustain, to varying degrees, prolonged resistance to ongoing capital accumulation strategies (Ellen Meiksins Wood 1999, 73). Industries and state policy co-opt or annihilate much of the basis for this free access to the means of life. The marketplace thus becomes more firmly entrenched as a disciplinary and regulatory force in the community, penetrating Ahousaht social relationships and undermining the material basis for Ahousaht culture. The erosion of the alternative social spaces that accompanied the commercial fisheries may weaken the opposition to industrial development inside claimed territories.

Social resistance is also managed by the utilization of legal and regulatory procedures – including the courts, treaty processes, and licensing and tenure permits – which act as vehicles for the enforcement of state sovereignty (Sullivan 2006, 45) and endorse and legitimize the ideological values underpinning capitalist accumulation. They attempt to direct the immediate outcomes of social movements along a narrow continuum of alternatives compatible with capitalist development. For example, a 2003 court ruling against the Heiltsuk First Nation's attempt to prevent an Atlantic salmon hatchery from operating in its territory chastised the band for a "zero-tolerance" policy on salmon farms and concluded that such a policy rendered its engagement in consultation processes disingenuous (Hume 2004, 42). The ruling effectively penalizes First Nations for refusing industrial development. In this way, legal and social exercises of (state) power aid capitalist enclosure.

Profoundly unemployed, surrounded by industrial salmon farming, unable to seriously influence state decision making, and intensely disappointed with regulatory prevention of environmental risks, the Ahousaht were left to use what political leverage they had in effecting a protocol agreement with Mainstream to secure some of their goals for a sustainable livelihood.

Struggles over natural resources are at the same time political struggles over rights of access and the nature of usage, as well as the forums in which political representation, authority, and domination are contested (Shiva 2004). This contestation provides opportunities to advance alternative, and more just, social formations, including Aboriginal sovereignty. The eventual decision to work with Mainstream demonstrates that, even in the face of policies and structural strategies advancing processes of commodification, a First Nation can use available venues to create spaces in which to influence capitalist development, exercise resistance to environmental degradation, and assert its sovereignty. These transgressive challenges to state and corporate power involve continuous engagement (Sullivan 2006, 60) and evoke ongoing regulatory and industry responses that attempt to contain, redirect, and eliminate such challenges. In these ways, the Ahousaht case study provides insight into the contingent and social nature of power, as well as the factors that advance or obstruct social and ecological resistance to capitalist relations of production.

The Ahousaht experience also foreshadows the forms of power that state and capital may unleash against assertions of ecological self-determination voiced by non-indigenous communities; as such, it should inspire reflexive study and, more importantly, strategies of solidarity between these communities and First Nations. Just as Friends found ways to continue working with the Ahousaht on issues of social need and economic justice (Vodden and Kuecks 2003), so, too, should this solidarity facilitate the profound and immediate material requirements of First Nations. This must include critical anti-racist public education to limit resistance in white settler society to the allocation of funds, services, and resources to First Nations, including the equitable allocation of Aboriginal commercial fisheries and the rights to commercially utilize and manage other territorial resources. Such solidarity must also materially support the clarification of Aboriginal treaty rights – considered Canada's biggest unresolved human rights issue (*Ha-Shilth-Sa* 2005, 3) – and the enforcement of ecological legislation. This requires demonstrable support for First Nations sovereignty and the preservation of cultural spaces that fall outside of, or are incompletely subsumed by, the requirements of capital.

Mainstream, and the salmon farm industry generally, must also be held accountable for the contractual obligations they enter into with First Nations. Specifically, the Ahousaht need jobs and the company must provide them. Secrecy surrounding financial compensation and employment figures prevents public scrutiny of the company's subsequent behaviour and isolates the Ahousaht when breaches occur. Agreements must therefore be widely accessible public documents. In addition, the establishment of contractual models such as that between the Kitasoo/Xaixais Nation and Marine Harvest Canada, in which the band retains substantial ownership and managerial control over the salmon farm facilities in its territory, will ensure greater Aboriginal influence over the terms of development (Sullivan 2006, 54; see also recommendations of Leggatt 2001). Ultimately, these negotiated ventures must redistribute profits into First Nations communities.

The Ahousaht's struggles to exercise autonomy in the interstices of capitalist market and neo-colonial state relationships demonstrate their desire for a more democratic and ecologically sustainable future. Their worldview pushes the boundaries of existing conceptualizations of citizenship rights and responsibilities, incorporating an understanding of community linked materially and epistemologically to ecological processes (Richard Atleo 2004). This worldview can inform non-Native struggles for an ecological and socially just future. In this future, as Robyn Eckersley (2004b, 246) puts it, material and social existence will have to be determined not by the priorities of capitalist accumulation, but by "the needs of the lifeworld." For these reasons, the success of struggles for a green state that facilitates ecological citizenship and democracy is intimately linked to the success of First Nations' struggles for meaningful self-determination.

Notes

1 In 2003, 31 percent of all fisheries-related jobs were occupied by First Nations, who at the time represented only 4.4 percent of the total population (James 2003, 26).

2 Salmon licences are property rights bestowed by the state upon fishers and attached to individual vessels. They require yearly payment (licence fees) to remain active.

3 The salmon fleet consists of seiners, which scoop salmon in a large net (a purse), gillnet vessels, which extend nets across river mouths, and troll vessels, which catch salmon with hooks and lines. Area licensing divided the coast into two areas for seine and three areas each for gillnet and troll vessels.

4 These include the A-I reduced-fee salmon licence held by individual ("status") First Nations, N licences held by the Northern Native Fishing Corporation, and F licences held communally by bands.

5 In the first half of 2005, for example, more than forty suicide attempts were recorded in the community (Wiwchar 2005a, 2).

6 The three are Marine Harvest Canada, part of Nutreco Holding N.V., based in the Netherlands and the world's largest salmon-farming company; Heritage Salmon, part of George Weston; and Cermaq ASA, a Norwegian corporation.

7 In 2000 it established a Program for Sustainable Aquaculture that includes a $20 million per year commitment over five years for research and development.

8 This firm is notorious for its whitewashing of the ecologically disastrous 1984 Union Carbide pesticide leak in Bhopal, India, which killed eight thousand instantaneously and many more afterwards. It has successfully defended the tobacco industry during a long-term campaign to counter evidence of the negative health consequences of smoking; it has helped to promote positive images of repression by Uganda, Turkey, and China; and it has been hired by the US government to promote the Gulf War, among other conflicts.

9 Pan Fish signed a thirty-year contract with the Kitkatla First Nation in April of 2004. Pan Fish negotiated similar contracts with the Kwakiutl and Gwa'SalaNakwaxda'xw. The Kitasoo/Xaixais Nation and the Kyuquot/Checkleset Nation have signed deals with Marine Harvest Canada, and the Ahousaht partnered with Mainstream in 2002.

10 Infectious hematopoetic necrosis (IHN) outbreaks swept through farm sites in March and June of 2002 and February 2003, resulting in the culling of more than 750,000 diseased fish, extended site fallowing, and a five-month closure of the company's Tofino processing plant, with a temporary loss of more than eighty jobs.

11 When nitrogen- and phosphorus-rich fecal waste is discharged into the ocean, it provides nutrients for plankton populations. The rapid growth of algae depletes oxygen in the surrounding aquatic area, suffocating fish.

12 As with the Mainstream protocol agreement, the Interfor contract details are confidential. Any ecological conditions, if such exist, have not been made public.

5

Fisheries Privatization versus Community-Based Management in Nova Scotia: Emerging Alliances between First Nations and Non-Native Fishers

Martha Stiegman

The legitimacy of the federal government's role in fisheries management is hotly contested in Maritime Canada. Following the collapse of northern cod stocks, inshore fishers doubt the stewardship abilities of Fisheries and Oceans Canada (DFO) and are frustrated by policy that has facilitated a dramatic concentration of corporate ownership in the industry, effectively privatizing marine resources and deregulating fisheries management (Apostle, McCay, and Mikalsen 2002). Empowered by the Supreme Court's 1999 *Marshall* decision, First Nations see their participation in and management of commercial fisheries as a treaty right and are reluctant to fish under DFO jurisdiction.[1] In Southwest Nova Scotia, inshore fishing communities are building community-based management (CBM) as a means of improving fisheries management and as a strategy of resistance against fisheries privatization. In Mi'kmaq communities, CBM is also part of a long-standing struggle for self-determination. Through these efforts, they are developing a common vision for ecologically sound and democratic self-governance of the fisheries and building the foundations for a united challenge to DFO's privatization and deregulation agenda.

In this chapter, the potential of CBM as a model of local participatory governance is considered. This is done by examining the particular processes that have brought CBM into being in both Mi'kmaq and non-Native fishing communities, the political issues that CBM has helped these communities to address, and the impacts of CBM organizing at the local level. It is argued that, though CBM has not been sufficient to address larger political issues, it has enhanced community capacity to deal with fisheries management in particular and local development issues in general – most notably, in the strengthening of relationships between the Mi'kmaq and non-Natives. The case study presented here is based on twenty key-informant interviews conducted over the winter of 2003 and field research carried out in the summer of 2005 involving three months of participant observation and in-depth interviews with two dozen community leaders and harvesters in

Bear River First Nation and the neighbouring town of Digby, Nova Scotia, and with First Nations and non-Native fisher organizations across the Atlantic region.[2]

What Is Community-Based Management?

A sense of urgency prevails in fisheries management circles as stock depletion, the decimation of fishing communities, and shrinking government budgets become global phenomena (Neis, Binkley, and Gerrard 2005). Many perceive top-down, state-centred resource management to have failed (Bryant and Wilson 1998). There is now a global trend toward the decentralization of resource management (Jentoft and McCay 1995; Andersson, Gibson, and Lehoucq 2004). Multiple currents exist within this trend: one advocates the privatization of fishing rights and the deregulation of fisheries management (Hannesson 2004); another features the devolution of resource management to local populations and interest groups through state-stakeholder partnership arrangements known as co-management. Within the co-management literature, CBM refers to institutional arrangements that feature a high degree of local control grounded in democratic community-based governance. Both co-management and CBM have gained acceptance among social and natural scientists as potential vehicles for sustainable development and equitable resource use (Pinkerton and Weinstein 1995; Berkes et al. 2001).

The 1990s witnessed the beginnings of a paradigm shift in resource management. Arguments stressing the positive relationship of effective resource management to participatory local governance (McCay and Acheson 1987; Berkes et al. 1989; Jentoft 2000) were reinforced by international declarations calling for public participation in resource management and recognition of the rights of indigenous peoples. International agencies from the World Bank to the Nature Conservancy championed community-based approaches (Li 1996; Agrawal and Gibson 1999). In Canada, DFO claims community consultation and public participation as hallmarks of such a new approach.[3] First Nations, especially in the Far North, have won a limited voice in resource governance through co-management systems – often established within comprehensive land claims settlements (Stevenson 1996).

A convincing case for devolution has been made; however, the continued frustrations of indigenous (Nadasdy 1999) and small-scale resource users (Kearney 2005), and the growing disenchantment of government and international agencies (Brechin et al. 2002), suggest these actors hold very different understandings of both the definition and utility of "community-based" approaches.

Growing attention focuses on the neo-liberal political climate that contextualizes the emergence of CBM (Peet and Watts 1993). Neo-liberal orthodoxy prescribes the rollback of government in direct service provision and an intensification of state intervention to facilitate the privatization of publicly owned resources and assets. Community organizations are not left out of this process – they are assigned a leading role in the social economy and in social service provision (Jessop 2002; Shragge 2003). In such a policy climate, CBM is in danger of legitimizing DFO downsizing and playing into state strategies to download the costs associated with management and responsibility for already decimated ecosystems, with little meaningful transfer of management authority or political power (Wiber et al. 2004). In an era marked by the globalization of economic power, the decentralization of administrative responsibilities risks becoming what J. Anderson (2000) has termed "devolution without empowerment" and presents communities with the challenge of naming and negotiating these new power dynamics.

As Michael Mascarenhas also documents in Chapter 14 of this volume, First Nations are well versed in the art of deciphering the double-speak used by the federal government to avoid genuine power transfer (Marshall, Denny, and Marshall 1989). Many commentators have noted the paternalistic approach and continued agenda of assimilation and rights extinguishment that make Canada's post-colonial age difficult to distinguish from its colonial past (Angus 1992; Rynard 2000). CBM and co-management arrangements have become hallmarks of Canadian post-colonialism. Yet, critics charge that token Aboriginal participation is the norm, with the end goal of management often predetermined by government partners to fit capitalist market imperatives (Stevenson 1996; Nadasdy 1999) – a pattern First Nations rightly associate with Canada's long-standing project of assimilation (Green 1995).

In Canada, CBM has developed in three contexts: as the result of cutbacks and state downloading of management costs onto resource users (Bradshaw 2003), as a form of community resistance to resource privatization (Kearney 2005), and as a way for First Nations to appropriate resource management arrangements (Wiber et al. 2004). Given the dialectical forces from which CBM emerges, it is important to defend a focused definition of the practice – one that makes explicit the end goals and underlying values guiding resource management (Berkes 2003), one that includes power sharing between community and state actors (Castro and Nielsen 2001), and one that looks beyond the scale and mechanism of governance to see how CBM is both embedded in a larger political economy and the result of local political processes.

The Area of Study

Fishing virtually defines the culture and economy of Southwest Nova Scotia. Although the area was not as hard hit by the collapse of northern cod as the rest of the Atlantic region, groundfish landings are a fraction of historic levels and continue to decline despite more than a decade of severe quota reductions.[4] Most harvesters now rely on the lobster fishery for 80 to 100 percent of their income (Kearney 2005).

The area under study is roughly contiguous with Kespukwitk (which comprises the southwest region of Nova Scotia) – one of the seven political districts of the Mi'kmaq, whose traditional territory, Mi'kma'kik, includes most of the Maritime provinces, the Gaspé Peninsula, and the southern coast of Newfoundland. Like the Nuu-chah-nulth, studied by Donna Harrison in Chapter 4, the Mi'kmaq are a coastal people, for whom fishing is of profound importance. It was a central component of their traditional migratory lifestyle and is foundational to their worldview (Ricker 1997). Systemic racism, matched with the large capital investments needed to enter the commercial fisheries, kept most First Nations people out of the industry, which has been dominated by Nova Scotians of Loyalist and Acadian descent. However, the Mi'kmaq have established a growing presence on the water since the Supreme Court's 1990 *Sparrow* decision opened an Aboriginal fishery for food and ceremonial purposes.

Privatization by Stealth: Resistance in Non-Native Fishing Communities

For those interested in the dynamics of capitalist development and neo-liberal globalization in Atlantic Canada, there is no better example than the fishing industry. DFO policy has long favoured the development of a centralized corporate-owned fleet capable of large-scale harvesting and processing for international trade, and has imposed industrial discipline on small-scale independent producers in order to integrate them into an ever-expanding and deepening capitalist market (Davis 1991; Fairley et al. 1990; Veltmeyer 1990).[5] The current policy thrust is consistent with a wider neo-liberal agenda: privatize rights to commonly held resources, downsize government services, and deregulate management (Neis, Binkley, and Gerrard 2005). This is being achieved primarily through the imposition of individual transferable quotas (ITQs), a market-based approach to fisheries management whereby significant control is transferred from publicly accountable government bodies to private corporations. With the establishment of an ITQ regime, the total allowable catch (TAC) of any given species is divided among existing licence holders who may then transfer their quota by selling or leasing it. This is intended to set off market competition for control of quota, ending in the survival of the most "efficient" and "competitive" fishers.[6] The approach assumes that a rationalization of the industry will allow for a reduced government role in regulation and that conservation can most effectively be achieved through the creation of private-property rights (Hannesson 2004).

Such a neo-liberal vision has guided DFO policy since the influential 1982 Kirby Report (Canada. Department of Fisheries and Oceans 1982).[7] It is a policy direction that has been criticized by many – including the Senate Standing Committee on Fisheries and Oceans – for its lack of transparency. Successive bureaucratic initiatives have put in place a de facto management regime that contradicts the public right to fish that exists in common law, as well as policies designed to keep ownership of the inshore fleet in the hands of independent fishers and to ensure that the benefits of that industry are distributed within coastal communities.[8] In short, DFO has effectively redefined its role as the steward of marine resources and is undermining public ownership of those resources, with no genuine public consultation or parliamentary debate (Canada. Senate Standing Committee on Fisheries 1998).

In Atlantic Canada, the result of this privatization by stealth has been a dramatic consolidation of corporate ownership and the near extinction of the small family-owned businesses that characterized the small-boat fisheries for generations (Kearney 2005). Related to this social disaster is an ecological one: the collapse of the northern cod and a sustained downward trend in other groundfish populations (Copes 1998). In the words of one fishing leader interviewed by the author in 2003, "Unless something dramatic is done we are witnessing the last generation of family fishermen." This sense of urgency is found throughout what is left of the independent small-boat fisheries; many are convinced the department is wilfully undermining their position within the industry.

The Emergence of CBM and the Fundy Fixed Gear Council (FFGC)

In Nova Scotia, CBM emerged in the mid-1990s as a community response to DFO downsizing.[9] In 1994 the federal department announced that it would stop enforcing trip limits, a regulation designed to ensure an equitable distribution of catch among fishers. After groundfish quota reductions of close to 75 percent over the two previous years, the move would have resulted in fatal losses for all but the largest vessels with the greatest catching power. In response, fish harvester organizations formed an alliance and negotiated with DFO to manage their sector of the fishery at the community level, developing management plans for the various geographic areas concerned (Kearney 1998, 2005).

The 1995 fishing season proceeded under an experimental community-based management, having won the tolerance but not the support of DFO. Department officials continued working toward implementing ITQs; fishing communities were outraged. Protesters occupied DFO offices across the province, and thousands demonstrated, demanding a moratorium on ITQs and government support for CBM. DFO conceded, halting its plans to implement ITQs in this sector and negotiating with harvester organizations to implement their proposed community management boards. On the Nova Scotian Fundy Coast, the focal point of this study, harvesters set up the Fundy Fixed Gear Council (FFGC) to coordinate the democratic self-governance of their fishery (ibid.; see Kerans and Kearney 2006 for a more detailed discussion).[10]

Viewed in the larger context of the Atlantic fishing industry, the impact of the FFGC has been marginal. The inshore groundfish sector is a fraction of what it once was; the quota at stake represents a negligible percentage of the fishing economy for the region; and, although the number of fishers involved in FFGC is significant, no powerful capital interests were displaced by the move. What's more, fishers got a raw deal: neither money nor institutional support were made available for capacity building, the costs associated with managing quota were downloaded from DFO to fish harvesters, and no legislation was enacted to recognize community authority.

Although this critical evaluation of the FFGC may be accurate, it misses the tremendous victory the organization represents. The FFGC is a result of popular mobilization in response to government mismanagement of marine resources and in opposition to a corporate takeover of the fisheries, facilitated by DFO policy and practice. The organization has played a crucial role in slowing and perhaps preventing the further corporate takeover of this sector and in providing a voice for independent fishers in resource management. The FFGC has, moreover, strengthened local democratic self-governance and has had ripple effects in the Digby area – most notably in terms of mediating conflict and building relationships between non-Native fishers and First Nations.

The Marshall Decision and CBM in Mi'kmaq Communities

On 17 September 1999, the Supreme Court of Canada's *Marshall* decision sent shock waves through an already struggling fishery. On trial had been the validity and interpretation of the 1760-61 Peace and Friendship Treaties, which were negotiated between the British and the First Nations of what are now the Atlantic provinces of Canada. The court ruled that, despite the Crown's claims to the contrary, the treaties were valid and that the rights they defended for the Mi'kmaq, Maliseet, and Passamaquoddy included a collective right to earn a "moderate livelihood" through participation in the commercial fisheries (*R. v. Marshall* 1999).

Marshall is one of a series of cases focusing on Aboriginal access to fishing, hunting, and logging that have been brought to the Supreme Court to win recognition of First Nations rights in the Maritimes. As editorial comments in the September 1989 edition of the *Micmac News* make clear, harvesting and management of natural resources have been defended as part of a larger vision for Mi'kmaq self-determination: "Micmacs base their hunting rights on a covenant chain of 18th century treaties ... To the native community, such agreements were binding, political compacts between two independent and sovereign nations which form the legal foundation of their self-determination and self-government" (quoted in Prins 1996, 13-14).

To the Atlantic Policy Congress of First Nations Chiefs (APCFNC), the *Marshall* decision was an opportunity to develop a First Nations–driven and –managed commercial fishery as part of a larger project toward self-government. Referring to the First Nations' inherent right of self-determination and the nation-to-nation spirit of the treaties, the APCFNC put forward a vision for community-managed fisheries tied to strengthened Aboriginal self-governance, local economic development, increased access to the traditional resource base, and the incorporation of traditional knowledge and values (APCFNC 2001a, 2001b). DFO's vision was very different.

Ottawa responded to *Marshall* with a two-pronged approach. Over the long term, the parameters of a treaty-based fishery would be established through the Made-in-Nova Scotia Process – formal negotiations involving First Nations and the federal and provincial governments as part of a larger process aiming to interpret and implement the historic Peace and Friendship Treaties in a modern context.[11] In the short term, a negotiator was appointed to establish interim access agreements on a band-by-band basis. These "MacKenzie agreements" (named for the federal negotiator James MacKenzie) provided bands with funds to access communal commercial licences, vessels, fishing gear, and training. In exchange, communities agreed to "shelve" (Milley and Charles 2001) their right to manage their fisheries for the duration of the agreements and to fish by DFO regulations.

The MacKenzie process echoed the paternalistic divide-and-conquer approach of the Aboriginal Fisheries Strategy (discussed in detail below) that had alienated Mi'kmaq communities. Many were dissatisfied with both the funds and the fishing access that DFO put on the negotiating table, its inflexible stance, and the rushed pace of negotiations that made a proactive approach difficult and gave DFO control of the negotiation agenda.

Critics of DFO's neo-liberal agenda hoped the Aboriginal commercial fishery created via the *Marshall* decision would challenge the status quo by incorporating a diversity of regulatory schemes within a broad strategy for conservation. This might create a precedent for non-Native forms of CBM as well as potential allies in the fight for sustainable fisheries policy (Stiegman 2003); what has transpired is quite the opposite.

Faced with high levels of poverty and unemployment on reserves, and fearing violence from both DFO officers and neighbouring non-Native fishers, most chiefs felt they had little choice but to enter into interim access agreements. At present, thirty-two of the thirty-four bands in the Maritimes have signed such MacKenzie deals. Some have been able to develop innovative and distinctive community-based fisheries, but the general orientations of most First Nations' fisheries do little to challenge the status quo.

A comprehensive evaluation of the MacKenzie process and an inventory of the Aboriginal fisheries it helped to create have yet to be undertaken (such tasks are complicated by the fact that, because no single policy guided the negotiation process, the resulting agreements were diverse in nature). However, it appears that, if the process was successful in helping bands to enter the commercial fisheries, it was equally successful at establishing DFO control over the orientation and management of this Aboriginal fishery. The federal department has justified its course of action by referring to the *Marshall* ruling, which acknowledges DFO's prerogative to regulate commercial Aboriginal fishing in the interests of conservation. But, to many commentators, it appears that the department's primary motivation has been to retain control over management in the interest of furthering an agenda of rationalized fisheries development (Davis and Jentoft 2001; Wiber and Kennedy 2001). It is an approach that has been criticized on numerous fronts.

First Nations complain that DFO's refusal to cede management authority and its insistence on containing Aboriginal fishing rights within established regulations are infringements of the rights outlined in both the *Marshall* and *Sparrow* decisions (Milley and Charles 2001). Brought to the Supreme Court by the Musqueam First Nation of what is now British Columbia, *R. v. Sparrow* won recognition of the Aboriginal right to fish for food and ceremonial purposes, opening an Aboriginal food fishery on all three coasts (McGraw 2003). Both rulings limit DFO's regulatory authority unless justified on the basis of conservation. Many are cynical regarding the department's approach. In the words of one Aboriginal leader, "We don't see any evidence of DFO supporting conservation; we see them supporting big business" (field research interview 2003).

Although in principle, MacKenzie agreements are without prejudice to the negotiation of treaty rights, some fear that the federal government will consider the agreements to be part of the larger treaty implementation process by referring to the MacKenzie negotiations as consultation regarding the infringement of Aboriginal rights and by deeming the funds made available through these deals compensation for the infringement of such rights.[12] Another concern is that interim agreements will undermine treaty rights by laying the foundations for an Aboriginal fishery within the dominant framework of the current colonial management regime. The danger perceived is that, once interests are created, they are difficult to uproot, and that interim agreements will determine the parameters of the treaty-based fishery being negotiated within the Made-in-Nova Scotia Process.

The MacKenzie process has also been criticized for its negative impacts in non-Native fishing communities. DFO made room for First Nations in the commercial fisheries through a voluntary licence buy-back program. Though the buy-back scheme did succeed in its intention of negotiating a peaceful transition and avoiding an overall increase in the fish harvest, its result has been to facilitate the concentration of corporate ownership in the small-boat sectors. This occurred because speculation caused by the licence buy-backs led to huge increases in the market value of fishing enterprises. In Southwest Nova Scotia, the value of a lobster licence and vessel more than doubled to over $1 million shortly after the *Marshall* decision.[13] At such a high price, it is corporations, not young fishers who can afford to buy a licence; and coastal communities fear that inshore lobster, the last independently owned sector of the fisheries, will "go corporate" in the coming decade as the next generation of fishers retires.

CBM and Conflict Resolution

Following the *Marshall* decision, First Nations across the Maritimes took to the water in celebration, exercising their treaty right to fish. Tensions flared, most notably in Esgenoôpetitj (also known as Burnt Church), where the community's determination to pursue a Mi'kmaq-managed fishery outside DFO jurisdiction made the village a target for government enforcement, non-Native reprisals, and media attention. Shocking images of RCMP officers beating Esgenoôpetitj fishers and DFO boats ramming Native dories made international news for two consecutive summers.

In Southwest Nova Scotia, more than six hundred inshore fishermen blockaded Yarmouth harbour in a show of force to keep Mi'kmaq would-be fishers off the water. The standoff was diffused in a dramatic behind-the-scenes meeting in which the chiefs of the two local First Nations and non-Native fishing leaders were able to look past the crisis of the moment to begin discussing the underlying issues: their more than four hundred years of shared history, the legacy and continuing impacts of Canadian colonization, and their common interest in working toward ecologically sound community-based fisheries management. "At some point," recalls one non-Native fishing leader, describing initial meetings with Mi'kmaq organizations, "we realized we both wanted the same thing: sustainable livelihoods developed through community management of the fisheries, based on democratic self-governance – or in the case of First Nations, self-government" (field research interview 2003). Though the experience was not as transformative for rank-and-file fishers, the relationships formed around the table were strong enough to make future joint initiatives possible. Such a remarkable example of conflict resolution cannot be explained without leaving substantial room for inspired leadership and exceptional personalities. However, the experience of both Mi'kmaq and non-Native communities in building CBM was crucial in preparing the terrain for such a coming together.

Through the popular mobilizing that led to the creation of CBM, non-Native fishers developed a political analysis of the privatization of the fisheries. Because they could identify DFO's bias toward corporate ownership as the greater threat to their livelihoods, they were able to see their Mi'kmaq neighbours as potential allies, not simply as competitors for marine resources. One fisher reflects on the lessons learned: "Personally, I didn't feel threatened by [the] *Marshall* [ruling], because I knew already that there was enough corruption in the policies, in the fisheries, and within DFO, and all through the system – without pointing fingers at Aboriginal people. I would never have known that if I were just a bystander, just listening to what I heard on the news" (field research interview 2003).

Also, engaging in the deliberative practice of allocating quotas and resolving conflicts among fishers from the various regional communities gave fishers practice in the art of managing a certain degree of cultural diversity. Finally, the FFGC and other NGOs created as a result of CBM were crucial in mediating conflict and facilitating dialogue with neighbouring Mi'kmaq communities in the wake of the *Marshall* decision (a similar momentum within Mi'kmaq communities is described below). Although the Yarmouth blockade and the overall climate of racial tension triggered by *Marshall* made such a mediation process critical, it was a dialogue that had begun years before.

DFO had responded to *R. v. Sparrow* by recognizing an Aboriginal food fishery through its Aboriginal Fisheries Strategy (AFS). The program established a federal licensing regime to regulate Native fishing, with financial support for economic development for bands that entered into AFS agreements (McGraw 2003). The DFO approach of negotiating agreements on a band-by-band basis sparked resentment in many Mi'kmaq communities, as did the conditions tied to the deals. AFS agreements allowed bands to access much-needed funds for job creation but at the price of reduced authority over Aboriginal fishing and with that, a lost opportunity to develop resource governance capacity (Milley and Charles 2001). Many First Nations saw the program as yet another example of a paternalistic Canadian state denying Aboriginal communities the opportunity to build their own model of development.

Several bands refused to sign AFS agreements and pursued their food fishery outside the federal regulatory scheme. The 1995 creation of the Mi'kmaq Fish and Wildlife Commission (MFWC) was part of these efforts. This organization was established by the Assembly of Nova Scotia Chiefs to help Mi'kmaq communities develop resource management capacity, partly in anticipation of the rights the *Marshall* case might secure. Instead of reacting to DFO's proposals, the MFWC helped bands take a proactive approach by developing community-based management plans that combined traditional values with contemporary needs (ibid.). As a vision and approach that found resonance with progressive non-Native communities working to build CBM, this played an important role in building bridges between these two communities.

In making space for Aboriginal fishing in the commercial industry, DFO has been notorious for its insistence on negotiating with First Nations behind closed doors. The exclusion of non-Native fishers from this process has contributed to tensions between them and their Native counterparts on both the Atlantic and Pacific coasts. In Southwest Nova Scotia, CBM has helped to diffuse these tensions by offering participatory structures to circumvent DFO's divisive approach.

In Defence of Our Treaties: Bear River First Nation's Stand

Given the exacerbation of divisions between Natives and non-Natives by the MacKenzie process and the corporate agenda advanced by DFO, Bear River First Nation has chosen not to sign an interim agreement until the department is willing to negotiate a deal that reflects its vision for the fisheries. One Bear River Nation member, interviewed by the author in 2005, summarized the community's position in this way:

> Marshall was based on a commercial fishing right ... but for us it was so much more than that. It was connected to our identity, our whole way of being ... It was so much bigger than fishing ... We were hoping to introduce a whole traditional approach – but in today's context – of bringing back some of our traditional values about conservation of the resources and respect for those resources and what fishing would mean to us ... But before we knew it, there was this "Marshall agreement" ... To put it quite frankly, it was a way of assimilating us into the commercial fisheries ... There was no talk about the food fishery or for ceremonial purposes or small-scale fishing; community-based resource management wasn't even a concept.

The community is unwilling to sign a deal that it feels will place it in yet another relationship of economic dependence on Ottawa, compromise Mi'kmaq treaty rights and self-determination, threaten the viability of marine resources, and force it into a model of fishing that goes against Mi'kmaq cultural and spiritual values. Its process of reflection has led it to embrace CBM as a model of resource governance that it feels is adaptable to its culture as well as more appropriate to the exercise of its treaty rights.

For Taiaiake Alfred (1999), First Nation self-government will be meaningless unless it is informed by indigenous principles, neither modelled on Western traditions and colonial institutions nor developed in reaction to them. Alfred advocates a self-conscious re-adoption of traditional values to address current political, economic, and social realities. Although such a vision does not reject modernization or participation in larger economies, it does challenge capitalism's insatiable desire to commodify everything. Most notably, Alfred identifies a spiritual connection to the land and stewardship responsibility as being at the heart of indigenous traditions.

Alfred's analysis describes Bear River's approach well. But, as Alfred observes and this example demonstrates, such an uncompromising vision for self-determination is difficult to realize from a position of economic dependency. International commentators are often puzzled by the fact that most indigenous organizing in Canada is funded by the very state from which First Nations seek greater autonomy (Long 1992). The practice is less confusing to those who realize that, if federal funding helps sustain Aboriginal community organizing, it also helps to solidify economic dependence and to structure both the scope and direction of local initiatives (Fiske 1990).

The experience of Bear River exemplifies the Catch-22 that contemporary politics presents to First Nations. Holding out on signing an interim agreement has meant foregoing the money and licences attached to such a deal, along with resources that could fund (or potentially co-opt) a community capacity-building process. As a result, Bear River has had a negligible presence in the commercial fisheries; and the frustration of waiting while other bands test their sea legs has caused divisions within the community. But, if negotiations with Ottawa have not offered Bear River the possibility of developing a fishery on its own terms, working with non-Native fishers organized along principles of CBM has provided an interesting alternative route.

When DFO negotiations proved unsuccessful, the band turned to neighbouring non-Native fishers working to build CBM. Recognizing that Bear River's stand strengthened their own movement, non-Native fishers leased the band a boat and a captain, and informed DFO that Bear River would be fishing with their support. DFO conceded, and a crew from the Bear River community fished lobster over the summer of 2003 (Stiegman 2003). One community member, interviewed by the author in 2005, recalls, "It was a historic moment that probably went unnoticed by the media, but it was historic because we were doing that in cooperation with non-Native fishermen – and DFO just basically had to toe the line."

Joining forces with this emerging movement to build CBM has helped Bear River to circumvent a DFO-led process it feels compromises its treaty rights and the long-term health of the fisheries. Instead, the band has embarked on an exceptional process that is strengthening relationships with non-Native fishers in its area and making a unique contribution to CBM.[14] These relationships have not been enough to get Bear River into the commercial fisheries on a permanent basis; nor have they increased its bargaining power with DFO enough to negotiate an acceptable interim access agreement. However, the common ground built by these two groups is a hopeful example of the benefits of local democratic governance; as such, it points toward the potential of alliances between First Nation and non-Native communities in their struggles to defend sustainable resource-based livelihoods.

Conclusion
That Ottawa does not represent their interests as indigenous peoples is hardly news to the Mi'kmaq, for whom the Peace and Friendship Treaties and the *Marshall* case are but two examples of a more than four-hundred-year struggle

for self-determination. In this period of neo-liberal transformations, small-scale non-Native fishing communities also perceive the state to be hostile to their way of life. Inshore fishing communities are permeated by the sense that a crisis is occurring not just in the fisheries, but in Canadian democracy as a whole. This view appears to be well founded. Public ownership of marine resources is becoming a legal fiction as resources and management are effectively privatized through DFO's imposition of ITQs. Policy designed to protect the independent character of the inshore fisheries is wilfully being undermined by the department (Canada. Senate Standing Committee on Fisheries 1998). Cutbacks matched with public-corporate co-management arrangements are quietly shifting control of management from publicly accountable government institutions to the largest corporate interests in the industry (Copes 1998). As one fisher, interviewed by the author in 2005, described the situation, "I don't even think DFO gives a hoot about the fish. I really don't. To them it's resource extraction, pure and simple. Once the fish are gone, there won't be any pesky inshore fishermen to get in the way of oil and gas exploration, and the companies can move on to aquaculture."

In Southwest Nova Scotia, CBM has emerged as the result of interrelated processes of popular mobilization, neo-liberal transformations, and Mi'kmaq struggles for self-determination. Non-Native inshore fishing communities see CBM as a strategy of resistance against an unstated DFO policy designed to push them out of the fisheries. Bear River First Nation is building CBM as a form of resource governance in line with Mi'kmaq culture and consistent with the exercise of its treaty rights. Although these two communities take up CBM with differing causes and concerns, they have found substantial overlap in their positions: both have interests in defending the place of independent small-scale harvesters within the fisheries, as well as in ecologically sound management grounded in local democratic governance.

Given the past failures of top-down, state-centred management, and the social and ecological crisis triggered by the neo-liberal policy proposed by DFO as a solution, the conclusion that CBM is a badly needed innovation in fishery management is obvious to those involved. But substantial political ground must be won if this model is to realize its potential.

Within CBM literature, it is generally accepted that communities should not be left to their own devices to manage resources. A number of issues cannot be dealt with at the local level. These include the migratory nature of fish stocks, the incapacity of communities to deal with large-scale environmental problems (Berkes et al. 2001; Bradshaw 2003), and – as demonstrated by the case of Southwest Nova Scotia – the fact that the regulation of capital lies beyond community control. Growing attention focuses on the need for cross-scale linkages – cooperation among institutions at various scales from local to international (Berkes 2002; Stern et al. 2002). Central governments have crucial roles to play in CBM. These include devolving political power, recognizing local authority, and providing funds and support for community capacity building, technical training, and scientific research, to name but a few (Berkes et al. 2001; Bradshaw 2003). However, in Southwest Nova Scotia, CBM is not yet a site of cross-scale linkages: it is a local strategy of resistance to DFO's devoutly neo-liberal approach – what Linda Christiansen Ruffman (2002) has referred to as "economic fundamentalism."

As other cases – some examined in this book – have demonstrated, alliances between environmental groups and indigenous peoples are often fraught with contradictions; often, environmentalists' support for indigenous peoples is grounded solely on the latter's ecological orientation (Head 1990). Yet, acknowledgment of past and present injustices – as well as respect for First Nations self-determination – are necessary steps toward reconciliation (Green 1995). In Southwest Nova Scotia, the legal ground won by the Mi'kmaq and the task of managing fish have forced non-Native small-scale harvesters to take the first step down such a path. It remains to be seen whether the movement to build CBM will succeed in deepening this alliance and mounting a substantial challenge to the now interrelated processes of neo-liberal transformations and (de)colonization.

Notes

1 *R. v. Marshall* recognized the rights of the descendants of the signatories to the 1760-61 Peace and Friendship Treaties to earn a "moderate livelihood" through participation in the commercial fisheries.

2 The author wishes to thank the communities participating in this study for their hospitality and generosity. Names of the interviewees have been withheld. Field research was supported in part by the Royal Canadian Geographical Society's Maxwell Studentship.

3 For an example, see "A Policy Framework for the Management of Fisheries on Canada's Atlantic Coast" (Canada. Department of Fisheries and Oceans 2004) http://www.dfo-mpo.gc.ca/afpr-rppa/.

4 Groundfish species are those that live and feed near the ocean floor. In Atlantic Canada, those caught commercially include cod, haddock, and pollock.

5 Many note a significant exception to this trend in Romeo LeBlanc's 1977-84 tenure as fisheries minister, suggesting that LeBlanc's Acadian origins gave him greater understanding of and sensitivity to inshore concerns (Williams and Theriault 1990).

6 In Atlantic Canada, ITQs were first implemented in the early 1980s in the offshore groundfish trawler fleet; they were subsequently extended across the midshore sectors and much of the inshore fishery. For a more detailed account of this progression, see Canada, Senate Standing Committee on Fisheries (1998) and Patrick Kerans and John Kearney (2006). For an analysis of the gendered impact of fisheries privatization, see M. MacDonald (2005).

7 A Task Force on Atlantic Fisheries was commissioned by DFO following an economic downturn and the near collapse of the corporate trawler and processing sectors. Its report, Navigating Troubled Waters: A New Policy for the Atlantic Fisheries, became known as the Kirby Report, after task force head Dr. Michael Kirby. The report signalled a major change in orientation within DFO that sought to maximize economic efficiency within the industry through fisheries privatization.

8 The fleet separation policy is designed to prevent vertical integration in the inshore fisheries by forbidding processors to own licences. Its owner/operator clause stipulates that licences must be owned by individuals who operate fishing vessels for their livelihood. See CCPFH (2001).

9 It should be noted that CBM is a local strategy that evolved in diverse ways across the Maritimes and had distinct impacts in each area.

10 "Fixed gear" refers to a category of fishing technologies that includes handline, longline, and gillnet. Generally more selective than other technologies and therefore ecological, they are used by small-scale inshore fishers.

11 The Made-in-Nova Scotia Process Framework Agreement aims to "negotiate the definition, recognition and implementation of Mi'kmaq rights and title" (Canada. Indian and Northern Affairs Canada 2007). Eleven of the thirteen Mi'kmaq chiefs have signed the framework agreement.

12 As explained in Chapter 4, governments have a "duty to consult" Aboriginal groups whenever a policy or development decision may infringe upon an Aboriginal treaty claim or right.

13 Southwest Nova Scotia is home to the most lucrative lobster fishery in the Maritimes. Although the value of its lobster enterprises is not as inflated as in other areas, the pattern is consistent throughout the region.

14 Although Bear River First Nation's collaborative approach has been exceptional, it is one of a growing handful of examples of cooperation between the Mi'kmaq and non-Natives. See P. McKintosh and J. Kearney (2002).

6

First Nations, ENGOs, and Ontario's Lands for Life Consultation Process

Patricia Ballamingie

From September 1997 until June 1998, the Ontario provincial Conservative government led by Premier Mike Harris sought public input into the development of regional land-use strategies for an area comprising 45 percent of the province, or almost half of Ontario's Crown lands. Coined the Lands for Life area, it included 39 million hectares of Crown land and a further 6 million hectares of private lands, federal lands, and Indian reserves. The Lands for Life area covers the traditional territories of the Nishnawbe Aski Nation (NAN) and the Treaty 3 and Treaty 5 nations. Most of the planning region falls within the 338,000 square kilometres of NAN traditional lands. Within these lands are forty-eight Aboriginal communities totalling over thirty thousand people.[1] The government framed Lands for Life – Ontario's first-ever comprehensive land-use planning exercise and Canada's largest-ever public consultation process – as an attempt to spur democratic debate. Various social actors – First Nations peoples, industrial representatives from the forestry and mining sectors, conservationists, and tourism advocates – hotly contested the protection and use of these lands. For the purposes of the process, the area was divided into three planning regions (see Figure 6.1). This research focuses on the southernmost one: the Great Lakes–St. Lawrence planning region.

This chapter is divided into three parts. The first part provides substantive background on the case study, exploring the ways in which the government orchestrated both the public consultation process and its outcomes. The second part examines the paradox faced by First Nations in the context of Lands for Life. In colloquial terms, this paradox would be described as a Catch-22, since the only acceptable outcomes for First Nations were denied by the government in the formulation of the problem to be solved. As a result, their engagement would have legitimated an outcome unfavourable to their interests. This section explores their concerns (which the round table members charged with the task of consulting the public lacked the

Figure 6.1 Lands for Life Planning Regions. Map drawn by Khyati Nagar, based on figure in Ontario, Ministry of Natural Resources (1998, 10).

mandate to truly address) and summarizes what First Nation participants felt *meaningful* consultation might have entailed. The third part focuses on the relationship between First Nations and environmental non-governmental organizations (ENGOs). There appeared to be potential for a convergence between these actors, which, in theory, could have challenged the entrenched power dynamic of government-industry collusion. However, attempts to form a First Nation–ENGO alliance failed. This research identifies the difficulties these groups faced in finding common cause, in hopes that these obstacles might be overcome in the future (recognizing that these differences may also prove irreconcilable).

This study adopts a methodological approach that is both genealogical and discursive. The former aims to reveal dissent, to deconstruct generalizations, to examine contradictory evidence, and to break apart globalizing hegemonic discourses. Michel Foucault (1977, 162) writes that a genealogical approach to research "seeks to make visible all of those discontinuities that cross us." Joan Scott (1992, 23) views the genealogical project as the process of *becoming visible;* its aim is "to reveal something that existed but that has been suppressed ... to write about and thus to render historical what has hitherto been hidden from history." Thus, a genealogical approach does not necessarily supplant the master narrative in its entirety; rather, it illustrates those dissenting counter-hegemonic voices that are missing and/or subjugated. Similarly, a discursive approach aims to critically interpret a "text" (broadly defined to include not just written documents, but also interview transcripts and visual representations), with a view to challenging truth claims by revealing inconsistency, contradiction, and uncertainty. Discourse analysis examines how language is used and deployed to construct political positions, how ideas are contested, and how certain phrases and concepts become normalized while others are silenced (Foucault 1980, 105).

To begin, this research systematically reviewed all government documents (such as press releases and reports) as well as popular press and scholarly literature related to Lands for Life. It also examined the minutes from the public hearings (including detailed testimony from a wide range of social actors) with a view to uncovering the dissent that – though part of the public record – was not evident in the government's final announcement. This evidence was then supplemented with the input of fourteen interviewees, nine of whom are cited here; these individuals were either *insiders* to the process (and as such, offered privileged insights into its inner workings) or were *outsiders* who were publicly critical of the process. Their identities have been coded with ENGO or FN (for First Nation) to ensure anonymity. The three individuals coded as $ENGO_1$, $ENGO_2$, and $ENGO_3$ spoke on behalf of conservation groups within the Partnership for Public Lands (a powerful coalition comprised of representatives from the World Wildlife Fund, the Federation of Ontario Naturalists, and the Wildlands League, hereafter referred to as the Partnership). $ENGO_4$, $ENGO_5$, and $ENGO_6$ spoke on behalf of other key environmental groups. FN_1 spoke on behalf of the Union of Ontario Indians (UOI), and FN_2 and FN_3 each spoke on behalf of two First Nations in the region, one as chief, the other as communications coordinator.

The Lands for Life Process

In July 1997, round tables comprised of twelve to fourteen citizens from supposedly diverse backgrounds were appointed within each of the planning areas by (then) Minister of Natural Resources John Snobelen. During Phase 1 (September-December 1997), the round tables sought input into the formulation of a Regional Land Use Strategy designed to inform broad decisions about land use and resource allocation within the planning area. During Phase 2 (February-March 1998), a second round of public consultations sought more detailed recommendations concerning the management of specific local resources as well as feedback on preliminary land-use scenarios. During Phase 3 (April-June 1998), the round tables submitted their recommendations, which were consolidated into one document by the Ministry of Natural Resources (MNR).

According to the MNR, the consultations aimed to achieve several ends. Specifically, they sought to "identify areas that represent the natural diversity of the region," to "protect new areas," to "foster sustainable land-use planning and management," to "retain and restore representative and special areas," and to "encourage partnerships that will lead to more protected areas on public and private land" (Ontario. Ministry of Natural Resources 1997). However, though the government's professed end was to reconcile the protection and use of Ontario's natural resources through a consensus-building approach (ibid.), both the process and the outcomes reveal an unacknowledged bias toward the latter of these goals. To start with, Crown lands had already been largely allocated to resource-extractive industries when the land-use consultation began. Thus, there were vested interests and questions of compensation to contend with, should any of the land be designated as off-limits for industrial or commercial exploitation. The National Round Table on the Environment and the Economy (NRTEE 2004) case study of Lands for Life reports, "Within the L4L region, 6.6 percent of the land was protected in federal or provincial parks before the process commenced. Virtually everything outside those areas was licensed for one or more extractive uses (forestry, mining, hydroelectric development or aggregate extraction). The land in the region had been allocated primarily to forestry on a piecemeal basis and with little public engagement. A 1995 judicial review determined that the government was not properly implementing forest planning and called into question 70 to 80 forest management plans."

While proposing to "complete" a system of provincial parks and conservation reserves – increasing Ontario's protected land base to 12 percent of the province's area – the provincial government also aimed to "secure a land base for the forest industry, for tourism, and for sport hunters and anglers" (ibid.). Thus, from the outset, the goals of ecological integrity and expanding industrial development were simply assumed to be reconcilable. Or, rather, the ecological goals were expressed as the protection of "representative" ecological areas, rather than as a more rigorous set of criteria drawing on conservation biology and such concepts as biodiversity or ecosystem integrity. The forest industry's interests were privileged in a number of ways. The land claim and treaty rights issues of the region's First Nations were, on the other hand, excluded from the agenda. First Nations were not represented during the negotiations that produced the Forest Accord (a three-way agreement between the forest industry, conservation organizations, and the Government of Ontario to achieve stability in the forest sector while advancing a scientifically sound system of protected areas) in 1999-2000; nor have they been included on the resulting Ontario Forest Accord Advisory Board (OFAAB). The OFAAB, which is comprised of representatives from the MNR, the Partnership, and the forest industry, reports to the MNR. In terms of outcomes, as this substantial tract of land was divided among competing uses (conservation, forestry, mining, hydroelectric development, and tourism), the primary resource-extraction industries benefited the most (Winfield and Jenish 1998, 141), whereas the treaty rights of First Nations have been jeopardized.

The *Consolidated Recommendations* report (Ontario. Ministry of Natural Resources 1998) of the three round tables proposed to protect 6-7 percent of Crown lands – only half of what the government had publicly promised. A public outcry ensued, with some fourteen thousand people writing to the government to protest "the inadequacy of parks and protected areas" (CELA 1999). The government then brought together key stakeholders at the Fern Resort, in Orillia, Ontario, to negotiate a different, more "politically palatable" (Cartwright 2003, 124) outcome. This shift in government approach to planning marks the end of the Lands for Life public consultations. These discussions, which took place "behind closed doors" (CELA 1999), were limited to representatives from the large forestry companies (namely, Bowater, Abitibi Consolidated, Domtar, and Tembec), from the Partnership, and from the MNR. In March 1999, the government announced a long-term land-use strategy for Ontario's northern and central Crown lands: Ontario's Living Legacy, including a commitment to protect 12 percent of the planning area, to create 378 new parks and protected areas, and to designate nine new "signature sites."[2]

On the surface (and in light of the 6-7 percent figure first suggested by the round tables), Ontario's Living Legacy might seem a coup for environmentalists. Indeed, most ENGOs – from the moderate representatives of the Partnership to the more radical Sierra Club of Canada and Earthroots – embraced the outcome. Certainly, the Harris government sought to gain political mileage from the announcement through the strategic distribution of a slick advertising poster, included as a supplement in Ontario's major newspapers. One interviewee described the insert as "propaganda" – "political advertising with public funds" (ENGO₅ 2001, 2). Inside the supplement, the Harris government touted Ontario's Living Legacy as "the biggest single expansion of parks and protected areas anywhere." Similarly, Monte Hummel, president of the World Wildlife Fund (WWF) Canada, described the accomplishment as "an unprecedented contribution to conservation on a global scale" (quoted in Ontario 2000).

However, others characterized the outcome as "pseudo-protectionism" (Weis and Krajnc 1999, 36), "not meaningful" (ENGO₅ 2001, 2), and ultimately insufficient to ensure ecological integrity over the long run. Mark Winfield of the Canadian Institute for Environmental Law and Policy described the announcement as no more than an "attempt by the Conservatives to shore up in advance of the coming election what has otherwise been a record of hostility to environmental protection" (quoted in Mittelstaedt 1999, A4).

Many dismissed the Lands for Life process as nothing more than "smoke and mirrors" – giving the public the illusion of a consultation while failing to address, and deflecting criticism from, more substantive environmental policy issues (Krajnc 2000, 124). One environmental representative characterized public consultations and round-table committees as "busy work," "window dressing," "allowing the government to be able to say it consulted broadly" (ENGO$_6$ 2001, 1). In fact, critics argued that Lands for Life failed to provide adequate protection for both the environment and the rights of First Nation peoples. The interests represented at the hearings reflected a rather narrow epistemic domain (Gibson 1999, 56); certain voices were legitimized, whereas others were silenced. Following Foucault, this analysis aims to "isolate, identify, and analyze" (Dreyfus and Rabinow 1983, 185) the unequal power relationships that emerged (or rather, that were reinforced) during the Lands for Life struggles.

Government Orchestration of the Lands for Life Process

Many charged that the Lands for Life public consultations were intentionally designed as an undemocratic process and that the outcome was predetermined. Certainly, the Harris government employed various means through which it orchestrated the process and thus limited democracy (see Richardson, Sherman, and Gismondi 1993). In so doing, the government ensured that the hegemonic discourse of economic rationality prevailed while counter-hegemonic discourses of resistance were silenced.

To begin, the government very deliberately selected *who would consult the public*. By and large, the MNR charged this task to a group of white men who possessed close ties to industry (or in some cases, directly represented it). A serious political bias resulted – *toward* industrial development and *against* conservation. For example, when Mike Barker (who represented forest industry companies) deliberated over proposed conservation reserves, he would urge other members to restrict their decisions to "proposals that do not have an economic impact" (GLSL 1998b), threatening job losses and mill closures. Thus, rather than identifying which environmental values needed to be protected, Barker influenced the round table to first consider forestry interests. His success in doing so narrowed the terms of reference and reduced complex land-use decisions to purely economic ones. Other round table members also seemed closed to the notion of protection, characterizing the visions put forward by environmentalists as paranoid and restrictive doomsday scenarios (ibid.).[3] Clearly, such sweeping generalizations reveal an attempt to dismiss environmental concerns altogether. In spite of their claims regarding objectivity, the proceedings were shaped by industry-dominated discourses.

Second, the government determined *who would be consulted* (and ultimately, it deemed some interests to be more legitimate than others). First and foremost, the government failed to adequately consult First Nations. Also, although northern residents have unique stakes in the outcomes of decisions affecting their region, the round-table meetings excessively limited the concerns voiced by groups based in the south. To this end, the Great Lakes–St. Lawrence Round Table spent eight months in the northern part of the area and only three weeks in the southern part.

Round-table members portrayed southern concern as pure self-interest – the desire to keep the north "pristine" so that it could be enjoyed for vacations and recreational purposes (ibid.). They further typified southerners as fundamentally out of touch with, and unsympathetic to, the economic reality faced by those living in the area. Conversely, people local to the planning area who were directly reliant on resource extraction for their livelihoods were characterized as genuine, their testimony as authentic, heartfelt, and informed (GLSL 1997b). Of course, the strategy of categorizing intervenors in regional (north/south) terms is itself a device to divide and conquer. The dichotomy reinforces existing and perceived tensions, and falsely – but effectively – homogenizes interests into north/south categories, erasing differences within both.

Third, the government also intentionally framed (and narrowed) the terms of reference by limiting *what questions could be asked*. The resolution of First Nations concerns and substantive critiques of land management practices were deemed to lie outside the boundaries of the Lands for Life mandate, or, in Foucauldian terms, outside the totalizing field of activity for this political process (see Dreyfus and Rabinow 1983, 198).

Next, the government exercised power in determining *how the public would be consulted*. Here, locations in which the consultations took place (upscale hotels or large sports venues), the dress code and demeanour of (most) round-table members, the time limits imposed on public input, and the public nature of speaking to an audience through a microphone all added to the formality of the hearings.[4] Moreover, audiences sometimes heckled those who supported protection, no doubt pre-empting (and silencing) others from making similar statements.

Foucault would call such disciplinary technologies "meticulous rituals of power" (ibid., 192). These rituals enable "the exercise of power with limited manpower" by involving all those who "come into contact with the apparatus" in its functioning (ibid.). Thus, hundreds of participants tacitly consented to the structure of the consultations and to the protocol involved in soliciting public testimony. By normalizing such protocols, a dozen (or so) round-table members were able to conduct consultations regarding a hotly contested question with a minimum of overt discord. To publicly question the format of the consultations would have been to expose oneself to ridicule, since most participants accepted the normalcy of engaging in the process in the prescribed ways.

Rationale for ENGO Engagement

Notwithstanding the critiques that have emerged regarding the Lands for Life process, there were several legitimate motivations for creating a formal land-use plan. One key environmental representative cited the following reasons: to resolve long-standing resource conflicts, to protect fast-disappearing wilderness areas, and to address the landscape in a more comprehensive and holistic way. These desires spurred a number of moderate environmental organizations to create the Partnership and to actively engage in both the public and the not-so-public consultation process (ENGO₁ 2001, 1).

Laurie Adkin (1998b, 63) argues that private-sector associations and environmental groups alike generally view "advisory bodies created to oversee development of regulations and to make recommendations to the ministers" as "crucial terrain in the battle for representation within the State." To this end, the Partnership positioned itself strategically in relation to both the round tables and the MNR. In fact, the Partnership achieved insider status (and thus political clout) by maintaining a constant presence throughout the consultation process, by generating its own alternative land-use scenario for the round tables to consider, and by representing a seemingly moderate (and obviously well-funded) "southern" contingent. Other, "more radical" ENGOs such as the Sierra Club of Canada, Animal Alliance, and Earthroots – a group one environmental representative characterized as "the pointy stick brigade" (ENGO₄ 2001, 1) – demanded the protection of no less than 30 percent of Ontario's Crown lands. In contrast, the Partnership's objective (of 15-20 percent protection) must have seemed infinitely reasonable. (For a detailed examination of the strategies employed by the Partnership, see Cartwright 2003.)

Certainly, the active participation of such a wide range of conservation interests speaks to what Raymond Bryant and Sinéad Bailey argue is the growing power and assertiveness of "civil society" (specifically, ENGOs) vis-à-vis the state. Bryant and Bailey (1997, 131) identify two underlying motives for this expanded role: first, the declining capacity of the state as an actor to ensure environmental and social well-being, and second, the "widespread public perception that states have contributed to – rather than mitigated – poverty and environmental degradation." This is a perception particularly held with regard to neo-liberal regimes – like that of the Harris Conservatives – in which government-industry collusion is deeply entrenched.

In fact, government-industry collusion was evident at every stage of this struggle; it characterized the role played by the Harris government in uncritically embracing an industrial model of development. The Harris government acted on behalf of Ontario's industrial interests as if those interests completely and utterly coincided with those of Ontario's citizens. However, the Conservatives seemed also to have become more adept at employing green discourse to pacify an increasingly critical segment of their own core constituency, the Lands for Life process being a case in point.

Certainly, when one evaluates the environmental motives of the Harris government, it is imperative to remember that, during its first term in power, it systematically weakened environmental policies, reduced Ontario Ministry of Environment budgets, devolved power to the private sector, and fundamentally restructured government (Krajnc 2000; Winfield and Jenish 1998). Moreover, as Mark Winfield and Greg Jenish (1998, 144) argue, some within government seemed to hold contempt for both "the concept of public goods" and "the idea of collective societal action, through government institutions, to protect these goods." In fact, as Tony Weis and Anita Krajnc (1999, 36) assert, the values underlying the Harris government's environmental policies and practices – embodied in its Common Sense Revolution – closely resemble those of the Wise Use Movement in the United States.[5] However, the government could not wholly dismiss the environmental concerns of its electorate. Thus, a high-profile public show in favour of protection – Lands for Life – proved just the ploy to distract the public from its otherwise hostile environmental record.

Effecting the Erasure of Ontario First Nations

From the outset, various Ontario First Nation organizations and community representatives participated warily in the round-table meetings, although they soon condemned the process, demanding government-to-government negotiations. Specifically, the Union of Ontario Indians (UOI) followed the process closely, since the planning region encompassed all of its treaty areas (FN$_1$ 2001, 1). On 10 July 1998, the NAN chiefs recalled their Aboriginal representatives from the Boreal East and Boreal West Round Table discussions. Soon after, Grand Chief Charles Fox, on behalf of the Chiefs of Ontario and Independent First Nations, formally rejected the Lands for Life initiative (Chiefs of Ontario 1998).[6] Fox lamented, "There was no mention whatsoever of our particular issues and we are profoundly and deeply disappointed by that because we agreed to participate in the process in good faith" (quoted in ibid.). Grand Chief Francis Kavanaugh also commented, "The Ontario government is trespassing on First Nations' rights every time it cuts a tree in Ontario" (quoted in ibid.). Others called a halt to the proceedings, recommending that the process be put on hold until all outstanding land claims had been settled. With these (and other) concerns in mind, the NAN applied to the Ontario Court, General Division, to have the process declared illegal (Mackie 1999, A13). Several political analysts shared the indigenous groups' outrage. Tony Weis and Anita Krajnc (1999, 36) remarked that "Reforming the land management regime over such a vast extent of land could ideally have been a vehicle to address the ongoing and debilitating colonial inequities faced by Ontario's First Nations."

Instead, the government employed various strategies to effect the "constitutive erasure" of Ontario First Nations (Braun 2002, 27). First, it proceeded with Lands for Life rather than settling outstanding land claims and treaty rights beforehand (that is, these issues were not deemed worthy of resolving). Second, the government charged the task of public consultation to a group of round-table members who, by and large, lacked experience in consulting Aboriginal people. Moreover, this group was given a huge responsibility but no corresponding authority; the government ultimately disregarded and overrode the round table's recommendations. As a result, in spite of some good intentions on the part of individual round-table members, they failed to adequately consult the First Nations and ultimately perceived them as merely one among a number of competing "interests" regarding land use. By representing Aboriginal peoples in this way, the government eliminated the need for a separate consultation process. Thus, it effected a legal and historical erasure, and "dissolved" sovereignty claims into a pool of interests.

The government also failed to represent the First Nations on the maps that were used to illustrate the various land-use scenarios. That is, the location of bands and the boundaries of unresolved land claims or of traditional treaty areas and sacred sites were all absent. Unfortunately, at the time the First Nations lacked the requisite geographic information systems (GIS) capacity to challenge such cartographic erasures (see Gibson 1999). At the very least, such a cartographic project might have enabled Ontario First Nations to assert their presence in the region – to make themselves *visible* – and to challenge the MNR in its public construct of Crown lands as devoid of Aboriginal peoples.

Erasure continued in the round table minutes, where a caveat claimed that the Aboriginal voices that did emerge could not be considered representative. Specifically, the testimony of John Beaucage (Wasauksing First Nation), Sam Kewaquado (Shawanaga First Nation), and Dwayne Nashkawa (UOI) was prefaced with the following comment: "Note that those in attendance do not necessarily represent the views of the communities they come from but are monitors of the process or speak only for themselves from their own experiences" (GLSL 1997a). Although this caveat may well have been intended to address the thorny ethical issues associated with indigenous representation, it also served to undermine, discount, and ultimately dismiss the messages that emerged. One First Nation interviewee characterized the caveat as a "deliberate misinterpretation" by the government (FN$_1$ 2001, 2), explaining that the comments of two chiefs and a UOI official represented more than the sum of the three individuals voicing them. Moreover, Beaucage, Kewaquado, and Nashkawa offered a potent critique of power and a call for justice.

The government attempted to appease First Nations by holding a few meetings in Aboriginal communities and by appointing a single First Nation representative (Holly Charnya) to the Great Lakes–St. Lawrence Round Table. Ultimately, the government portrayed as *meaningful* consultation that which First Nation representatives characterized as *limited* involvement. In this context, it is ironic that the round tables advocated what they called an "Aboriginal principle of long term stewardship" (GLSL 1998a) and made specific recommendations concerning Aboriginal peoples in their *Consolidated Recommendations* report (Ontario. Ministry of Natural Resources 1998). Because First Nations felt that the round tables failed to address their concerns in a meaningful way, the rhetorical adoption of such a principle was perceived as an act of cultural appropriation.

In effect, the Ontario government precluded certain development options for First Nations, and in so doing, it limited and narrowed their possible futures – a particularly insidious outcome that legitimates differential access to resources and power. This is why one First Nation proponent viewed the exclusion of Aboriginal rights and claims from the Ontario's Living Legacy policy as so disastrous: "If those rights and claims are never realized, First Nations will have virtually nothing left in the north on which to assert their authority and with which to practise their rights to hunt and fish" (FN₃ 2001, 2).

First Nation peoples identified several shortcomings in the Lands for Life process. To begin, participants cited various legal precedents, reports, and decisions (ranging from international to national to provincial and territorial in scale) that they argued ought to have informed the Ontario government in any land-use planning process of this nature. Specifically, they cited the Draft Declaration on the Rights of Indigenous Peoples, the *Report of the Royal Commission on Aboriginal Peoples,* three recent Supreme Court decisions *(Sundown, Delgamuukw,* and *Sparrow),* and, finally, the Ontario Environmental Bill of Rights. All of these affirm First Nation rights to self-determination, the need for an adequate land base, and the protection of Aboriginal treaty rights. Thus, from a First Nations perspective, ideally, the government would first have engaged in a separate nation-to-nation negotiation process aimed at resolving outstanding land claims, the exercise of Aboriginal treaty rights (in the context of Crown lands), and past grievances.

First Nation peoples further argued the need for sufficient resources (both financial and technical) to enable meaningful engagement. Here, Foucault (1983, 223) acknowledges that individuals are differentiated in their ability to act – by their social or economic status, by "linguistic or cultural differences," by "differences in know-how," and by technological capacities. He (ibid.) summarizes: "Every operation of power puts into operation differentiations which are at the same time its conditions and its results." Thus, differential abilities to act self-perpetuate and reinforce those differentiations.

Clearly, access to resources would have allowed First Nations to participate on a more equitable playing field with other actors (such as industry or ENGO representatives) and would have addressed the differential ability to act (making the entire process more inherently democratic).

Finally, First Nations stressed the importance of consulting the whole of the community, including elders and children, in a way that is mindful of the significance of oral culture and relationships of trust. At a minimum, consultative bodies (as quasi- research entities) ought to engage in an ethical review of their proposed projects. Especially when Aboriginal peoples are consulted, special consideration must be given to their unique "rights and interests" (Tri-Council 1998, 6.1), including (but not limited to) informed consent, representation, cultural differences, linguistic barriers, and historical context.

Earth Allies? First Nations and ENGOs

At least in theory, the potential convergence between First Nations and ENGOs could have proved a powerful and strategic alliance to counter government-industry collusion. For some time now, post-structural theorists have deliberated over political cooperation among diverse subjects. To this end, Donna Haraway (1990, 197, 199) reflects on *coalitions* based on *affinity* rather than *unity*, suggesting that different groups (such as environmental associations, feminist organizations, unions, and church groups) may identify commonalties that enable political alliance without reducing specific identities to a single overarching meaning. Indeed, she views such alliances as a critical political imperative. Arturo Escobar (1995, 227) also reflects on the significance and limitations of such counter-hegemonic alliances: "They certainly do and must occur ... But social movements are not ruled by the logic of all or nothing; they must consider the contradictory and multiple voices present in such experiences without reducing them to a unitary logic." Finally, Jennifer Barron (2000, 102) argues for an "articulatory politics" whereby an alliance (or connection) between various political actors occurs in a variety of ways, depending upon specific constellations of subject positions and interests. Rather than resolving each and every point of dissent, organizations join to form temporary negotiated partnerships. Ultimately, Barron (ibid., 111) views such alliances as "moment[s] of discursive articulation [that] do not necessarily persist for all time, nor under all circumstances."

Unfortunately, attempts to form a political alliance – either durable or momentary – in the context of this struggle failed. Interviews with key representatives from the Partnership and from First Nation communities in the Great Lakes–St. Lawrence region help to elucidate why. First, they reveal that the potential for an alliance did in fact exist. Second, they uncover at least in part why such an alliance failed. And finally, they identify the obstacles that need be overcome if a future alliance is ever to be forged.

I should state at the outset that I assumed that Ontario's left-leaning environmental constituency would be sympathetic toward First Nation struggles, an assumption that largely proved accurate. Moreover, I generally found that both Partnership and First Nation representatives were receptive to the idea of working together insofar as it might strengthen their respective positions.

First Nations' Perspectives

The UOI has struggled for years to join forces with ENGOs to create what one representative argued ought to be "a natural partnership or advocacy relationship" (FN$_1$ 2001, 4). He further stated, "I think, by and large, we have entirely the same goals and aspirations and concerns about protection of Mother Earth" (ibid.). One First Nation chief reflected on the generally positive support that First Nations have received from Ontario environmental organizations, but he also cited several instances in which environmentalists directly opposed First Nation development projects (FN$_2$ 2001, 3). So, clearly, support has been qualified and conditional.

According to the UOI representative, members of the Partnership were initially critical of the Lands for Life process, characterizing it as inherently flawed, and, as a result, they were keen to establish a working relationship with First Nations to bolster their political power. However, First Nations perceived a lack of long-term commitment and concluded that ENGO objectives were largely instrumental. The UOI representative reported bitterly that, although ENGOs claimed to want a substantive relationship, "The only interest that we really found ... was to add colour to their campaigns" (FN$_1$ 2001, 4). He further argued that ENGOs often include "flowery principles in their mandate" to (among other things) "respect Aboriginal rights" (ibid.). But if those words were more than mere rhetoric, how then, he asked, could the Partnership justify signing onto the Forest Accord in the absence of First Nations? For First Nations, this was a complete betrayal of trust.

Partnership Perspective

Certainly, members of the Partnership attempted to cooperate (and find common cause) with First Nations. One Partnership representative explained, "As a group, we would have liked them to be allies, and we did some exploratory work on trying to create more of an alliance with them, and it just didn't lead to anything" (ENGO$_2$ 2001, 7). Partnership representatives attributed part of the challenge to the fact that each of the groups has fundamentally (and legitimately) different "starting points": for First Nations, this is economic survival and development; for ENGOs, it is preservation. Thus, although it seems reasonable to think that ENGOs might show support for First Nations in the Lands for Life process (as they attempted to), the Partnership representative did not think it realistic to expect these groups to place First Nation interests *above* their own. Nor would it be acceptable for environmentalists to speak *on behalf* of First Nation peoples (as they themselves recognized).

The Partnership representative felt that the voice of First Nations was, to a certain extent, silenced because they withdrew from the process. He explained, "The Partnership *chose* to participate; the First Nations community *chose* not to" (ibid., emphasis added). Thus, members of the Partnership made pragmatic choices, and in some cases, compromises, to achieve what they perceived to be the best possible environmental outcome.

Obstacles to Convergence

Both ENGOs and First Nations felt that they had tried to join forces but blamed the other side for the failure. These points of dissent are useful as they flag the very obstacles that need be overcome for a successful convergence.

First, although there are, of course, internal differences within these broad categories, environmentalists and First Nations often hold fundamentally different (and seemingly incompatible) notions of nature (and thus protection). As one would expect, the raison d'être of conservation organizations is *protection*. To this end, nature is represented as *distinct from* human community – something in need of protection and restoration, something with which we should seek to harmoniously co-exist, something we are not fundamentally *a part of,* but rather, *apart from.* Such a construction serves to perpetuate a nature/society dichotomy (see Castree and Braun 1998, 2001; Braun 2002). Representing nature in this way has proven a tremendously successful strategy in garnering support from urban populations to secure protected areas. But it is also a representation in fundamental conflict with First Nations' constructions of nature and self. Thus, the UOI representative asserted, "In the long run, who we are, who the Anishnabek people are, is really quite closely related to our relationship to the land" (FN₁ 2001, 5). According to the regional First Nation chief, "What is our relationship to the land? We say the people and the land are one. We say we are self-sustaining. And I have seen and visited communities that did not need government handouts to sustain them. They built their houses out of logs, they hunted and they fished" (FN₂ 2001, 4).

Competing constructions of nature result in differing understandings of what constitutes protection. One First Nation proponent felt that a major stumbling block to forging an alliance may well have been the Partnership's objective of protecting only 15-20 percent of the land base – seen by many First Nations as inadequate to maintain ecological integrity (FN₃ 2001, 4). Similarly, the UOI representative argued, "You can't just say '12 percent is a good number, 10 percent isn't, 14 percent is better.' You should protect all of it to the best of your ability" (FN₁ 2001, 6). However, he also provided this important caveat: "That doesn't mean setting 100 percent of it aside, either. Part of protecting the land is protecting the people who make a living from it and enabling people to make a living from it" (ibid.). Dividing the land into parcels (earmarked for either protection or exploitation) is incompatible with the perspectives offered by First Nations in this process. The UOI representative explained First Nations reticence in this regard: "It doesn't make a lot of sense to our elders, or to me personally ... to say we have to save this, and preserve it, but we'll do whatever we want over here ... It's like we're creating a museum" (ibid.). And herein lies, perhaps, the greatest point of contention: whether protection is a philosophy that ought to be applied to the entire landscape or whether only certain ecologically significant (and ideally, representative) areas ought to be designated as protected.

When First Nations question the split between protection and exploitation, they radically challenge the assumption that progress necessitates environmental destruction and the displacement of people from the land – predominant characteristics of the existing economic order. When ENGOs do not question this split, they implicitly accept a hegemonic model of capitalist growth. Instead, what is needed is a conception of sustainability that encompasses both ecosystem integrity and sustainable livelihoods. To be fair, the Partnership promoted an alternative development strategy that combined ecological protection and the economic health of northern communities (though the vision it offered did not explicitly target First Nations, nor did it resolve the protection/exploitation dichotomy).[7] This issue will no doubt lie at the heart of future negotiations between First Nations and ENGOs, and its resolution will prove critical to alliance building.

One First Nation proponent argued that ENGOs must temporarily set aside the environmental agenda (FN$_3$ 2001, 4). However, he further reflected that environmentalists must first be satisfied that "to recognize First Nations' rights is to protect the environment" (ibid.). What remained unclear in the context of this struggle was the extent to which environmentalists perceived First Nation land stewardship as better for the earth than (and fundamentally different from) industrial capitalism. Certainly, the risk of essentializing and/ or romanticizing the Aboriginal relationship to the land would loom ever present (see Jennifer Barron 2000; Hornborg 1998). But ultimately, the interviewee described the task of finding common cause with First Nations ("people who have first claim to the land") as difficult but satisfying work (FN$_3$ 2001, 5).

Conservation initiatives (such as the designation of park or another protected status) can also threaten Aboriginal rights of access and thus limit First Nation rights to self-determination. The Great Lakes Heritage Coast – one of the signature sites designated by Minister Snobelen in 2000 under Ontario's Living Legacy – highlights this tension. In particular, one regional chief felt that the rights of First Nations to "hunt, fish, trap, and gather on those sites" were being jeopardized (FN₂ 2001, 1). He further insisted that "There had to be recognition of our treaty right of access to resources on those islands. Certainly we were supportive of not privatizing the islands but also not supportive of making them into parks as well" (ibid.). Aboriginal subsistence practices (in both historical and contemporary contexts) offer evidence of a fundamentally different conceptualization of self and genuinely closer ties to the land than are embodied by mainstream Canadian culture. Moreover, in asserting their collective right to a traditional subsistence livelihood, First Nations challenge the juridical categories of private and public property established by the colonial state. Certainly, they are at odds with the individual ownership, commodification, or even outright protection of nature (if this means the exclusion of any human presence).

ENGOs may be unwilling to take positions on First Nation constitutional issues. James Morrison, author of a 1993 World Wildlife Fund discussion paper titled *Protected Areas and Aboriginal Interests in Canada* (1993, 2), explains that resolving such questions of jurisdiction and title may delay governments from creating new protected areas. Similarly, Mac Chapin (2004, 21) argues that ENGO reticence stems from the fact that ENGOs consider such actions to be too overtly political and ultimately "outside their conservationist mandate."[8] According to one First Nation proponent, even the Canadian Environmental Law Association – a group of environmental lawyers attuned to First Nations struggles – was reluctant to become fully involved in the treaty claims issue, since it understood (perhaps better than others) "the extent of the problem" (FN₃ 2001, 5). Moreover, if and when First Nations are successful in asserting their constitutional right to resources, ENGOs might eventually find themselves in conflict with some First Nations over conservation issues (ibid., 6).

Yet another challenge lies in the fact that First Nations, far from monolithic, represent a diverse range of values and perspectives. Clearly, the aspirations of elders, band development officers, organizational representatives, and other community members may not always align. So, deciding exactly *who* represents a given community and *whom* to engage with politically can be difficult. Here Jennifer Barron's (2000, 107) insights apply: "As a result, political unity ... is not assumed, but achieved with some effort (and invariably some dissent)." Of course, the same difficulty applies equally to coalitions of environmental groups (agreement among the Partnership organizations notwithstanding). Although the government endorsed the Partnership as broadly representative of environmental interests, the group represented only the view of "moderate" conservation organizations.

One Partnership representative also cited logistical challenges in developing a meaningful relationship with a spatially diffuse and linguistically diverse community: "Communications and misunderstandings are a real problem since we rarely can afford to get to northern communities" (ENGO₁ 2001, 4). To the problems created by linguistic barriers and physical distance, Chapin (2004, 21) adds "the difficulty of reconciling cultural differences between industrialized and indigenous ways of viewing the world, deliberating, negotiating, and making decisions." Certainly, these differences came to the fore in First Nation criticisms of the process.

Conclusion

A counter-hegemonic alliance, or "articulatory politics" (Jennifer Barron 2000), between First Nations and ENGOs will probably remain tenuous and elusive at best. Adkin (1998b, 10, 18) flags one of the limitations of convergence among diverse groups: bringing together different social subjects to form a common front will not necessarily "result in their actually engaging in a discussion about both their conflicts and their commonalities, and hence, in the modification of their particular identities." She adds that "equivalences must be established among these different subject positions [e.g., gender, race, class, ethnicity] which acknowledge what is unique about each experience while demonstrating their common stakes in a counter-hegemonic project." In the context of conservationists and Aboriginal peoples, Morrison (1993, 29) cites two points of commonality: a shared opposition to large-scale forestry and mining operations, and "a commitment to the ethos expressed recently by naturalist Ron Reid – that, while humans have become the dominant species on earth, 'we still have a cardinal responsibility to share our whole planet with all other living creatures, plant and animal, that evolved here.'" Both groups seem to embrace a sense of connectedness to non-human species, although they may do so from different cosmological starting points.

Clearly, First Nations and ENGOs require a forum through which critical reconciliation of both differences and past grievances can be achieved, a relationship of trust can be built, and overlapping (or at least complementary) interests can be identified and brokered, if only temporarily. For each conflict, representatives from these groups need to sit down and work through the difficult process of negotiating elements of a shared vision, including the means through which it will be achieved. This will require a real conversation about power relations, privileges, and compromises. They also need to explicitly acknowledge differences, including the contradictory and multiple voices present in both communities, so that these very tensions do not serve to unravel any hard-earned gains.

In spite of the daunting obstacles, an alliance between First Nations and ENGOs would better serve both sets of interests. However, the path toward convergence seems fraught with potential misunderstandings, risks of cultural appropriation or homogenization, and, indeed, the potential for succumbing to the divide-and-conquer strategy of the state. But even the dim hope of a potential convergence warrants a whole-hearted attempt to resolve the obstacles.

Democracy will never be served so long as powerful industrial interests dominate public consultation processes. In the case of Lands for Life, power imbalances were reinforced by government-industry collusion. Although the Partnership managed to engage these interests with a degree of success, it alienated First Nations in the process. This chapter has sought to identify the causes of this failure, in the hope that such a tactical relationship may eventually be forged to their mutual benefit.

The active role played by civil society vis-à-vis the liberal democratic state is an increasingly critical aspect of contemporary democracy. Here, engagement in, and critique of, quasi- or pseudo-democratic processes represent important attempts to challenge existing power dynamics and to shed light on unequal power relationships. Counter-hegemonic discourses (put forward by First Nations and environmentalists) are essentially democratizing discourses. These actors offer potent critiques of process and outcome, and envision alternative ways of understanding and relating to the earth, to non-human species, and ultimately to one another. Only by being obliged, on democratic grounds, to embrace differences will political institutions facilitate movement toward what Bruce Braun (2002, 5) describes as a "radical environmental democracy."

In the case of Lands for Life, neither First Nation nor environmental discourses proved powerful enough on their own to supplant the logic of industrial capitalism. As a result, only 12 percent of Ontario's Crown land received official protected status (of course, the corollary to this figure is that 88 percent has been slated for some degree of industrial development). Had First Nations and ENGOs been able to form a strategic alliance, perhaps their negotiated vision might have resulted in an entirely different material outcome for Ontario's Crown land. Specifically, had Lands for Life been framed as an opportunity to redress colonial (and post-colonial) inequities, as well as to ensure social and ecological sustainability, greater gains in these regards might have been achieved.

Notes

1 A summary and critique of the 1999 Forest Accord and the Government of Ontario's policy document, *Ontario's Living Legacy: A Proposed Land Use Strategy,* is available from the Canadian Environmental Law Association. See Paul Muldoon, Theresa McClenaghan, Ramani Nadarajah, Laura Shaw and D. Zabelishensky, *The 1999 Ontario Forest Accord Commitments by Members of the Forest Industry, the Partnership For Public Lands and the Ministry of Natural Resources; and Ontario's Living Legacy: A Proposed Land Use Strategy; and Ministry of Northern Development and Mines, March 1999 Minerals Industry Announcements: An Analysis by the Canadian Environmental Law Association,* publication no. 373, May 1999, CELA (Toronto), http://www.cela.ca/files/uploads/373land_life.pdf.

2 The nine signature sites were Great Lakes Heritage Coast, Nipigon Basin, Algoma Headwaters, Spanish River, Kawartha Highlands, St. Raphael, Nagagamisi, and additional sections of

Woodland Caribou and Killarney Provincial Parks. Each site received special protection and promotion under Ontario's Living Legacy.

3 According to the NRTEE (2004) account, "the forest industry countered [environmentalists' demands for protection of 20 percent of the planning area] with its own predictions of job losses and mill closures. Forest industry workers responded in droves, packing community meetings with busloads (1,600 people in one case) of sometimes very angry people. This factor tipped the balance of round table recommendations away from protected areas goals."

4 Mary Richardson, Joan Sherman, and Michael Gismondi (1993, 11) similarly posit (in the context of hearings related to Alberta-Pacific Forest Industries), "There are other, more subtle ways that governments and project proponents can limit debate and legitimate the desired outcome of the hearing, including dress, demeanor, and a choice of words intended to indicate expertise, trustworthiness, objectivity, and concern for the well-being of society."

5 The Wise Use Movement argues that humans must use natural resources in order to survive, that the earth is inherently tough and resilient, that technology and ingenuity will allow us to surpass finite carrying capacity, and that human modification of the earth is an ultimately benevolent part of our evolution as a species (Arnold 1996).

6 This rejection resulted from Resolution 98/14, determined at the All Ontario Chiefs Conference (AOCC) in West Bay, Ontario, 2-4 June 1998.

7 For more information on the Partnership's vision for northern Ontario, see Partnership for Public Lands (1998, 2000) and Wildlands League (1998a, 1998b).

8 Chapin's "A Challenge to Conservationists" (2004) is particularly relevant. He criticizes the World Wildlife Fund, Conservation International, and the Nature Conservancy for not adequately involving indigenous and traditional peoples living in the very territories these groups attempt to protect.

7
Participation, Information, and Forest Conflict in the Slocan Valley of British Columbia

Darren R. Bardati

In July 1997, on a road above the village of New Denver in the Slocan Valley of south-central British Columbia, an enormous forest company bulldozer made its way up into the forested mountainside past a blockade of about 375 peaceful protesters – artists, teachers, priests, politicians, sawmill workers, grandparents, children, and others (see Figure 7.1). The bulldozer's progress was assisted by a dozen or so RCMP officers who had calmly arrested and carried off those who did not step behind the police tape they had installed along the edges of the road.[1]

I was an observer to these events, having arrived in the valley just a few days earlier on a visit to the region. On that misty morning, seeing the sombre faces of everybody involved – the protesters, the RCMP officers, even the bulldozer operator – I sensed that I was witnessing something deeply undemocratic and unjust. But I had no knowledge of the history of the local residents' road blockade or the company's decision to conduct operations in that particular watershed. In the coming years, I sought to understand what had led to the blockade and what the conflict I witnessed means. This chapter describes in part what I have found.

The War in the Woods and the Commission on Resources and Environment (CORE)

For several decades, conflict over the use of forest lands has marred the peace of the Slocan Valley. On the one hand, the valley's forests provide a host of both timber and non-timber values deemed important by local residents, including small-scale eco-forestry, drinking-water supply, nature-based recreation and tourism, visual quality, and biodiversity conservation. On the other hand, these same forests are coveted for their merchantable timber by an industrial forest company that has been granted long-term timber harvesting rights by the provincial government.

The events taking place in July 1997 in the Slocan Valley were similar to what was happening in many areas of the province during the late 1980s

Figure 7.1 Protesters in the Slocan Valley, July 1997.
Courtesy of Valhalla Wilderness Society.

and early 1990s – what was popularly referred to as the "war in the woods." This war was not fought with guns and bullets (although some incidents of violence did occur) but rather with words and public relations campaigns that appeared in the daily media, on the steps of the provincial legislature, and in the regional offices of the Ministry of Forests (MoF). Its front lines included road blockades and large public demonstrations in the Carmanah Valley, Clayoquot Sound, the South Moresby archipelago, and the Stein Valley (Drushka, Nixon, and Travers 1993).

This clash between combatants, motivated by opposing interests and worldviews, was forcing the government of British Columbia to rethink the way in which the province's vast Crown (or public) forest lands were being allocated and managed. For over a century, forestry has been the driving force of the provincial economy. Land-use planning had traditionally been the domain of the MoF, whose mandate focused on timber production and harvesting – what has been referred to as the "timber bias" (Draper 1998). The challenge facing MoF managers increased in the 1980s, when the need to include the public's input and a host of non-timber values such as wildlife, nature tourism, and drinking-water protection were included in the MoF mandate without any clear policy guidance for the staff to perform these new tasks (CORE 1995). As a result, public input was limited to consultation near the end of each planning process. By the late 1980s, the continued emphasis on timber values over all others had caused a significant public outcry regarding the lack of meaningful public participation in forest land-use decisions as well as the apparent lack of concern over the long-term sustainability of the forests.

Acknowledging the "social dysfunction" in this approach to planning the uses of the province's forests (Owen 1993), the provincial New Democratic Party (NDP) government created the Commission on Resources and Environment (CORE), whose mandate was to develop a comprehensive land-use strategy integrating the "new" non-timber resource values into land-use plans.[2] These plans were to be based on principles of economic, environmental, and social sustainability. CORE used round-table-style consensus-driven deliberations, in which all affected parties participated in the development of land-use plans. CORE provided logistical support, mediation services, and training in interest-based deliberations. When the CORE process began, it was the most innovative forest planning process the province had seen because of its unprecedented levels of public participation and its stated goals of sustainability.

However, after two years of intensive (often very heated) deliberations, none of the regional planning tables – among them, the community table in the Slocan Valley – had reached an agreement on all issues in the time provided. After the deliberations adjourned, CORE staff drafted a series of recommendations and incorporated them into regional land-use plans. Subsequent negotiations between the government and affected groups adjusted the CORE recommendations and positioned the government to make decisions that would affect land use and resource management for years to come. Although the new regime allowed for the formal protection of some wilderness areas and consideration of forestry practices that were more ecological in nature, the dominant paradigm of industrial forest exploitation remained unchanged. Thus, the status quo of traditional forest management was hidden behind what Michael M'Gonigle (1997) has called "a green curtain."

Research Approach
The record of the CORE process provides an ideal opportunity for research into public participation in resource decisions and for evaluating such consultation processes as instances of deliberative democracy. Both CORE and British Columbia forest policy have been well studied in recent years, especially from a political-economy perspective (see, for example, Barnes and Hayter 1997; Cashore et al. 2001; Gunton 1998; Hayter 2000; Salazar and Alper 2000). Although these studies explore the macro-scale implications of

conflict within British Columbia forest policy and make suggestions for improvements through alternative approaches to forest planning, there is a relative dearth of research into environmental conflict on a micro scale – at the level of the communities – to examine the roles that local residents play in shaping conflict and its potential resolution.

This chapter provides an overview of the local history that is important to understanding the particular conflict in the Slocan Valley. It then assesses the Slocan Valley CORE land-use planning process to elucidate its democratic strengths and weaknesses. Knowledge of the historical context and of the process provides insight into why the conflict remained unresolved. The analysis merges the critical communicative theory of Jürgen Habermas (1984, 1987) with public participation literature (Webler 1995). Particularly, the analysis makes use of two meta-criteria: *fairness* and *competence* (which are explained later) to assess how well CORE permitted a full airing of views and the consideration of alternatives with regard to forest uses.

This analysis forms the basis for a discussion of the role of civic science, which has emerged recently in resource planning (Bäckstrand 2003; Fischer 2000; Funtowicz and Ravetz 1993; Krajnc 2002). Briefly, "civic science" refers to constructive interaction between experts and citizens that includes information exchange and creative dialogue for the generation of new knowledge and understandings. I offer some conclusions regarding both the limitations of liberal democratic institutions as a framework for ecological democracy and the potential for their reform through civic science. The outcome of the CORE process for the Slocan Valley reveals the substantial obstacles to a deliberative model of decision making in the context of liberal-capitalist regimes. At the same time, the sustained activism of the local residents demonstrates the demand for procedural democracy as well as the capacity of communities to generate alternative visions of ecological democracy.

Overview of the Conflict

Understanding the conflict in the Slocan Valley requires an overview of its environmental conditions, its history of resource use, and its unique cultural and political dynamics. This overview paints a picture of a community that is familiar with conflict and that – in the 1990s – had done its homework in exploring alternative paths of development.

Environmental Conditions

The Slocan Valley is the drainage basin of the Slocan River, located in the Selkirk Range of the Columbia Mountains, at the western edge of the Rocky Mountains. The valley, which lies in a north-south orientation, is roughly a hundred kilometres long and about thirty-four kilometres wide. The dominant features of the landscape are the rugged, perennially snow-capped peaks rising beyond three thousand metres above sea level and steep forested mountainsides. The Interior Cedar-Hemlock Forest zone, which characterizes most of the Slocan Valley forests, has the highest diversity of tree species of any zone in the province (British Columbia. Ministry of Forests n.d.).

Slocan Lake, forty-five kilometres long, sits at the bottom of the valley, with the Slocan River flowing southward as its only outlet. The headwater tributaries that feed both lake and river originate in very small drainage basins, which are less than ten square kilometres, in the flatter mid-elevation areas. Over 80 percent of the Slocan Valley's fifty-six hundred residents are rural and rely solely on the runoff flowing in these small tributaries for their drinking water. These gentle headwater areas are, however, particularly attractive to forest companies because of their large, commercially valuable trees and relatively easy access. Even the most careful logging and road-building activities in these headwater areas remove significant portions of biomass, causing openings of the forest canopy and changes in micro-climate and soil moisture conditions that affect invertebrate, plant, and wildlife habitat, as well as compacting the ground and increasing surface runoff. Forestry activities also exacerbate erosion on steeper mountainsides and put the drinking-water supply of local residents at risk (Voller and Harrison 1998).

Resource Use

For over a century, the region's resource-based economy was built on three main pillars: mining, agriculture, and lumber milling. In 1890 the discovery of rich silver-bearing ores brought an onslaught of some six thousand people into the area, creating a veritable "mining boom"; the towns of New Denver, Silverton, and Slocan City grew almost overnight. By the end of the First World War, mining in the area had significantly diminished. The small operations that remain are quite unsteady due to rising extraction costs and fluctuating metal markets. The flat, fertile valley bottom in the valley's southern portion, though very small, attracted pioneering homesteaders. But, by the 1960s, the little commercial agriculture that had existed in the first half of the century had fallen victim to the construction of a provincial highway through the southern portion of the province, which made produce from the Slocan Valley less competitive with cheap imports into the valley. Lumbering of the valley's vast Crown timber resources became the dominant economic activity. By the 1970s, small-scale logging operations and local family owned sawmills had been replaced by one company – Slocan Forest Products (SFP) – which enjoyed almost exclusive ownership of harvesting rights. By the 1980s, the company's sawmill in the village of Slocan was the largest single source of employment in the Slocan Valley.

Cultural Dynamics

In the nineteen hundreds, waves of very different communities arrived in the Slocan Valley, joining the miners and loggers to make a living from the land. A diversity of values held by various groups has caused conflict to flare at times between neighbours, and a dominant theme of this cultural history is resistance to government authority. Painful moments in the collective memory of valley residents include the internment of seven thousand Japanese Canadians from the West Coast during the Second World War and, during the 1950s, burnings and bombings over the forced residential schooling of the children of the Sons of Freedom, a religious sect of Doukhobors. During the Vietnam War years, the Slocan Valley saw the influx of many young Americans, some of whom were draft resisters. Many young "back-to-the-landers" also settled there, hoping to escape the perceived ills of life in California, Vancouver, Toronto, and other parts of Canada, bringing with them their social and environmental values and alternative lifestyles (Gordon 2004). By the mid-1970s, these new residents comprised only about 15 percent of the valley's total population, but their presence was strongly felt (Gower 1990). Most of these young people advocated local control of natural resources and the preservation of wilderness areas. In the 1980s and 1990s, the Slocan Valley attracted mainly semi-retired financially independent professionals, whose ideological perspective also largely favoured conservation and community-based management of natural resources.

These post-1960s residents, in particular, promoted a vision of democracy and ecological sustainability that challenged existing policy-making processes and inequalities of power. The Slocan Valley is the home of the Valhalla Wilderness Society, a prominent environmental organization in the province with an international reputation, and the Silva Forest Foundation, a consulting firm whose president, Herb Hammond, is one of Canada's leading authorities on eco-forestry. As well, a grassroots organization of local water users, the Slocan Valley Watershed Alliance (SVWA), functions as an umbrella organization for roughly a dozen watershed groups operating throughout the valley that monitor water supply and advocate scientific studies on water quality, quantity, and timing of flow.

Resource Planning Battles

Out of local residents' efforts came the 1974 Slocan Valley Community Forest Management Project (Slocan Valley Resource Society 1974), which suggested that authority for managing the area's resources be transferred from the provincial Ministry of Forests to a local resource committee made up of representatives from both the local public and government agencies. Although the proposal was not accepted by the government, it was one of Canada's earliest community-based forest management plans, bringing together both public participation and ecological sustainability goals (Jeremy Wilson 1998). Another notable achievement was the establishment in 1983, after a long and fractious battle between the Valhalla Wilderness Society and the provincial government, of Valhalla Provincial Park, a fifty-thousand-hectare area in the Valhalla Mountains on the western slopes of Slocan Lake. However, these events had created a strong polarization between pro-industry supporters, who lobbied against the establishment of Valhalla Park, and the environmental groups, who lobbied for its creation. This polarization of interests remained evident throughout subsequent land-use planning initiatives.

During the early 1980s, the government recognized the ecological boundaries of the Slocan Valley drainage basin as a suitable unit for forest land-use planning and proposed a strategy for "economic diversification," with protection for visual quality and watershed areas. However, this challenge to the status quo was met with intense opposition from the pro-industry supporters, who – with their "Can the Plan" campaign – argued that the plan would result in extensive job losses for the forestry sector. Consequently, the plan's guidelines were never implemented and were eventually abandoned.

In the late 1980s, a newly created Integrated Watershed Management Planning (IWMP) Process attempted to address the conflict over SFP's logging in watersheds, which the Slocan Valley Watershed Alliance (SVWA) strongly opposed. According to SVWA field observations, SFP plans would have had negative impacts on several watercourses, including unnamed rivulets and creeks that were not identified on SFP or Ministry of Forests maps. This claim sparked a debate about the validity of the technical information used by the government to support resource decision making.

Following the IWMP consultations, the Ministry of Forests decided to approve forest road construction in Hasty Creek, above the town of Silverton. In the fall of 1991, a company crew assigned to commence road-building operations was met by over a hundred protesters on-site. Eighty-three people were arrested for refusing to abide by an enforcement order obtained by SFP. The war in the woods had reached the Slocan Valley.

CORE Comes to the Slocan Valley

The creation of CORE appeared to signal the government's commitment to the change in forest development planning for which many Slocan Valley residents had been calling since the mid-1970s. As part of its proposed activities, the CORE commissioner announced that he wanted to investigate options for developing what he called "community resource boards" at the sub-regional (or community) level. A community resource board was defined as "a group of people structured in such a way as to represent a full range of the resource interests and values in a community and who come together to develop advice and recommendations on land use and resource management issues ... Community resource boards serve as a valuable touchstone for government agencies who are seeking advice on complex decisions. They provide a forum where community representatives can identify and develop objectives and strategies for resource management which balance environmental, social, and economic interests" (CORE 1995, 69).

Amid the ongoing conflict at Hasty Creek, the water users asked the CORE commissioner to implement a "pilot project" to assess the feasibility of implementing a community resource board for the Slocan Valley, along with a moratorium on forest development activities pending the CORE deliberations. A letter to the commissioner from a director of the Slocan Valley Watershed Alliance, dated 12 February 1992 (Rutkowsky 1992), states,

> As you are aware, current planning, manifested in "Integrated Watershed Management Plans" is considered inadequate for protecting water by Slocan Valley residents, including those in the Hasty Creek area, for reasons thoroughly detailed [already] ... Last summer's events in Hasty Creek, including the mass arrest of 83 people, are a symptom of government's failure to meaningfully involve people in decision-making that seriously affects them. Rural people value their water highly and are profoundly frustrated with government's apparent unwillingness to take the steps necessary to protect water. Faced with plans for logging and road building activities this spring and summer, residents in Hasty Creek and throughout the Slocan Valley feel an urgent need for a moratorium on resource extraction activities in all consumptive-use watersheds in order to give your commission time to adequately address the issues and propose a strategy ... We urge you to support the notion of a moratorium as an essential ingredient in a fair, effective planning process. Such a moratorium with reasonable time limits, coupled with the dedication of adequate government resources, will hopefully serve as a catalyst for efficient dispute resolution.

Within weeks, a logging moratorium was implemented in the most contentious areas pending preparation of the CORE process. The Slocan Valley CORE process, which ran between 1993 and 1994, is outlined below (A more detailed version of this information was published in Bardati [2003]).

SUMMARY OF THE SLOCAN VALLEY CORE PROCESS

Purpose	To facilitate community participation in developing and advocating the implementation of land and resource management plans that are environmentally, economically, and socially sustainable.
Funding	$100,000 by CORE.
Boundary	All Crown lands in the drainage basin for the Slocan River.

Interest-sector representatives (stakeholders)

Agriculture, forest independents, IWA/labour, local enterprise, local government, mining, outdoor recreation, Slocan Forest Products, tourism, watershed, wildcraft (the harvesting of berries and mushrooms), wilderness. A representative for the provincial government also sat at the round table but was not considered a stakeholder. A representative of the Siniixt (Arrow Lakes) First Nation was an observer at the table.

Constituencies

Interest-sector representatives were responsible for representing, informing, and being accountable to their constituencies.

Timeline

March 1993 to June 1994. The group met seventeen times over sixteen months.

Meeting logistics

Two-day meetings held in various locations in the Slocan Valley.

Procedure

Process design
Participants deliberated a "ground rules" document (approved October 1993), which defined a code of ethics, participants' roles and responsibilities, deliberating principles, meeting procedures, the formation of working groups, the use of technical and substantive information, and the role of the media. The participants also jointly developed a "terms of reference" document (approved February 1994), which defined the purpose and scope of the CORE process, and land-use and resource-management issues and interests, and provided a protocol for liaison with the regional CORE process.

Building agreement toward a final plan
1. Preliminary organization (June-July 1993)
2. Terms of reference (July-August 1993)
3. Information assembly (August-September 1993)
4. Planning zones to define specific objectives for each (September-October 1993)
5. Scenario development of specific management objectives for each planning zone (September-October 1993)
6. Scenario evaluation (November-December 1993)
7. Scenario selection for producing a consensus plan (no date); preparation of plan, implementation, and monitoring (no date)

Implementation and monitoring
These stages were never reached due to delays in discussing process design issues.

Information base	The group had the assistance of a team of government staff from the Ministry of Forests and the Ministry of Environment, Lands and Parks, known as the Technical Working Group (TWG).
Decision making and authority	All decisions were based on consensus by verbal polling. If the participants were unable to reach agreement on a land-use plan, responsibility for delivering a plan to government defaulted to CORE. The province retained authority for approving, rejecting, or modifying the final land-use plan.

Assessing the Slocan Valley CORE Process

The strongest rationales for public participation are based in deliberative democratic theory (Bocking 2006; Fischer 2000), and critical theory (Habermas 1984, 1987), which have focused attention on the social and cultural functions that are served by public participation experiments (Calhoun 1992; Forester 1980). As J. Parkins and R. Mitchell (2005) observe, democratic theorists have synergistically incorporated the ideas of critical theory into political science, creating a robust new approach to assessing public participation.

My analysis of the Slocan Valley experience draws upon T. Webler's (1995) normative theory of public participation, which is inspired by Habermas' (1984, 1987) theory of communicative action. This approach suggests that it is not enough to evaluate a single public participation exercise "objectively" (according to outcome), because of the impossibility of knowing the generalized will. Nor is it sufficient to evaluate it "subjectively" (according to the participant assessments), because of the problem of aggregating individual preferences. Aggregating individual preferences can be very misleading because, though all participants will have their own interests, they may have shared interests that may be more important (to each and every one) than individual interests. Rather, building on the universal principles of communication found in Habermasian theory, Webler's evaluation framework uses criteria that are meant to collectively depict a normative ideal. Two meta-criteria, *fairness* and *competence,* are the central pillars of this normative evaluation.[3] The fairness criterion holds that the participation process should provide every individual an equal and fair chance to defend his or her interests and to contribute to the development of the collective will. Assessing a participatory process for fairness involves asking such indicator questions as follows: Do the deliberations involve all those parties who consider that they have a stake in the outcome? Do participants have the opportunity to set the agenda, discuss, and decide the issues?

The competence criterion means that the process should allow for the construction of the most valid understandings and agreements possible, given what is reasonably knowable at the time. The needs of a competent participatory process are two-fold: participants must have access to information and its interpretation (knowledge), and be able to use the best available procedures for resolving disputes about this knowledge. Applying the competence criterion to the participatory process involves asking such indicator questions as these: Is knowledge accessible to all? Do the deliberations use terms that all understand? Is information true and correct? Are all participants able to meaningfully debate and decide on technical issues? Do the deliberations recognize and address social needs and values?

Like Habermas' work, Webler's approach advances the goals of human emancipation. A key difference is that Habermas defines competence in terms of individual capabilities, whereas Webler defines it in a procedural sense, that is, the use of the best available techniques for resolving validity claim disputes (which are disputes about correctness, appropriateness, and truthfulness of assertions). In performing this interpretive-evaluative exercise, I am aware that these normative criteria measure actual processes against a perfect, or "ideal," deliberative process, where everyone aspires to produce political actions without any hierarchy and no one comes to the deliberations with any "hidden agendas." Habermas (1987) himself admits that egalitarian discourse is nearly impossible to achieve, even in a functional democracy. Therefore, the main question is not so much whether the Slocan Valley process perfectly met these ideal, perhaps unrealistic, criteria, but rather where the gaps are most significant and how they explain the outcomes.

Fairness Criterion

The Slocan Valley CORE process ensured that everyone could have equal opportunity to put concerns on the agenda and to make the process rules. The independent mediator, who was approved by the planning table, ensured that rules were enforced to the satisfaction of all participants. The discussion was open, and everyone had an equal chance to put forth and criticize anyone else's claims about everything under discussion. The only area that appeared to require major improvement was the creation of a pre-approved method to resolve disputes surrounding which claims (about anything under discussion) the participants would consider valid in the process discussions. The Habermasian ideal states that, during the discussion of any agenda item, all participants potentially affected by the problem must have an opportunity to participate and, more importantly, to have a say in how any potential disputes will be resolved. Neither of these conditions was fully possible in the Slocan Valley process, and these failures remained a significant problem throughout its duration. For example, at the sixteenth meeting – less than one month before CORE concluded the process – some factual information was still contested by opposing parties at the deliberation table. The minutes of the 15 May 1994 meeting report that "there was discussion about technical information and decisions where certain sector representatives felt out of their depth, whereas other representatives felt that the challenge was to present good information in a way that enables people to make good decisions." This difficulty in interpreting technical information appeared to inhibit effective discussion participation for some people, which violates the fairness principle. Not everyone had access to the resources and expertise needed to verify claims made about the state of the valley's forest and watercourses. Thus, it became difficult for some participants to agree to a method to resolve this conflict.

To remedy this situation, according to Webler's normative model, a pre-approved consensus method for validity claim disputes resolution should be required not just for the achievement of procedural fairness, but for effective functioning of the discussions. However, the criterion of fairness on this aspect demands a deferral to the competence of others, which may undercut the ability of the deliberation to arrive at a solution to the conflict. Asking participants to agree beforehand regarding which data are acceptable undercuts the fairness criterion about their role of bringing new information and developing new interpretations of data in the process deliberations. The participants' inability to arrive at such an agreement means that the conflict revolved around more than mere access to information and interpretations of data. This problem is discussed further by exploring the competence criterion.

Competence Criterion
In accordance with the conditions of Habermas' competent ideal speech situation, Webler's model requires that the competence criterion meet two basic needs: access to information and its interpretation, and use of the best available procedures for knowledge selection. Access to information is a matter of time, effort, and cost. Although it is difficult to make determinations about how accessible information and knowledge must be, unreasonable inaccessibility is grounds for criticizing the deliberative process.

The budget for the Slocan Valley CORE pilot project was $100,000 and was initially scheduled to last until 31 December 1993. In the minutes of the 25 July 1993 meeting, the CORE staff person states, "If we assume modest expenses, one meeting per month, and that technical information needed is restricted somewhat and done by Ministry of Forests as opposed to contracting out, and that mapping domestic watersheds is not in great detail, then the pilot project could have money until March 31, 1994." The explicit trade-off between the duration of the project and data acquisition, as a consequence of limited funding, is made in this statement. The comment about the lack of detail in mapping domestic watersheds is particularly problematic, given the controversial nature of these areas and their significance in the resolution of the conflict. Although the evidence suggests that the CORE staff did not aim to limit mapping of these areas, the reason given – lack of sufficient funds – was not well received by most participants and proved to be a stumbling block toward successful collective deliberations.

On the matter of information interpretation and use of the best practices for knowledge selection, the Slocan Valley process encountered its most significant pitfall, despite CORE's laudable attempt to level the playing field among its participants. CORE staff encouraged all participants at the table to produce an "interest statement" that defined not only their interests in the process, but specific terms and concepts associated with them. Participants were encouraged to question each other if any terms needed clarification. In addition, CORE's model explicitly required that consensus had to be achieved on everything produced by the deliberating group, including definitions. As a result, the Slocan Valley CORE process scored rather highly on most indicators regarding references made to the participants' use of language and of definitions, references made to social needs, and references made about the subjectivity of participants, even though a few participants insisted that they were not being subjective in their discussions.

The main exception was the diverging characterizations of existing environmental conditions in the valley's forests. Slocan Valley CORE process participants needed access to relevant information, usually technically oriented types of data found in forest cover maps, forest development plans, and hydrological maps, to deliberate about substantial issues related to land-use planning. A persistent dispute surrounded the mapped data of the area under the so-called interim measures. The term "interim measures" was used by the process participants to refer to the temporary logging moratorium placed on the most controversial areas in the Slocan Valley, such as watersheds, while the deliberations proceeded. Throughout the CORE process, the participants from the environmental-interest sector were aggravated that both the government and the industrial interests were proceeding with logging plans, while at the same time deliberating about the fate of the valley's forests. For example, the minutes of 2 March 1993 state, "The table agreed more information was required to address interim measures, and arrangements were made for a meeting to be attended by interested sectors where Slocan Forest Products and the Ministry of Forests could show and explain maps outlining planned harvesting, so that at the next meeting participants could identify to the planning table areas of low, possible or high concern, together with options."

At the meeting, the environmental and watershed sectors questioned the omission of rivulets and creeks from MoF and SFP maps. This issue, which had been raised in the late 1980s in the failed Integrated Watershed Management Planning Process by the Slocan Valley Watershed Alliance, again resurfaced as a key flashpoint in the deliberations. SFP logging plans were being approved by the MoF for areas that – in the opinion of environmental participants – were too sensitive ecologically. The latter claimed that their data and understanding of the local conditions were superior to the inaccurate maps produced by the government ministry and the forest company. Attempting to address the issue, CORE agreed to build a common information database and to form a Technical Working Group (TWG) of experts enlisted from various government resource agencies to coordinate it. Several technical presentations were made by the TWG and outside experts to explain the relevance of the local environmental information to the deliberating participants.

Although CORE staff attempted to create fair exchanges of information by compiling this common information base, the forest company and the government resource ministries proved reluctant to contribute to it. Issues of confidentiality of technical information had to be explicitly negotiated in the process design phase. For example, the government representative agreed to provide all information requested of it, except for that covered by Cabinet confidentiality, "within the limits of government budget and staffing power" (Item 54 of the Ground Rules). The Slocan Forest Products representative requested that the process participants recognize the company's right to limit the amount of information available to the planning table: "Information which is both proprietary and confidential will be withheld where its disclosure would significantly harm a competitive position or result in undue financial or cultural loss or a conflict of interest" (Item 55, ibid.). The other participants countered by deliberating a clause stating that "information will not be withheld from the table for tactical advantage" (Item 58, ibid.). Deliberations over these points diminished the time available for discussion of the more substantial issues about the land-use plan, and the conflict of interpretation remained. In the end, time ran out before participants could agree on a final land-use plan.

Ideally, to meet the normative framework, the process should promote independent verification of data from various sources to check whether factual claims were consistent with the prevailing opinion in the expert community. However, the empirical experience in the Slocan Valley case revealed that this criterion could not be met, demonstrating a deeper underlying problem about the role of information and expertise to support decision making.

First, though CORE's attempt to provide a common information base for all participants is laudable in the interest of fairness, it reinforced the perspective that only professionals, speaking a common language of technical and managerial expertise and sharing similar assumptions concerning knowledge, values, and priorities, have the competence to participate. This assumption is problematic because it tends to reinforce established power influences instead of democratizing information and its interpretation in the deliberating process.

Second, relying on so-called independent verification to determine whether sources are consistent with prevailing scientific opinion may, in fact, be counter-productive to meeting the competence criterion. It may simply reinforce perceived biases in the system. For example, in the case of forest mapping accuracy, one may ask who might be sufficiently "independent" to perform the "verification" of information. Perhaps the registered professional foresters of the province who work for the MoF or the timber companies have the competence to make such a judgment? (Participants from the environmental sector disagreed.) Perhaps the maps should be verified by consultants with expertise in forestry-related issues but who have no affiliation with government or industry? (The industry and government sectors disagreed.) Social actors such as the Slocan Valley process participants can no more agree on the objectivity of experts than they can on the objectivity of information. Hence, the conflict is not resolved, only perpetuated, by the assumption of independent verification.

Third, the fact that expert opinion may be "prevailing" is no proof of its validity. The concept of objectivity – including scientific knowledge in resource planning and management – has been profoundly undermined by epistemological critiques (Ludwig, Hilborn, and Walters 1993; Lee 1993). Our knowledge of the natural world – including the impacts of logging activities – is incomplete, uncertain, and riddled with ignorance (Bocking 2006, 81). For instance, some have suggested that forest inventories, including maps, are "more often than not ... outdated, speculative, or just plain wrong" (May 1998, 33). The Slocan Valley case leads one to conclude that a new approach is required in generating scientific information for use in participatory planning forums: a civic science approach.

Implications: Linking Participation, the Use of Information, and the Need for Civic Science

In democratic societies, government authority derives from the consent of the governed, and public participation is seen as both morally and functionally integral to the fundamental democratic values of political equality and legitimacy of the government's authority (Renn, Webler, and Wiedemann 1995).

The Slocan Valley case suggests that local residents are increasingly less content to let distant governments and private corporations make decisions that will inevitably impact their day-to-day lives without having some sort of input into those decisions. The traditional approach to resource planning, dubbed the "decide-announce-defend" scenario, in which the public is consulted only after a course of action is determined, is no longer acceptable (Connor 1996). As William Hipwell also observes, in Chapter 9 of this volume, decisions to log a particular forest are often made in corporate head offices – head offices that may not even be located in the same region as the forest. Government policies on a range of topics, from timber supply determinations to land-use allocations, are made from distant areas, with little input from or reference to the communities in which the impacts are felt. In many communities, the futures that local constituencies desire for their immediate environment may differ markedly from those sought by industrial giants and government officials. This struggle is not solely one for access to, or development of, rural public lands and resources. It is also a contest for control of the decision-making power that will determine the future of the locality. In many cases, this contest is played out in the very public forum of resource and land-use planning debates, a forum where alternative visions for the future will clash (see also Mason 1999).

Despite repeated statements by government (even following the CORE process) that lauded CORE's unprecedented levels of community input, true control over the land base was never removed from the firm grip of the forest industry. A similar outcome occurred in many British Columbia regions during the 1990s. As A. Gill and M. Reed (1999, 167) state, "Despite the emphasis on improving local autonomy and well-being, local communities still do not have the capacity to make effective contributions to land-use decisions concerning (re)allocation and management. More specifically, existing policy initiatives and institutional arrangements operating within local communities as well as those operating at larger regional and national scales render the contribution of local decision makers marginal."

As seen in the case of the Slocan Valley CORE process, the prevailing standards of knowledge legitimacy play an important role in this marginalization. Although the Slocan Valley residents had been active participants in the many land-use planning forums since the 1970s, their own sources of information and knowledge about forest science, watercourses, and other ecological matters were systematically considered unacceptable for decision making. Only data generated by the forest company and government technical experts appeared to find their way into the final decisions regarding forest planning.

K. Hanna (2000) states that participation presents a paradox: although it is commonly viewed as an essential and influential part of any planning success, public participation may not be as significant to outcomes as the acceptability of knowledge claims rendering certain kinds of information legitimate and others not. According to Hanna (ibid., 399), "planning runs the risk of becoming a reinforcing exercise for determined decisions, where participants serve as window dressing, lending credibility to decisions rather than actually helping to construct them." There is power in being able to determine the definitions of terms, especially when the terms involve technical or scientific complexity – not familiar to the "average" citizen. As seen in the Slocan Valley deliberations, some people were quite unwilling to have knowledge claims forced upon them by the government-industry technical experts. There, scientific and technical information within a planning process played a role in undermining public participation in democratic decisions.

Thus, participation by local residents in the Slocan Valley CORE process was merely tokenistic (Arnstein 1969), while at the same time the planning exercise served to legitimize the provincial government's public forest policy. Situations such as this are all too familiar across Canada, as other chapters in this book attest. Michael Mascarenhas and R. Scarce (2004, 34) state that "a legitimate plan will also attempt to balance regional and provincial concerns with local interests, and expert knowledge with local knowledge." In the Slocan Valley case, the outcomes suggest that the balance seems to have been very heavily tipped against local interests and local knowledge. Many analysts have commented that the BC government's intention was good, in convening the CORE planning forums, but questions of legitimacy remain unanswered (ibid.; Halseth and Booth 2003).

The Slocan Valley case confirms, once again, that public participation mechanisms need to overcome such barriers to information flow and to participation, problems that have existed for decades (Meredith 1997; Grima 1985). What is needed is an information flow that is two-way, as opposed to simply the one way, top down provision of information. Hanna (2000, 400) states that "in some settings the relevance of participation may lie in its ability to contribute to information development." This is an area that may hold promise for the resolution of some resource conflicts. K. Lee (1993, 161) states that managing large ecosystems (such as forests) should rely not merely on science but on *civic* science.

The term "civic science" has been used interchangeably with "civil," "participatory," "citizen," "stakeholder," and "democratic science," and with "lay knowledge" (Bäckstrand 2003, 28). On the one hand, it has been employed to refer to efforts by scientists to reach out to the public, communicate results effectively, and encourage scientific literacy (Clark and Illman 2001). On the other hand, it denotes a science that is developed and enacted by citizens, who may or may not be trained as conventional scientists (Irwin 1995). Not surprisingly, the term remains ambiguous. Yet, civic science serves as an umbrella for various attempts to increase public participation in the production and use of scientific knowledge. Civic science is a pluralistic model of science that provides a more flexible approach to integrating diverse epistemologies, knowledge sources, beliefs, and values (Borchers and Kusel 2003).

Karin Bäckstrand (2003, 25) contends that the science-politics interface needs to be reframed to include the triangular interaction between scientific experts, policy makers, and citizens: "The citizen is not simply the recipient of policy but an actor in the science-policy nexus." This approach serves to question the assumptions of the traditional mode of top-down scientific expert knowledge used in many resource planning efforts, even highly participatory ones such as CORE.

When applied to forestry, civic science engages local communities and workers in the production and use of knowledge for adaptive management of the forest systems on which they depend – a knowledge based in practice. It actively engages all parties with interests in and knowledge of the forest and devises means to share the information for the generation of new knowledge in partnerships. Precisely because it contains these qualities, a civic science approach may offer communities and governments a means to resolve conflicts and find solutions that can be sustained in the long run.

Conclusion

The implications of the Slocan Valley case outlined above suggest that successful public participation in a forest planning process is more complex than staff from CORE and the BC government might have imagined. The case suggests that what lie at the root of the conflict are differing understandings of the environmental and social consequences of development choices, and more fundamentally, differing visions for the future of the locality. A group of local residents, having done their homework and having produced an alternative forest plan, suggested a vision for the future of their valley that they claimed would better balance economic, social, and ecological interests than the plan generated by the provincial government. Their frustration lay in the fact that their vision could not be put in place because the government would not support it. Meanwhile, the same government claimed that its own plan, heavily influenced by timber interests, best served the public good. That the conflict remains is not necessarily a failure of democracy. However, that the industrial-government resource hegemony over British Columbia forest use remains intact, even as it claims full and transparent involvement of interested parties and community groups, is evidence of the need for a democratization of societal decision making. As Commissioner Stephen Owen (1998, 15) stated after he resigned, "[CORE] was created by the provincial government as a challenge to itself and to the public to embrace uncomfortable but essential change. Having unsettled established practices and influences, and mobilized a broader range of interests in the decision-making process, CORE provoked strong reactions ... It was this catalytic role, and the emotions it stirred, that led, perhaps inevitably, to CORE being discontinued in 1996."

The Slocan Valley case demonstrates that good intentions for deliberative planning need to be supported by tools to incorporate varying constructions and interpretations of local environmental information. Habermasian normative theory suggests that the resource planning process is ideal when it is open, when the rules are clear and consistent, and when the element of mystery (an open door for abuse) can be eliminated. When the decision quality becomes a central aspect of the process design, decisions that are most favourable to the plurality of interests are most likely to be encouraged if one interest group does not dominate the outcome (Webler 1995). Unfortunately, this assessment of CORE revealed that domination of outcomes is still possible even when a participation process is fairly close to the Habermasian criterion of procedural fairness. Although the willingness of the provincial government to become more inclusive by consulting with various interest groups has proven necessary, it is insufficient to reform its forest policy.

At the root of the problem are data sources, information quality, and questions of competence. Competent participation requires answers to the following questions: Which data sources should be used? What level of information quality is appropriate? And who is competent enough to interpret these data sources and information types to resolve forest planning conflicts? Again, in an ideal planning process, the deliberating group should decide these things. This is where the Slocan Valley process failed. Participation failed because the competence of local residents – along with their alternative proposals for change – was consistently ignored.

The Slocan Valley case points to a need for the development of a new way of incorporating public participation in resource planning circles, one that remains inclusive and fair but that also supports and promotes the competence of the public. This new development is the emerging civic science approach to resource and environmental planning. Lee (1993, 161) suggests that "civic science is a political activity; its spirit and value depend upon the players, who make up, modify, implement, and perhaps subvert the rules." Because civic science integrates rather than separates both the technical and the social in the production of information for deliberation, it may hold some promise for resolving conflict and for addressing pressing environmental problems in a manner that is more socially acceptable to the general public in a democracy. Unfortunately, this promise has yet to be tested.

The main lesson to be learned from the Slocan Valley experience is that, in some communities, agents for the emerging civic science are already in the making; they need their chance to be tested. The local people, having suggested alternative information for use in CORE's highly deliberative process, were denied the opportunity to have it validated. The Slocan Valley case provides a useful model to chart the possibilities of practical civic science working its way into the realm of public participation processes for local forest management.

This chapter began with a quest to explain the meaning of the Slocan Valley road blockade I witnessed in July 1997. The analysis of the antecedents of the conflict and the assessment of the strengths and weaknesses of the CORE land-use planning process leave me with answers to some questions, though not to all. The solutions are not easy to come by. Short of completely overhauling the entire provincial forest policy system, including the diversification of forest ownership away from corporate control, as some propose (Drushka, Nixon, and Travers 1993; Hammond 1991; M'Gonigle 1997), participation by communities in policy making must be more than procedurally fair. There must also be authentic deliberations on substantive matters such as conflicting data and information used for resource planning purposes. Promoting authentic deliberations means exploring and testing a civic science approach, which is a more inclusive way of generating information and knowledge about environmental conditions and the impacts of various resource planning alternatives.

Notes

1 For photos and details, see "Watershed Logging Protests: Slocan Valley – 1997," http://www. watertalk.org/news/archives/slocanvalley97-8/slocanvalley97.html.
2 The Commission on Resources and Environment (CORE) Act, which created the advisory commission, was passed unanimously by the British Columbia legislature in 1992 to assist the transition to sustainability through the development of an overall provincial strategy, regional strategic land-use plans, increased public participation and Aboriginal involvement, improved government coordination, and dispute resolution processes (see CORE 1995). Stephen Owen, the former provincial ombudsperson, was appointed as its commissioner. Building on experiences of the BC Round Table on the Environment and the Economy, and the work of the Land Use Coordination Office and the Environmental Assessment Office, CORE convened several round-table-style interest-based mediated negotiation forums. These included representation from various interest parties, such as environmental groups, water users, forestry workers, Aboriginals, recreation industry, wildlife, and others.
3 For a more detailed explanation of the evaluative framework, see Webler (1995).

8

The Limits of Integrated Resource Management in Alberta for Aboriginal and Environmental Groups: The Northern East Slopes Sustainable Resource and Environmental Management Strategy

Colette Fluet and Naomi Krogman

The province of Alberta is rich in natural resources including oil, gas, bitumen, and thousands of hectares of merchantable forests. Alberta's boreal forest has come under increasing pressure in recent years as the cumulative impacts of decades of industrial development become apparent and tensions between multiple users of the land base continue to intensify.

One attempt by the provincial government to address the growing complexities of resource management is the Northern East Slopes (NES) Sustainable Resource and Environmental Management Strategy, commonly called the NES Strategy. It applies to Alberta's Northern East Slopes region, a vast area of 40,405 square kilometres on the northern east slopes of the Rocky Mountains. The NES experience provides insights into the limitations of consultative processes in Alberta and has served as a "practice run" for the Alberta government with regard to subsequent land-use consultations. The stated aims of this strategy, which was launched in 1999, were to "enhance communication, cooperation, and consultation among industries, communities, governments and Aboriginal peoples" and to "balance use between diverse interests" (RSG 2001, 4, 6). Terms of reference for the strategy were developed by the Northern East Slopes Environmental Resource Committee (ERC) and "key stakeholders" (ibid.).[1] The NES Strategy was expected to produce values, goals, indicators, prioritized issues, strategies to address those issues and to achieve stated goals, and a process to monitor, evaluate, report on, and improve the consultative process (RSG 2000).

This chapter describes the findings of a study in which we explored the complexities of involving environmental non-governmental organizations (ENGOs) and Aboriginal groups in Alberta's environmental policy network, specifically in the NES Strategy.[2] We compared the involvement of these two groups to that of the forestry and oil and gas industries. We examined the process and the views of the participants to determine how exercises such as the NES Strategy function as opportunities for democratic debate and citizen involvement in policy making. This case study adds to the relatively small body of research on environmental policy formation in Alberta and draws some conclusions regarding the ways in which consultation processes may be democratized. In addition, it contributes to the theorization of the nature of state-society relations in the province; we situate the Alberta case within a number of characterizations of state-society relations.

In the summers of 2001 and 2002, we interviewed nine representatives from ENGOs, five from Aboriginal groups, eight from the forestry industry, eleven from the oil and gas industry, seventeen from various government departments, and two people who were otherwise involved in the NES Strategy. All of these fifty-two individuals had either participated in the strategy in some way or had opted not to engage in it. We selected interviewees through purposive snowball sampling – that is, we generated our initial list of contacts from website publications, newspaper articles, and NES documents, and "snowballed" subsequent interviews from interviewees' suggestions about other persons who were knowledgeable and/or concerned regarding integrated resource management in the region and who did not necessarily hold opinions similar to their own. Interview questions focused on perceptions of the relationships between the provincial government and three key stakeholders in land and environmental management: environmental organizations, Aboriginal groups, and the forestry and oil and gas industries; they also examined the degree and nature of respondents' involvement in the NES Strategy and their perceptions of the involvement of other groups.

Our case study also draws on analysis of NES Strategy documents, including *Final Terms of Reference* (RSG 2000), *The NES Strategy: Interim Report* (RSG 2001), minutes from public consultations (Abells and Henry 2001, 2002), and the strategy's draft recommendations (RSG 2003). Transcripts from interviews and strategy documents were analyzed using NVivo qualitative analysis software.

Environmental Policy and State-Society Relations in Alberta

In Canada, environmental organizations and Aboriginal groups have typically been part of the forest policy *community*, a diffuse group of many actors who share common policy knowledge, but they tend to be excluded from existing forest policy *networks*, a subset of the larger policy community in which members are bound by common material interests (Howlett and Rayner 1995; Hessing and Howlett 1997).

Canada's policy networks have tended to be corporatist, with only the state and the major economic players of industry and labour involved in policy making (though the role of labour is much less significant than that of industry) (Howlett and Rayner 1995). In corporatist arrangements, participants enjoy "representational monopoly within their respective categories in exchange for observing certain controls on their selection of leaders and articulation of demands and supports" (Philippe C. Schmitter, quoted in Downes 1996, paragraph 7). Trevor Harrison (1995, 118) argues that corporatism is an "apt description of how State-civil society relations are today structured in Alberta" but asserts that right-wing corporatism in the province involves only the relationships between the state and the private sector, excluding labour and other "interest" groups.

There are several reasons why "non-economic" actors, such as Aboriginal groups and environmental organizations, have been excluded from policy networks or have had limited ability to be effectively involved in them. Generally speaking, "networks exist because the members have something to offer each other" (Howlett and Rayner 1995, 401). In the past, environmental organizations and Aboriginal groups have not had the economic or political leverage to warrant their inclusion in policy networks. Unlike industry – and, for the most part, labour – environmental organizations and Aboriginal groups have not been sufficiently organized and hierarchically structured to enter into neo-corporatist bargaining or to ensure the compliance of their members with the outcome of any agreement reached.[3] Nor have they been in a position to implement and enforce negotiated policies (Downes 1996, paragraph 19). As well, it should be noted that resource industry associations "have resisted any effort to expand subsystem membership, contesting the legitimacy of environmental groups, for example, to participate in allocating resource lands and organizing local industry supporters to counter grassroots environmentalism" (Hessing and Howlett 1997, 111). On the other hand, as we see in other cases examined in this book – for example, opposition to mega-hog-farming in Quebec (Chapter 3) or the mass protests against the logging of BC's old-growth forests during the 1990s (Chapter 7) – non-economic bodies such as environmental organizations and Aboriginal groups may exercise political power because of their ability to disrupt or prevent development projects or to mobilize public opinion (Downes 1996; Bombay 1996). The following sections situate Aboriginal and environmental actors in Alberta within these general theoretical characterizations of policy networks.

Aboriginal Groups
Although, historically, Aboriginal groups have had relatively little influence in Canada's policy networks, their influence in land and environmental management decision making may be increasing. On both the global and the national scale, there is a growing recognition of Aboriginal rights to share in the economic benefits and decision-making processes of resource development (Peggy Smith 1998; Bombay 1996; Robinson and Ross 1997). In the 1960s, Aboriginal groups began to challenge the "restricted interpretations of Native rights contained in Canadian jurisprudence" (Hessing and Howlett 1997, 58). As Melody Hessing and Michael Howlett observe, through

various strategies including demonstrations, court challenges, and media campaigns, Aboriginal groups are occasionally successful in placing their issues on the political agenda in Canada.

Hessing and Howlett (ibid.) have advanced three reasons to explain why the involvement of non-urban Aboriginal people in policy processes regarding resource and environmental decision making has been strengthened: these include their claim to prior ownership and use of resources, their expertise in aspects of resource use, especially related to hunting, trapping, and fishing, and the impacts of resource policy on their lives. But Aboriginal people, particularly First Nations people, are asking for "a role in collaborative planning beyond general 'public involvement' practices, such as development-by-development notification, open houses, public meetings ... and information sharing workshops" (Treseder and Krogman 2001, 211). As in the case of the Lands for Life land-use consultation in Ontario (Chapter 6), First Nations in Alberta are asking for involvement that is long-term and more fundamental than most public involvement processes – such as co-management arrangements, where management of natural resources is shared in some manner between Aboriginal people, the provincial and/or federal governments, and possibly industry and other stakeholders (Treseder, Krogman, and Tough 2005).

In the case of Aboriginal groups in Alberta, a number of factors make their inclusion in land and environmental policy processes such as the NES Strategy complex. It is Ottawa's responsibility to manage "Indian Affairs"; however, the natural resources that provide the basis for Aboriginal ways of life are managed by the province (Peggy Smith 1995). This split jurisdiction over Aboriginal issues related to natural resources has allowed the two levels of government to "pass the buck" and to avoid directly addressing the concerns of Aboriginal people (ibid.). Also, numerous unsettled specific claims and treaty entitlement claims exist in Alberta, as well as the Lubicon Lake Cree land claim (Canadian Broadcasting Corporation 2008).[4] Ongoing claims consume valuable resources of Aboriginal communities and complicate land-use decisions for all stakeholders. Finally, Aboriginal participation in Alberta policy making is further complicated by the fact that some groups, and/or their leaders, actively pursue resource development for their communities in an attempt to increase employment levels and strengthen their local economies, whereas others are concerned about the impacts of industrial development on the environment and their traditional ways of life.

Environmental Organizations
Non-governmental organizations such as environmental groups are increasingly being incorporated into governance frameworks, and their influence is growing. For example, through networking directly with industry, the Clean Air Strategic Alliance of Alberta produced recommendations – which

were accepted by government – to cut mercury emissions and significantly reduce three other pollutants (Brookymans 2004). As one Strategic Alliance member commented, the gains of the alliance were worth ten years of lobbying (ibid.). Similarly, the US Environmental Defense Fund was able to reach an agreement with McDonald's Restaurant on the design of new packaging that would reduce the amount of waste produced by the fast-food chain (Moffet and Bergha 1999), and Greenpeace has succeeded in lobbying both industry and government to halt oil drilling near Australia's Great Barrier Reef and to eliminate toxic compounds from children's toys (Greenpeace International 2008a, 2008b). Numerous ENGOs have successfully lobbied home improvement stores, such as Home Depot and IKEA, to sell only those wood products that are certified under a sustainable forest management certification system (Cashore, Auld, and Newsom 2004). These are only a few of many examples of the growing ability of NGOs to achieve their policy objectives.

Although environmental organizations have made gains in their ability to affect policy, small ENGOs working at the regional or community level still face obstacles to meaningful involvement in policy processes: these include uncertain funding, their temporary and issue-specific nature, their organizational instability, which may be exacerbated by struggles between and within ENGOs, and the inconstant levels of public support (Hessing and Howlett 1997).

Through the 1980s and 1990s, Alberta was the site of a number of heated land and environmental management controversies, such as the Alberta-Pacific, or Al-Pac, pulp mill public review hearings, the Cheviot Mine proposal, the Alberta Forest Conservation Strategy, and Special Places 2000. Space does not permit detailed discussion of these cases here, but suffice it to say that environmental organizations were ultimately dissatisfied with the outcomes of all of them.[5] In spite of the hearings and public involvement processes, ENGOs felt that their concerns were not addressed and that their work in multi-stakeholder processes was ignored. Economic development seemed to override environmental concerns in all cases, resulting in the ENGO decision to boycott government-led environmental management processes.

This overview of the involvement of Aboriginal groups and ENGOs in Alberta's environmental policy networks in the past few decades sets the stage for the NES Strategy process and helps explain the results of the public involvement processes in this recent case of environmental policy development.

Fairness of Representation on the Regional Steering Group

A Regional Steering Group (RSG) was assembled to function as the lead group in developing the NES Strategy. Membership in the RSG emerged as a major issue for the stakeholders interviewed in this study. The RSG was to receive input from the public consultation forums and to develop recommendations that would be presented to the ministers responsible for sustainable resource development and environmental management (RSG 2001). Advertisements soliciting applications from the public for RSG positions were posted in weekly newspapers in the region and in Edmonton in April 2000 (Abells and Henry 2001). The Northern East Slopes Environmental Resource Committee (ERC) received thirty-four applications from people wanting to sit on the RSG, interviewed twelve persons, and recommended seven, all of whom were appointed by the minister of environment. These seven individuals, plus one person who represented Aboriginal perspectives, made up the "citizens-at-large" portion of the RSG. Although they were linked with various industries and associations, they were not considered to "represent" them in their work on the RSG (see Table 8.1). The chair of the RSG himself (who later shared the position with a government member) was employed by Luscar Ltd.'s Cardinal River Coals and had acted as project manager for the controversial Cheviot Mine Project (RSG 2001).

Aboriginal representatives were recruited through a separate process (Abells and Henry 2001), but only one Aboriginal individual was consistently involved in the RSG. At one point, another Aboriginal person participated, but he did not attend meetings regularly and consequently was asked to resign his seat. At an early stage of NES Strategy development, a third Aboriginal seat was offered to an Aboriginal association that had insisted on having representation in the RSG. This group, however, declined the seat when it was eventually offered.

Six government representatives sat on the RSG: one from local government, one from Ottawa, and one each from the provincial departments of Environment; Sustainable Resource Development; Agriculture, Food and Rural Development; and Economic Development.

The RSG membership was based on a non-representative model. Applicants for citizens-at-large positions on the RSG were reportedly selected based on their leadership qualities, knowledge, skills, and experience in resource, community, and environmental management (ibid.). A government respondent whom we interviewed stated that applicants were also evaluated on their ability to take a regional perspective, regardless of whether a decision affecting the region would have a direct impact on them, and on their willingness to work in a consensus-based decision-making process.

Our interviews and the strategy documents themselves revealed that some saw the RSG as having a "clear and direct bias towards industrial development" (Abells and Henry 2002, 58). This perception arose because an industry person was selected as chair, because many of the so-called citizens-at-large were in some way affiliated with industries in the region, and because many of the government members were seen as being pro-industry.

Table 8.1

Regional Steering Group members for the NES Strategy

Category	Employment	Community
Citizens-at-large	Family farm operator, worked in oil field industry, logging, trucking, heavy construction business	Edson
	Burlington Resources Canada Energy	Calgary
	Petro-Canada	Calgary
	Weldwood Hinton Forest Resources	Hinton
	Forestry Training Centre	Hinton
	Environmental program management with local industry, family-owned guide and outfitting business	Grand Cache
(Chair/co-chair)	Luscar Ltd. Cardinal River Coals	Hinton
	Millar Western Forest Products	Whitecourt/ St. Albert
Non-status Aboriginal	President of Aseniwuche Winewak Nation	Grand Cache
Local government	Mayor, teacher, and principal	Edson
Federal government	Superintendent of Jasper National Park	Jasper
Provincial government		
(Co-chair)	Alberta Environment, NES director for the Land and Forest Service of Alberta Sustainable Resource Development	
	Alberta Resources Development, Mineral Access and Development (replaced by a representative from Alberta Resources and Development, Land Access and Development)	
	Alberta Agriculture, Food and Rural Development (replaced by a representative from Public Lands of Alberta Sustainable Resource Development)	
	Alberta Economic Development, Tourism and Regional Development	

Sources: RSG (2001, 2003); author interviews.

Representation of the environmental community on the RSG was weak to non-existent. Although three RSG members had educational backgrounds related to environmental studies, and two had work experience with industry in environmental management, none were connected to an ENGO. Likewise, none appeared to have experience in environmental advocacy centred on protecting natural systems in their own right. One ENGO respondent, who was interviewed for an RSG seat but was not selected, attributed her exclusion to her refusal to pre-approve the committee's final decisions: "Within the actual interview, I was asked 'Will you support and endorse this process within Alberta to the environmental community?' And I said to them, I will endorse the process after it's finished if we feel that it has been a fair process. But I can't tell you now whether I'm going to do that or not ... You are asking me to say right now that I will carte blanche just go ahead and endorse the process with the environmental community within Alberta, and I said I can't do that."

Aboriginal groups were also discontented with their representation on the RSG. Viewing it as the only legitimate form of involvement in the NES Strategy, they were dissatisfied with the two seats reserved for Aboriginal members (Four Winds and Associates 2000, 2002). Some Aboriginal groups expressed skepticism concerning the legitimacy of the Aboriginal RSG members. Discussing one of these, an Aboriginal interviewee remarked, "I don't know him. He was never commissioned by any of the chiefs or the tribes I know around here. I've been asking around. So they probably just recruited anybody off the street just to say there's Aboriginal involvement."

When asked about the selection of RSG members, government respondents and other members of the RSG explained that it was not intended to be a *representative* body. A government representative commented, "They weren't necessarily looking for representatives of various sectors, or interests necessarily, although the Regional Steering Group is made up of individuals that, not coincidentally, come from those various sectors ... What they were looking for was people who had proven skills and abilities, the ability to sort of look at the bigger picture ... who had a broader range of experience, and in working with these sort of multi-stakeholders processes."

Thus, it was suggested that the dearth of ENGO members on the RSG had come about simply by chance. However, one government respondent offered this explanation as to why ENGO representatives were not selected: "We didn't want somebody that halfway through the process said, 'Well, I don't like the way this is going,' and start lobbying government outside the process ... We wanted to ensure that members participating on the steering committee dealt with their issues on the process and then the issues locally here within the steering committee and not lobby government outside of the process."

Quality of Public Consultations

The development of the NES Strategy involved three phases of public consultations to collect feedback on the work of the RSG as the strategy progressed (for details of the three phases, see Fluet 2003; RSG 2001, 2003). Open public meetings and targeted meetings were held, as were separate meetings organized for Aboriginal communities. The RSG submitted its final report in 2003, and although many government departments supported some of the NES Strategy recommendations, the minister of environment did not accept them (Alberta Environment policy analyst, telephone conversation with Naomi Krogman, 1 June 2007).

Environmental Organizations

Although environmental organizations were aware of the strategy and the opportunities for participation, they had minimal involvement in these public consultation phases. Twelve persons attended the first targeted consultation meeting for environmental organizations, and only two attended the second such meeting (Abells and Henry 2001, 2002). In general, provincially based environmental organizations made a conscious, strategic choice to remain outside the NES Strategy, saying that their decision resulted from frustrating experiences with similar processes in the past. One ENGO representative stated, "We feel we have more influence on the decision making related to land in that area externally ... than we do inside the process as it currently stands."

ENGO respondents mentioned a variety of reasons for opting out of the NES Strategy. They wanted to retain their status as critics, or their autonomy (Chess and Purcell 1999), and they feared that government would construe their participation as endorsing the legitimacy and outcomes of the process. Because the consultation phase was separate from the policy formation and approval phases, they believed they would have minimal influence over the final outcome (especially given that government had failed to follow through with recommendations generated by past processes). Ultimately – as proved to be the case – ministers can exercise discretion over resource allocations, regardless of what regional planning organizations might recommend.

Similarly, some respondents remarked that the political playing field was not level and that the real decisions were made, not around the table with all groups present, but behind closed doors where only government and industry met. ENGO respondents rejected the government's decision not to apply any recommendations that might emerge from the process to existing resource allocations. They felt the strategy lacked attention to protected areas and that the involvement of local people in the NES process was overemphasized, whereas the concerns of Albertans who did not live in the region were marginalized (MacKenzie and Krogman 2005; Parkins 2002; Stefanick and Wells 2000). Finally, some respondents felt the framing of issues within the terms of reference was too narrow to meaningfully address the intrinsic value of landscapes and of nature, requiring environmentalists to frame their concerns in anthropocentric and utilitarian terms.

A 2002 press release signed by eleven ENGOs set out the conditions under which environmental organizations would participate in future processes: that ecosystem-based management be applied, as it had been in the Alberta Forest Conservation Strategy; that participants be self-selected within the major sectors; that funding be provided to cover costs of participation; that consensus decision making, not majority voting, be used; that adequate resources be provided for necessary scientific and economic research; that the terms of reference be defined by all participants at the outset; that an external facilitator be used; and that full access to government information be allowed (Semenchuk and Wallis 2002).

Aboriginal Groups

The process to involve Aboriginal groups in the NES Strategy evolved over time as strategy staff adapted their methods to engage them. Some communities had little or no involvement in the strategy; a few had more. The transcripts of the Aboriginal consultation meetings of the NES Strategy show that some communities saw development of the strategy as a threat to their treaty rights and/or the environment, whereas others saw integrated resource management as an opportunity to improve their economic, social, and cultural well-being (Four Winds and Associates 2000, 2002).

The first round of consultations (which had been designed without Aboriginal input) was regarded as a dismal failure, according to one government respondent, with only six of fourteen identified Aboriginal groups meeting with NES staff (Four Winds and Associates 2000). In the second phase of consultations, a new approach was developed that incorporated Aboriginal concerns. This involved the formation of an Aboriginal Task Team and the creation of a communications tool kit featuring "aboriginalized" terminology, which was specifically intended for use in Aboriginal consultations. Of note, the term "consultation" was removed from documents to address the concern of Aboriginal respondents that their participation would be taken as consent. According to one individual involved with the Aboriginal consultation process, dozens of small meetings were held with people from twenty-one Aboriginal groups.

Comments from Aboriginal groups in the NES Strategy's interim report reveal that Aboriginal communities did want to be involved in the strategy, "but not just in the *consultation process*. They also want[ed] to be involved in the *decision-making process*" (RSG 2001, 31, emphasis added). Like other studies, this one found that Aboriginal groups want to be involved in land and environmental decision making as more than just another stakeholder

(Peggy Smith 1995; Ross and Smith 2002; Treseder and Krogman 2001). Aboriginal communities are bombarded with requests for their involvement in various processes, and their limited resources and capacity require that they prioritize participation – often in federal- and provincial-level projects, such as the Aboriginal Policy Framework, that have far-reaching implications for their continued legal access to the forest.[6]

Aboriginal respondents did not support the RSG format in which three Aboriginal people represented the general interests of all Aboriginal communities in the NES area, preferring a structure that would allow all Aboriginal groups with interests in the region to be represented in the NES Strategy. The strategy's conceptualization of Aboriginal peoples was misguided because it failed to see them as tied to specific places and as sovereign polities who were unable or unwilling to abstract their concerns in a way that could encompass a general "Aboriginal perspective." An Aboriginal Steering Group to function parallel to the Regional Steering Group was one suggestion for more effective Aboriginal representation.

Aboriginal participation in the strategy was also hampered by the Alberta government's apparent lack of preparation, training, and/or commitment to carry out the Aboriginal component of the strategy in an appropriate manner (for example, conducting the first phase of consultations for a new strategy in a short time frame with consultants who were unknown to Aboriginal communities). Cross-cultural communication barriers also affected Aboriginal participation. For instance, English was used in the participation process even though some Aboriginal participants did not speak it as a first language. Procedural norms, such as the short time allotted for decisions and actions, and the process of presenting information and soliciting a response all in one meeting, caused difficulties as well (Treseder and Krogman 2001; Natcher 2001).

For all of these reasons, the participation of Aboriginal groups in the NES Strategy was less than NES staff had hoped, and – for most Aboriginal communities in the strategy area – it could not be called meaningful involvement.

The Forestry and Oil and Gas Industries

Participation of the forestry and oil and gas industries in the NES Strategy was substantial, certainly as compared to that of ENGOs and Aboriginal groups. According to one government respondent, industry was the only stakeholder group involved in developing the terms of reference for the strategy. As well, five of the nine non-government members of the Regional Steering Group were directly employed by the forestry and oil and gas industries. Government staff from various departments were also seen by some respondents across stakeholder groups as representing the interests of the industries they managed. "Targeted" consultation sessions were held with various industry groups throughout the NES process. As one government respondent explained, "We're working directly with each of those industries ... because they're the ones most affected. And they also have the most valid information ... I think we're doing what we need to for alleviating the concerns for their [the industries'] future use."

Unlike their ENGO counterparts, industry respondents clearly felt an incentive to participate in the strategy and that they would have some influence over its outcome. As one put it, "If you're involved up front, you can really help form the process or regulation that's going to come out. That's a big thing because then you can kind of build it how it will work for you." Most respondents indicated that few difficulties were encountered in efforts to involve the forestry and oil and gas industries operating in the region in the NES Strategy. One issue, mentioned by three government respondents, was industry concern over the cumulative effects assessments undertaken by the strategy: "Industry was pretty upbeat until we started to do cumulative effects assessments. And I understand why there is a bit of a shift there because all of a sudden ... there was a possible threat because ... if some bad things come out of there and some red flags come up, especially if they make one industry in particular look bad, that's not good. So, we took measures to deal with that."

Specifically, industry was concerned regarding the "credibility" of the information being put into the cumulative effects model – a concern that NES staff addressed by using industry's own data to populate the model.

Insiders and Outsiders in Environmental Policy Making in Alberta

Taken together, these findings paint a clear picture of the distinct positions of these three stakeholder groups. Industry obviously held a privileged position with regard to involvement in the NES Strategy and had an incentive for participation as it was confident in its ability to influence the strategy outcome. Although a sincere effort was made to engage Aboriginal groups in the process, as evidenced by the revision of approaches along the way, their level of involvement remained limited. Some Aboriginal groups felt the need to be involved but were dissatisfied with the mechanisms for participation, and due to limited resources and time, they were unable to take part in a meaningful way. Attempts to engage environmental organizations in the NES Strategy were weak, with little effort being made to address their underlying concerns. In light of their past experiences in similar government-led processes in which they had had little influence on the outcomes, ENGOs generally chose to stay outside the NES Strategy process.

In the case of the NES Strategy, it appears that, as expected, Aboriginal groups and ENGOs were part of the resource and environmental management policy community but not part of the policy network (Hessing and Howlett 1997). In fact, the network consisted solely of the state and industry (ibid.). Typically, such networks are defined in terms of clientelism or capture – those dominated by the state are perceived as clientelistic, whereas those dominated by non-state bodies such as industry are considered to be captured (see Chapters 4, 5, and 11 for instances of the industry capture of government networks). However, in the case of the NES Strategy, state and industry shared control; thus, the network structure can be described as both clientelistic and captured. The reason for this dual character is that government and industry actors perceived their interests to be virtually identical. The Alberta government has a direct share in the financial benefits of resource exploitation on its public lands and sees the province's prosperity as depending upon private investment in the resource-extraction sector (Walther 1987).

The fact that the efforts made to include Aboriginal groups in the strategy were greater than those soliciting ENGO participation may be attributed largely to the legal status of First Nations claims. Should the provincial government fail to adequately address such issues, costly legal ramifications are a very real possibility, as evidenced by numerous court cases in recent years, including the *Delgamuukw* and *Sparrow* (see Chapter 5) decisions (Peggy Smith 1995; Natcher 2001).

Environmental organizations in the province, on the other hand – although they also use litigation as a strategy to achieve their objectives – have much less legal weight. Their ability to challenge resource developments is hampered by what they see as inadequate legislation for protected areas and land-use policies falling under provincial jurisdiction (Kennett and Ross 1998; Kennett 1999). Without a legal basis for their claims, environmentalists can be dismissed as simply another "special-interest" group.

Another reason for the greater efforts to involve Aboriginals is that the goals of some Aboriginal communities align more with those of industry and government than do those of environmental or conservation groups. Any pro-development position will be easier for government and industry to work with than an eco-centric position.[7]

Narrowly Framing the Debate

The rhetoric around the selection of Regional Steering Group members raises some important questions regarding methods of representation of interests in public input processes. Government respondents and RSG members were insistent that the RSG was not intended to be representative. Thus, although no individuals from environmental organizations sat on the RSG, these respondents asserted that environmental views were adequately represented there. To make this claim is to employ a very limited definition of an "environmental perspective." The concerns of environmental organizations regarding non-use and/or conservation values were given superficial treatment in the NES Strategy, which was unlikely to inspire the participation of these groups.

Indeed, in a response to the draft recommendations of the strategy, the Alberta Wilderness Association stated that the plan, a "summary of long-known land-use conflicts," did little to clarify the process by which these conflicts would be resolved and land uses would be prioritized. Because no clear commitment had been made to achieving the goals of the strategy, this ENGO feared that it would "join the long line of Alberta Government policies that are unenforceable and unfulfilled" (Alberta Wilderness Association 2003).

The narrow framing of environmental values within the dominant paradigm of resource use, excluding eco-centric views, may have been key to garnering industry support but had the effect of discouraging any involvement by environmental organizations as, from the beginning, it ruled out certain courses of action for the strategy. Experts recognized by government do not typically espouse passive use or existence values of the forest. Indeed, industry's resource management experts and the bureaucrats who manage the policy process might be characterized as what Porter and Mirsky (2003) call "pragmatic capitalists," or key players committed to market values over other values, who would prefer that planning is carried out by local elites and professionals with a similar perspective. The selection of "pragmatic capitalists" for the RSG limited the agenda for integrated resource management in the Northern East Slopes. Moreover, in instructing selected members that they must buy into the process and not lobby outside it, the RSG effectively limited public involvement and closer outside scrutiny.

The rationale for omitting environmentalists from the RSG revealed a double standard in attitudes toward industry and ENGOs. The fear of the selection committee, according to several respondents, was that a representative from an environmental organization would leave the NES process to lobby government externally should the organization become dissatisfied with the process – such as did occur in the Special Places 2000 process where the eventual withdrawal of environmental groups undermined the legitimacy of the initiative. The set-up of the NES Strategy and the selection of RSG members effectively prevented ENGOs from employing the dissatisfied exit as a political tool and sought to place the blame for their exclusion on the ENGOs themselves. With environmental advocates absent from the RSG, discussions regarding land management could proceed within a narrow scope that excluded eco-centric perspectives.

Only one government respondent voiced concerns that industry might publicly criticize the process, and no one suggested that industry representatives should be excluded from the RSG because of the threat of exit. Interestingly, industry representatives' objections to the data used in the cumulative effects model were accepted and acted upon by government officials. Industry was clearly granted more licence than were the environmental organizations to oppose or question the terms of reference and the knowledge that would be considered credible. This, in addition to the significant degree of involvement of forestry and oil and gas companies in the strategy, made it unlikely that they would ever need to walk away from the process in protest.

Evidently, the RSG requirement that, at the outset, members agree to accept the final decisions of the strategy was significantly more risky for environmental organizations than it was for industry, given the preponderance of industry interests in the RSG. Furthermore, though the RSG itself may have functioned on a consensus model, the ultimate outcome of the strategy was to be decided by the minister. Several ENGO respondents expressed a willingness to participate in a consensus process in which the outcome of negotiations among the members at the table would not be overruled by higher-order government executives or altered as a result of behind-doors lobbying by industry.

The democratic quality of the Alberta government's consultation process continues to be tested. The province is currently working on a Land-Use Framework, which is intended to become policy for land-use decisions for all public and private lands in Alberta. It has also undertaken background analytical and process design work for the Southern Alberta Landscapes (SAL) project, a regional strategy for the southern part of the province (Alberta. Ministry of Environment 2006). A variety of public involvement processes have been employed in both cases, but a few key differences between the NES and SAL can be noted. First, the SAL steering committee that oversaw the analytical and design phase consisted entirely of provincial and federal government personnel, with plans for the planning and delivery phase to be undertaken by a "strategy development team" comprised of "southern Albertans with an interest in the sustainable development of the region" (ibid.). Design for the planning and delivery phase envisages a phased collaborative process with extensive public and stakeholder engagement throughout. SAL is currently on hold, pending release of the final report and direction from the Land-Use Framework. The province has a new policy on consultations with First Nations regarding land-use management and resource development (Alberta 2005), which may affect Aboriginal involvement in SAL, and it is possible that the SAL process will develop a "parallel Aboriginal process" to move forward with SAL planning (Alberta government employee with responsibilities for SAL, e-mail to Naomi Krogman, 28 February 2008). Opportunities for ENGO involvement remain vague.

We are cautiously optimistic that this process will lead to on-the-ground changes in resource allocations that include ENGO and Aboriginal priorities. The test of our hope will be whether, after two or three years, any tangible policy reform has occurred to incorporate ecosystem values.

Conclusion

This study sheds light on the procedural ways in which the forestry and oil and gas industries exercise a disproportionate influence in government-led integrated resource management efforts, and, more generally, on the ways in which public consultation processes are constructed to proceed without challenging the privileged position of the private sector. (Chapters 6, 7, 11 and 17, in this volume, reach similar conclusions concerning public consultation processes.) The ideological orientation of the government and the economic weight of the oil and gas industry in Alberta have effectively narrowed the debate about land and environmental management to exclude any points of view that might challenge the status quo of this industry. By failing to include all perspectives in land-planning processes, the Alberta government's approach to regional planning is likely to result in more conflict rather than less.

This case study contributes to our understanding of the ways in which public consultation processes may be represented as democratic but simultaneously maintain the power relations that produce the current model of economic development. Although consultative processes in Alberta have increased in number (Alberta. Aboriginal Relations 2007; Alberta. Ministry of Environment 2006; Alberta. Energy Resources Conservation Board 2007), citizen influence is curtailed by the structure of these processes. The lessons learned from the Aboriginal experience with the NES, as well as from the conditions set out by ENGOs for future participation in public consultation processes, point to the need for broader dialogue and agreement concerning the specific goals of the process, the terms of reference for participants, the process of decision making by the committee, the information brought to bear on the learning process of committee members, the authority the recommendations will carry for policy making, and the need for a parallel, culturally and legally appropriate, Aboriginal process. Documentation and analysis of these processes can play an important role in revealing their serious biases and in generating citizen pressure for meaningful participation in decision making and developmental choices.

Notes

1 The ERC consisted of four directors from Alberta Environment, one director each from Resource Development, Agriculture, Food and Rural Development, and Economic Development, as well as representatives of the Energy and Utilities Board, Jasper National Park, and Environment Canada (Francis 2001). Interview data suggest that representatives from the forestry and oil and gas industries were involved in creating the terms of reference for the strategy.
2 We use the terms "Aboriginal groups" and "Aboriginal communities" as a shorthand to refer to First Nations, Metis, and non-status Indian groups. Their use is not meant to disrespect the distinct status of these groups in any way.

3 Neo-corporatism, according to David Downes (1996), refers to corporatist arrangements within modern liberal democracies and does not assume dominance of the state. This is in contrast to the corporatism of the fascist states of Italy and Germany.

4 Treaty entitlement claims arise when the Crown has failed to fulfill its treaty obligations, whereas specific claims occur when the Crown has failed to fulfill its obligations to Aboriginal people as set out in the Indian Act (Alberta. Aboriginal Affairs and Northern Development 2000b).

5 For details, see Colette Fluet (2003); Larry Pratt and Ian Urquhart (1994); Paul Edwards (1990); Mary Richardson, Joan Sherman, and Michael Gismondi (1993); John McInnis and Ian Urquhart (1995); Ian Urquhart (2001); Sierra Club of Canada (2005); Glen Semenchuk and Cliff Wallis (2002); and Lorna Stefanick and Kathleen Wells (2000).

6 The Aboriginal Policy Framework, released in 2000, "sets out the basic structure for existing and new Government of Alberta policies to address First Nation, Métis and other Aboriginal issues in Alberta" (Alberta. Aboriginal Affairs and Northern Development 2000a, 1).

7 As Robyn Eckersley (1992, 2004a) explains, there are at least two schools of eco-centric philosophy. The first asserts that all entities that demonstrate self-production or self-renewal have "intrinsic value" and deserve moral consideration. Examples of these include species, ecosystems, or even the ecosphere. Arne Naess (1983) and William Devall and George Sessions (1985), among others, argue for the principle of biocentric equality, as opposed to anthropocentrism. The second tradition, transpersonal ecology (see Warwick Fox 1990), advocates the cultivation of a sense of self that extends beyond the individual ego to include all beings.

9
Environmental Conflict and Democracy in Bella Coola: Political Ecology on the Margins of Industria
William T. Hipwell

In 1995 members of the Forest Action Network (FAN) travelled to the remote Bella Coola Valley in northern British Columbia to assist in the defence of old-growth forests threatened by industrial logging. At Bella Coola, the Vancouver-based FAN signed a protocol of cooperation with the House of Smayusta (literally, House of Stories), a radical-sovereigntist faction of the indigenous Nuxalk Nation that was embroiled in a complex political dispute with the elected Nuxalk Nation Band Council. Together, FAN and the House of Smayusta (HoS) mounted civil disobedience actions in the region and a broader public relations campaign featuring Nuxalk dressed in ceremonial garb. During blockades of logging roads at Ista (Fog Creek, King Island), FAN activists were arrested alongside hereditary chiefs, elders, and other Nuxalkmc (Nuxalk people) associated with the HoS. The relationship between FAN and the HoS was characterized as an alliance between environmentalists and the "traditional leadership" of the Nuxalk people. This relationship reached its zenith in 1996 but persisted until as late as 2001.

The FAN-HoS alliance arose in response to a constellation of ecological and economic factors. A succession of transnational forestry and fisheries corporations have been exploiting the resource wealth of the Bella Coola region for several decades, using ecologically damaging and unsustainable harvesting practices. The companies involved have primarily employed non-residents and similarly have seldom sourced supplies in the area. Yet, as the following pages will show, the relationships between FAN, the Nuxalk, and non-Aboriginal residents ("Bella Coolans") were problematic and in many ways sabotaged local efforts to strengthen ecological democracy.[1]

My research into this conflict was undertaken from 1995 to 1997, with fieldwork performed during the summer of 1996 in the Bella Coola Valley. As a radical environmentalist myself, I initially set out to prove that radical environmentalists and indigenous sovereigntists were natural allies – a view widely held in the movement. This supposition was fractured by field data, and what emerged was what B.G. Glaser and A.L. Strauss (1967) have described as "grounded theory," where field data are allowed to guide and shift a researcher's conceptual framework. I conducted semi-structured interviews and focus groups with thirty-eight persons involved in or knowledgeable about the conflict. This group included nine Nuxalk elders, five Nuxalk hereditary chiefs, one elected Nuxalk band councillor, various Nuxalkmc who self-identified as supporters of either the band council or the HoS, several Bella Coolans including one elder, eight Forest Action Network activists, and three government employees with the BC Ministry of Forests (MoF) and the federal Department of Indian Affairs and Northern Development. This selection of respondents, undertaken by identifying spokespeople in media stories and by "snowball sampling," ensured that key opinion leaders and community activists represented all of the major competing viewpoints in the valley (Babbie 1995). All interview respondents are identified loosely by category (such as Nuxalk hereditary chief, Bella Coolan, environmentalist, and so on), as well as by an additional identifier of their own choosing (such as former logger, FAN activist, HoS supporter) in cases where they wanted to be identified more specifically. Only one respondent agreed to be identified by name. Respondents were encouraged to speak about any issues they considered important for the Bella Coola Valley. Interviews were conducted until I felt sure that no new themes or perspectives were emerging (consistent with the grounded theory methodology of Glaser and Strauss 1967). I subsequently coded interview transcripts according to themes that had recurred, including forestry/forest, fish/fishing/fisheries, Nuxalk, House of Smayusta, band council, and so on. The story I tell here is that which emerged from the interviews.

Secondary data analysis included a comprehensive review of local newspaper coverage of natural-resource-use management and community politics over a five-year period, extensive Internet research, and access to the archives of the Bella Coola offices of the BC Ministry of Forests, the Forest Action Network, and the Central Coast Economic Development Commission. This archival work was used to triangulate interview findings.

Using these data, this chapter seeks to uncover the "politics of difference" in the Bella Coola region (Deleuze and Guattari 1987). My underlying questions include the following: What lessons may be learned from this case about the potential for both neo-colonialism and solidarity in non-Aboriginal environmentalists' relationships with indigenous peoples? How do indigenous and non-Aboriginal communities use claims to the right of self-governance to resist and to negotiate with the corporate and governmental agents of "Industria," the global urban-based socio-politico-economic complex of industrial capital (Hipwell 2004a, 2007)? What are the implications of linking rights to decision making (in this case, regarding the terms of resource exploitation and the nature of the local economy) to one's belonging to a place?

Few studies have examined the relationship between environmentalists and indigenous peoples in Canada, and of these, fewer still pay attention to the theoretical sophistication offered by post-structuralism. Patricia Ballamingie (Chapter 6, this volume) makes important headway here, using a genealogical and discursive approach to deconstruct some of the generalizations that continue to dog critical scholarship. As she notes, "First Nations, far from monolithic, represent a diverse range of values and perspectives" (99). My research reveals that this is equally true of, and absolutely essential to understanding, the Nuxalk.

The nature of environmentalist-indigenous interactions varies considerably across cases. At times, the relationship is an intensely antagonistic one, such as that typified by the long-standing conflict between animal rights activists and indigenous trapping cultures. At other times, there is relatively unproblematic cooperation, as occurred in South Moresby on the Queen Charlotte Islands (May 1990). At yet other times, racial animosity between locally based environmental or conservation groups and local indigenous nations renders cooperation difficult, despite broad areas of common interest. More frequently, the interaction is ambiguous, combining a mixture of these elements, as in the relationship between the Ahousaht Nation and the Friends of Clayoquot Sound (Chapter 4, this volume). This was also the case in Bella Coola.

Local and "Glocal"

Noel Castree (2004) has aptly shown that critical scholars should not necessarily view localism, even where it erects "strong" boundaries, as regressive. He argues that, in light of the multiple ways in which local identities, ecologies, and economies are at once rooted in particular places and yet simultaneously transected and penetrated by global forces, a degree of local defensiveness is both justifiable and pragmatic. He follows W. Magnusson and K. Shaw (2003) in dubbing "glocal" this multiplicitous intersection that means that particular places are in many important ways simultaneously global and local. In many regards, the situation in Bella Coola was a manifestation of the age-old tension between translocal universalisms, on the one hand, and localisms of various definitions and degrees of militancy, on the other. The universalisms appeared in the guise of the multiplicitous Industrian actors, forces, and moral and legal codes that penetrate the margins.[2] Localisms will be discussed at greater length below.

Industrian universalisms were especially manifest in the discourses and actions of industrial forestry multinational corporations (MNCs) and their collective nemesis the Forest Action Network. Both FAN and these corporations told stories about what should take place in the Bella Coola region that were based on quite different but nonetheless universalist visions of "the public good." Industrial forestry MNCs typically attempt to reify their tenure, boasting of their technological and scientific ability to "manage" forests – with (preferably minimal) regulation by the provincial government – for the benefit of a mythical and undifferentiated "public" (Willems-Braun 1997). These corporations also routinely ignore the unsettled issue of legal title to the lands of BC. Variations in the social and political topography are flattened out in favour of an illusory homogeneity masking the effective dictatorship of the urban majority and the non-Aboriginal settler population. These discursive strategies played out starkly in Bella Coola.

In conventional democracy, majorities rule, and areas with the greatest population wield more political clout than sparsely populated rural areas. Politicians and civil servants allocate natural resources from remote regions to corporations that can promise to provide the greatest increase in provincial GDP over the short term (typically a politician's four years in office). Although the need for rural employment is sometimes the overt political justification for resource allocation decisions, this is belied by decisions in which large industrial extractors offering large production volumes but a decreasing number of jobs per unit of resource are favoured at the expense of small resource users offering lower rates of extraction but longer-term sustainability at higher levels of employment by volume. Wood supplies are directed to mills employing the greatest number of electors at the provincial level to create constituencies for resource exploitation. However, local employment is not a deciding factor in the investment decisions of companies. Automation of harvesting and processing operations has meant that rural unemployment continues to grow, provoking an out-migration of youth to the cities (Brownson 1995). In this neo-colonial arrangement, natural resources and the wealth created by their exploitation flow steadily from the rural "resource periphery" to the urban "core" (Barker 1986; McCann 1987; Murray 2005).

FAN, on the other hand, wished to universalize its vision of nature as a place where the activities of industrial humanity are at best unwelcome. The Great Bear Rainforest – especially the parts of it that could be designated "old-growth" – represented a living heritage, the protection of which could be entrusted to neither the Bella Coolans nor the Nuxalk Nation Band Council (NNBC). FAN had concluded that the HoS was the legitimate voice of the "traditional" Nuxalk Nation and that it was more culturally authentic – and more ecologically responsible – than the NNBC. This decision was typical of a larger alignment among "radical environmentalists" and "indigenous traditionalists" in Canada during the 1990s (Hipwell 1997). Moreover, citing international law and the history of colonization, FAN's implicit interpretation was that the Nuxalk claim to the Bella Coola Valley was superior to that of the descendants of later settlers.

Localisms at play were similarly multiplicitous, ranging from the militant indigenous sovereigntism of the HoS to the more diplomatic arguments for local control over natural-resource-use management emerging from groups such as the Central Coast Economic Development Commission and the local chapter of the Sierra Club of Canada. However, even the politics of engagement favoured by the NNBC paid scant attention to the needs or wants of people outside the region.

As with the inhabitants of the Slocan Valley or the Oak Ridges Moraine (Chapters 7 and 17, this volume), local people lay claim to a privileged position in decision making especially as it pertains to nearby lands and resources. Part of this may be a pragmatic recognition of the primacy of usufructuary rights in common law. Or perhaps this claim is predicated on the fact that local people are *physically comprised of, and animated by, local elements* – notably, the water they drink and the air they breathe. These sentiments – of belonging and the right to authority vis-à-vis a place – play out frequently in the development of various local resource-use or conservation plans. When local people perceive that their inherent right to input has not been adequately regarded, they frequently react with protests and blockades, coupled with a distrust of outside intervention.

In response, governments sometimes grant local control over some non-threatening part of the natural resources sector. As Donna Harrison (Chapter 4, this volume) argues, the Ahousaht First Nation was offered some participation in the development of local aquaculture, while being simultaneously driven out of the more lucrative and contentious offshore fishery. This act of redirecting or co-opting local energies into realms less threatening to power relations may also be at play in government-sponsored processes such as the Bella Coola Local Resource Use Plan (LRUP) discussed below.

Although these categories (universalism and localism) in Bella Coola were interpenetrating and fluid, we can discuss the penetration of the region by Industria as *neo-colonialism* and efforts by Bella Coolans and Nuxalk to stake out some version of local control over resource extraction as *resistance*. For all its isolationist characteristics, this localism may – as it was in the case of Bella Coola – "be founded on an explicit and conscious engagement with extra-local forces" (Castree 2004, 163). Places such as Clayoquot Sound or Bella Coola can be seen as functioning as "nodal sites" – locations where lateral forms of resistance to the verticality of Industria can meet and re-inforce one another (M'Gonigle 2003).

The Bella Coola Region and Its People

Concealed by Pacific fjords to the west, hemmed by the fifteen-hundred-metre-high Ts'ilhqot'in Plateau to the east and the towering peaks of the Coast Mountains to the north and south, the Bella Coola Valley is the centre of the ancestral territory of the Nuxalk indigenous people. It is home to about twelve hundred Nuxalkmc as well as a slightly larger number of Bella Coolans, primarily descendants of Norwegian settlers who arrived in 1894, along with more recent arrivals. As with many small, geographically isolated communities, local "geographs" (Dalby 2002) in the Bella Coola Valley contain a sharp distinction between "Inside" and "Outside," to the extent that these words are frequently capitalized in local newspaper stories.

The valley is the centre of the heavily treed region that the BC Ministry of Forests has designated the "Mid-coast Forest District." It is a significant source of the rapidly dwindling natural resources upon which Industria must feed. The valley is geographically isolated. By road, the journey from Vancouver is 1,100 kilometres, the final 460 kilometres of which is primarily gravel. From Anahim Lake, at the top of the Ts'ilhqot'in Plateau, in a stretch known in local parlance as "The Hill," the road narrows to one lane and descends down the face of a cliff from the top of the plateau to the floor of the Bella Coola Valley in fewer than nineteen linear kilometres of switchbacks, with grades reaching 18 percent. The valley may also be reached by small aircraft or an infrequent ferry service.

The Biotic Community

Local forests are made up of western red cedar (the primary building material used in traditional Nuxalk culture), western hemlock, Sitka spruce, Douglas fir, amabilis fir, yellow cypress, mountain hemlock, alpine fir, and black cottonwood. The Nuxalk have also used innumerable edible and medicinal plants in the forests for centuries.

At least twenty-three "rare or uncommon" animals are found in the region (G.E. Bridges and Associates 1994, 15). Large mammals include grizzly bears, black-tailed deer, moose, cougars, timber wolves, and mountain goats. In every interview in which any discussion of logging occurred, mention was made of the damage to wildlife habitat caused by industrial forestry (twenty-eight of thirty-eight interviewees addressed this topic). In the words of a Nuxalk hereditary chief, "I was concerned about the needless clear-cutting. It affects everything ... We aren't opposed to logging, but they don't take care; it's hurting the salmon" (Nuxalk hereditary chief 1, HoS supporter, interview June 1996).

The tidal estuary and saltwater fjords are similarly abundant in life. Numerous important fish species and harbour seals are found in or near the Bella Coola River estuary. The streams that run throughout the region provide some of the most important habitat in North America for five species of salmon and two species of trout. Overhead, the sky is dotted with hunting bald eagles.

The People – Nuxalk

The Nuxalk Nation has occupied the Bella Coola region since time immemorial, operating a natural-resource-based economy and engaging in trade with adjacent nations. Shortly after first contact with Europeans in the late eighteenth century, successive waves of European-introduced diseases decimated the population, which is now a small fraction of its historical size. The Nuxalk have also been significantly deterritorialized; the Bella Coola Indian reserves total a mere 20.24 square kilometres – only 0.13 percent of the approximately 15,600 square kilometres occupied by Nuxalk family groups in 1850. The Nuxalk unemployment rate in 1996 was almost 80 percent. The rundown houses and unpaved streets on the Nuxalk side of Bella Coola village create a stark visual contrast to the better-served non-Aboriginal side of town. Alcoholism and other poverty-born social and health problems continue to plague the Nuxalk.

Traditional Nuxalk democracy differed greatly from the putatively representative democratic institutions and mechanisms developed by European cultures. Decisions in Nuxalk communities were made by consensus. Members of a village, led by hereditary chiefs representing various clans, would meet to discuss possible courses of action on an issue of importance, "and if unanimity of opinion was reached, that course was decided on" (McIlwraith 1992, 340). The principle of collective consensus, guided by opinion leaders, was paramount.

During the 1990s, two distinct political factions divided the Nuxalk. The NNBC had been established by the Canadian government under the Indian Act of 1870 as part of the band council system imposed on indigenous nations across Canada. In 1996 it comprised eleven councillors and a chief elected every two years by Nuxalk aged eighteen or older, in conventional secret-ballot elections. In 1995, under the leadership of Chief-in-Council Archie Pootlass, the NNBC began holding public meetings (with a quorum of 230) to decide on issues of importance. Several interviewees noted that these meetings had revitalized the community participation that had been intrinsic to the traditional hereditary system. Ironically, FAN's intervention in the valley had the unanticipated side effect of promoting Nuxalk democracy, as the first such meeting was called to decide whether the HoS should be allowed to bury the body of FAN activist Corey Duncan, who had died in a hiking accident, in the Nuxalk community graveyard.[3]

As the headquarters of the putatively traditional "Nuxalk Nation Government," the Nuxalk House of Smayusta functioned as a cultural and political centre for Nuxalkmc interested in learning about the "old ways." The HoS claimed to represent the hereditary chiefs of the Nuxalk and generally those disaffected by the band council system. Claiming to operate according to traditional political principles of consensual decision making led by hereditary chiefs, it has long pursued a radical-sovereigntist policy, rejecting the authority of Canadian governments over unceded Nuxalk territory.

The HoS assertion of legitimacy was largely based on the support of Nuxalk elders and hereditary chiefs. Moreover, the HoS claimed that its radical-sovereigntist position, in which it refused to acknowledge any jurisdiction by Canadian governments or agencies, was more legitimate than the position of the NNBC, which maintained relations with Canadian government agencies and is actually constituted under Canadian law. On this basis, the HoS called into question or explicitly denied the legitimacy of the NNBC. At the time of field research in 1996, the HoS had the open allegiance of six of the eighteen Nuxalk hereditary chiefs (at least three hereditary chiefs were openly opposed to the HoS, and the rest remained publicly neutral). Of the twelve elected councillors including the chief-in-council, six at least nominally supported the HoS and its sovereigntist stand at Ista.

The People – Bella Coolans

Socio-economically, the Bella Coolans were only slightly better off than the Nuxalk. In 1996 the Bella Coolan unemployment rate neared 50 percent. Interview respondents spoke of resultant social problems including alcoholism, fetal alcohol syndrome, and domestic violence. Local municipalities were impoverished, and there were serious concerns about potable water supplies due to antiquated sewage and plumbing systems.

Bella Coolans were active in numerous conservation groups and local planning initiatives. The Central Coast Economic Development Commission was formed by valley residents to address the unsustainable nature of industrial forestry and to search for economic alternatives. It conducted research and produced several fact sheets (such as CCEDC 1995). Most significant for forestry matters, the Bella Coola Local Resource Use Plan Committee (the LRUP Committee) was initiated by the BC Ministry of Forests in early 1990 in response to concerns expressed by Bella Coolans about unsustainable levels of cutting in the region. Its members were mandated to assess forest resources in the planning area and to recommend a sustainable harvest rate to the government (LRUP Committee 1996). In August 1996, the committee released its report, which was titled the *Bella Coola Local Resource Use Plan* (LRUP Committee 1996). The LRUP itself was opposed by key industry players and ultimately shelved by the provincial government. For their part, both the Nuxalk and the neighbouring Heiltsuk indigenous nations complained that the LRUP report had been written "by non-native communities for the benefit of non-native people" (Arlene Wilson 1996).

Neo-colonialism?

Today the marginalization of native voices can be found, despite important differences, in the rhetorics and practices of both extractive capital and environmentalism (Willems-Braun 1997, 25, emphasis in original).

The ENGOs came in like environmental warlords. They are colonialistic. It reminds me of Indian Agents at the turn of the century (Nuxalk Nation band councillor, interview June 1996).

Figure 9.1   Industrial forestry at Clayton Falls Creek, July 1996.
Photo William Hipwell.

Industrial Forestry in Bella Coola

Forests in the Bella Coola region have been severely degraded by industrial forestry. Given that the main offices of both governments and corporations are located in major cities and that timber exports account for more than half of provincial export revenues (59 percent in 1992) (Brown 1995), resource-extraction corporations have greater access to and exert greater influence on senior government decision makers than do residents of the areas being logged.

Until the mid-1990s, International Forest Products Limited (Interfor), TimberWest Forest Industries Limited (a subsidiary of Fletcher Challenge Canada), MacMillan Bloedel, and Dean Channel Forest Products, known locally as "the majors," carried out most local forestry. Their logging practices, typified by what can only be termed the obliteration of the forest in Clayton Falls Creek valley by clear-cut logging in 1991, form an inescapable context for any assessment of industrial forestry in the Bella Coola region (see Figure 9.1).

According to independent studies based on BC Ministry of Forests (MoF) data, these companies had, both individually and collectively, been logging at unsustainable levels (between two and two-and-a-half times the sustainable

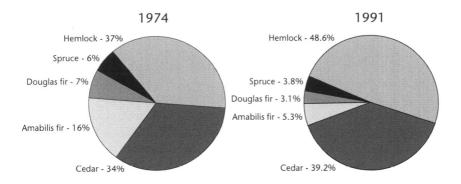

Figure 9.2   Mid-coast forest composition, 1974 versus 1991.
Based on statistics from CCEDC (1995); additional graphics work by Khyati Nagar.

long-term harvest level determined by the MoF) throughout the Bella Coola region since at least as early as 1983 (G.E. Bridges and Associates 1994; CCEDC 1995). This represents the deforestation of as much as three thousand hectares every year.

Selective deforestation and frequently unsuccessful efforts at reforestation with single species have progressively degraded the forests. The biodiversity of the Mid-coast Forest District has been significantly altered as a result of unsustainable forestry practices. Between 1974 and 1991, as shown in Figure 9.2, the percentage of the most commercially valuable species including amabilis and Douglas fir declined sharply (CCEDC 1995).

As they changed the overall forest composition, the companies cut stands with the biggest, strongest trees and replanted the logged area with imported seedlings. As a result, the gene pool in these areas deteriorated because the regenerated forest (such as it was) did not carry the DNA of the large trees that had been removed.

Between 1984 and 1993, this harvest earned forestry companies in excess of $3 billion after costs (ibid.). Yet in 1992, it generated only 191 jobs in the entire Mid-coast Forest District, including harvesting, transportation, processing, silviculture, and provincial government employment (G.E. Bridges and Associates 1994). The workers who played a major role in producing this wealth for the companies earned roughly $40 million only (ibid.). Most of them were shipped in by logging companies from outside the region, prompting one Nuxalk hereditary chief to comment, "Interfor brings in all these people from out of town and steal the Nuxalk's work" (Nuxalk hereditary chief 3, HoS supporter, interview June 1996). Nor did even the maintenance of logging camps benefit local communities: $50 million worth of supplies and services for the logging camps were being imported from southern British Columbia each year (CCEDC 1995).

The response of TimberWest to the LRUP report is telling. The LRUP Committee member representing TimberWest did not sign the plan and indicated in a letter to the committee that "economic realities dictate that the majority of the harvest must continue to flow to the [southern] conversion plants best able to extract the highest value from each log. To propose controls which would direct the harvest to local plants incapable of extracting the highest value constitutes a subsidy and [therefore] a cost to the public treasury. It may not be in the overall public interest and it could spark countervail [sic] actions" (Brennenstuhl 1996, 17). In the end, no forestry companies endorsed the plan.

As noted earlier, this invocation of the "overall public interest" in effect privileges the interests of people outside planning area communities and associates private interest in capital accumulation with a public interest in economic growth. This amorphous, ghost-like "public interest" is a key component of the modernist telling of the story of political life, used frequently to disguise the operations of small groups of elites. Nor is it clear that industrial plants can indeed extract the "highest value" from each log. Local arguments for value-added secondary production imply the opposite.

It is revealing to see the active hostility of TimberWest, as manifested in its threat of "countervail actions," toward the idea of local control. The subsidy argument is interesting in at least one additional respect. It renders invisible not only the cost of road building and other government expenditures made for the benefit of forestry corporations, but also the cost to the public treasury of unemployment insurance, welfare, and other social assistance to the unemployed population in the territory that is providing the trees. This is a price that the government and the community have to pay when companies are allowed to export the trees from the region rather than transforming them into commodities locally.

Enter the Forest Action Network
Formed in the summer of 1993 during logging protests at Vancouver Island's Clayoquot Sound, FAN was initially a loose collection of activists. In 1995, these activists coalesced into a more permanent group, dedicated to changing forest policy in British Columbia and promoting alternatives such as eco-forestry and value-added manufacturing through public education and non-violent direct action. In the late spring of 1995, with an invitation from the HoS, FAN arrived in the Bella Coola Valley.

A FAN newsletter lists three primary goals: to preserve the "Great Coast Wild Bioregion," to reduce the AAC (annual allowable cut) in British Columbia, and to work toward ending clear-cut logging and promoting ecologically sustainable community-based eco-forestry (Forest Action Network 1995). As a FAN member explained in a 1997 e-mail to the author, "The bottom line is that our industrial relationship with the forests has to change from one of ownership to one of stewardship ... This is why we promote alternatives such as eco-forestry and value-added manufacturing."

Claiming a unique position of legitimacy among environmental nongovernmental organizations (ENGOs) as a defender of both the environment and indigenous title, FAN alleged that other ENGOs – including the local chapter of the Sierra Club of Canada – implicitly endorsed industrial forestry and harboured inadequate respect for Nuxalk Nation sovereignty. Moreover, in interviews, several FAN members opined that Bella Coolans lacked adequate knowledge of ecology to render credible judgments about sustainability and were complicit in the illegitimate appropriation of indigenous territory and the impoverishment of the Nuxalk Nation.

Local Reactions

These representations provoked anger in the Bella Coolan community and among those Nuxalkmc not associated with the HoS. In February 1996, the NNBC held a community general meeting to discuss FAN's activities in the region and its physical presence on the Bella Coola Indian Reserve (FAN activists had been staying in the house of an HoS supporter). After hours of debate, a large majority of adult Nuxalkmc voted by show of hands to evict FAN from Nuxalk territory. After being served the eviction notice by the NNBC, in a display of disrespect that flew in the face of its publicly articulated defence of Nuxalk sovereignty, FAN simply moved across the road and relocated in Bella Coola village (Feinberg 1996). FAN thereby disregarded the spirit of the Nuxalk band council's democratically expressed wishes by not removing itself from Nuxalk territory altogether.

In the Nuxalk community, anger toward FAN continued to grow. On 4 May 1996, two Nuxalk families organized a so-called yellow-ribbon protest, which attracted dozens of Nuxalkmc wearing yellow ribbons (a common pro-logging symbol in BC during the 1990s) to a march against FAN. Although the protest was to a certain degree organized by Interfor, which supplied machines and personnel for the march, it did attract Bella Coolans who supported the Nuxalk organizers' desire to evict the FAN activists (Bella Coola resident 3, conservationist, interview June 1996). One anti-FAN placard, visible on the FAN video of the event, made a play on the LRUP acronym, reading "Local Responsibility, not Urban Pretension" (Forest Action Network 1996). This slogan highlighted the sense in the community that FAN members were Outsiders – urban people trying to dictate what took place in the region. The march eventually turned ugly, and a Bella Coolan protester assaulted at least two FAN members (Feinberg 1996; Forest Action Network 1996).

Prior to the internal conflict that came to a head during the FAN/HoS protests, the Nuxalk Nation had been a formidable local political force. Successful protests launched unilaterally by the Nuxalk had halted proposed developments, including a deep-sea port, a resort, and planned logging in Thorsen Creek valley (Nuxalk Nation band councillor, interview June 1996).

FAN's relationship with the HoS was, without question, excellent political capital in the group's global fight against industrial forestry. Nuxalk heredi- tary chiefs in ceremonial dress constituted powerful material for the inter- national media, as did rhetoric about working side-by-side with "Nuxalk traditional leadership" (a claim made in several FAN press releases). As B.A. Conklin and L.R. Graham (1995, 696) put it, "The new politics of the eco- Indian middle ground is primarily a symbolic politics; ideas and images, not common identity or economic interests, mobilize political actions ... Indians possess ... 'symbolic capital,' and positive ideas about Indians and their rela- tions to nature have become a potent symbolic resource in transnational politics."

However, FAN's effective "use" of the Nuxalk for their politico-symbolic value appears to have, if not caused, then at least exacerbated conflict within the Nuxalk Nation (Feinberg 1996). It must be stressed that FAN's relation- ship with the HoS may not have been purely instrumental; indeed, interviews with FAN activists revealed an apparently sincere, if not particularly well- informed, belief in Nuxalk sovereignty, the comparative legitimacy of the HoS versus the elected band council, and the notion that the achievement of Nuxalk independence would have desirable environmental outcomes. Nonetheless, the local impacts of FAN involvement were profound.

One FAN activist acknowledged the problems resulting from FAN's role in the Nuxalk conflict, saying that internal divisions among the Nuxalk "have at times made it difficult because you want to come in and you want to effect positive change, but you don't want to be a cause for division. I mean there was already a large schism in the community over logging, but to come in and then find yourself maybe exacerbating [it] or at least creating a focal point ..." (environmentalist 4, environmentalist-at-large, FAN sup- porter, interview June 1996).

Local residents, both Nuxalkmc and Bella Coolan, complained in inter- views and letters to the editor of the *Hagensborg Coast Mountain News (CMN)* that, although FAN had been invited to the valley by *one faction* of the Nuxalk, it represented itself and its allies as "the voice of the Nuxalk people" (Bella Coola resident 6, local naturalist, interview June 1996). FAN seemed to become aware of this and eventually stopped making general references to the Nuxalk Nation in its press releases. However, the nature of the Internet is such that once articulated, representations are often repeated.

FAN's use of the phrase "traditional Nuxalk leadership" to describe the minority of hereditary chiefs who had invited it to Bella Coola was as prob- lematic as employing the term "Nuxalk Nation" to describe the HoS. Inter- views with FAN activists revealed that they knew virtually nothing about Nuxalk history and culture. Despite their references to "the majority of elders" or "the head hereditary chief," respondents associated with FAN – including those in de facto leadership positions – displayed complete ignorance of such critical information as the total number of Nuxalk elders (and even of hereditary chiefs), of the method of choosing a "head" hereditary chief, and even of the process by which a Nuxalkmc becomes a hereditary chief. As one FAN activist commented, "Sixteen hereditary chiefs? Wow. I didn't know that. Hmm. We have only kept in contact with five" (environmentalist 1, FAN activist, interview June 1996). When asked about these subjects, FAN respondents described them as matters of internal Nuxalk politics or said that they "did not know" or were not mandated to speak about them. Yet FAN press releases and newsletters had consistently referred to cooperation with the "hereditary leadership" of the Nuxalk, which served to add legitimacy to contested HoS claims.

A common complaint focused on FAN's use of the Internet to publicize its campaign. The nearly universal lack of Internet access in the Bella Coola Valley in 1996 reinforced an impression that local people were being unfairly written out of FAN's representations. To be fair, though, FAN's Internet use constituted its strategic response to the anti-environmentalist bias it perceived in the "mainstream" media. An analysis of articles in the *CMN* archives validated FAN's concerns regarding media bias. In at least one case, the *CMN* published an Interfor press release verbatim without acknowledging its source, while simultaneously calling into question the veracity of a parallel press release from FAN. Other articles and editorials were sharply critical of FAN and the HoS, and sympathetic to the economic and political arguments of large industrial forestry companies (see, for example, *Hagensborg Coast Mountain News* 1995a, 1995b, 1995c). FAN's mistake was implying that its statements were representative of the Nuxalk Nation as a whole.

Another complaint concerning FAN related to its perceived attitude of superiority: "[Local people] consider it an urban arrogance that people are coming here and claiming to speak for them and tell them what's best for their community and how they should be operating in the community. And the assumption is that they are not capable of recognizing what's really good for their community, or the resource, or the environment. And there's a lot of resentment about that" (journalist familiar with the Bella Coola region, interview June 1996). This allegation was confirmed by comments of FAN activists, including one who said, "*They don't understand that once these trees are gone they're gone.* A lot of the areas around here that have been replanted haven't regrown very well. And it's going to take another thousand years to replace some of the old-growth. If they cut all the forests down, then they're not going to have jobs anyway ... *If we can show people what is going on, we can make them realize how much they are being controlled* by a corporation that employs only a really small percentage of people in the valley" (environmentalist 1, FAN activist, interview June 1996, emphasis added).

Field research corroborated what was already evident from the work of the LRUP Committee and the Central Coast Economic Development Commission: Bella Coolans possessed sophisticated knowledge of the local environment and of the implications of unsustainable logging practices.

Local Resistance: "They Have No Stake in Where They're At"

> We need the loggers themselves and the people themselves to have more voice in what's going on. Because they're out in the field and they see the devastation. People in boardrooms don't [see it]; they make deals. They make deals on paper (Bella Coola resident 1, business person and Sierra Club member, interview June 1996).

One of the most fundamental and long-lived claims to political legitimacy is that predicated on locality or "having stakes" in an area. S. Dalby and F. Mackenzie (1997) have pointed out that such phenomena as NIMBYism (not in my backyard) depend upon an acceptance that people who live in a local area and who will bear the burdens of pollution or other environmental disturbances have a special and prior claim to make decisions about such matters. Bella Coolan and Nuxalkmc respondents made similar claims, and the rights of locality underlay the Bella Coola Local Resource Use Plan.

Many saw the fact that the majority of Interfor employees were "shipped in from outside" as an explanation for both the degree of ecological damage being wrought by Interfor logging practices and the relative local economic disempowerment. As one Bella Coolan resident remarked, "A lot of these [logging] camps that I've been to, these guys are from god-knows-where just going into a camp. They don't have families there. They don't live there ... They work ten [days] on, four [days] off. And then they come back again ... and make their money and leave again. They have no stake in where they're at! They don't think 'Holy Christ, I can't show this to my kids in the next ten, fifteen years.' They have no stake in it!" (Bella Coola resident 6, local naturalist, interview June 1996).

The Nuxalk claim to legitimacy is also wedded firmly to place: "Our *smayustas*, our land, and our territory speaks for itself because we have names and we have stories and we have songs and dances of the different places within our territory ... They remind us what's there. A name may tell us ... a name of the river which means 'where the steelhead run,' that's what the name tells us" (Nuxalk hereditary chief 2, HoS supporter, interview June 1996).

Nuxalkmc Buddy Mack, an HoS supporter and former logger, reiterated this sentiment: "The culture is our land, I guess. It's like ... about the land, taking a piece of our culture away" (interview June 1996). Most Nuxalkmc and Bella Coolan interviewees made similar assertions about commitment to place and concern about interference from Outside.

Conclusion

It is rarely possible to mention indigenous land rights or forestry on the northwest coast of North America without triggering heated – even angry – debate. Few issues have so polarized communities in these resource peripheries as that of indigenous title. Similarly, the centrality of forestry in the region has ensured that most actors will have an abiding interest in discussions concerning it. They may be local residents, frequently divided along racial or other lines, or they may be interlopers from Industria. The latter can include global capital (multinational corporations or investors), imported labour, police or other security agencies, non-governmental organizations (NGOs), government administrators, militants of various stripes, or even, it must be added, academic researchers!

The multiplicitous set of actors that comprises Industria has penetrated resource-producing regions in Canada in a variety of ways. Federal and provincial governments claim territorial sovereignty, which in turn legitimates the presence of resource-extraction corporations. Mandated by government, these effectively plunder the regions of their natural-resource wealth by extracting it at unsustainable rates and exporting it to the global Industrian economy.[4] Urban-based NGOs and corporations use the Internet, other media, or lobbyists to assert their often diametrically opposed interests. Finally, local economies are penetrated by – and made peripheral parts of – the global economy through the local consumption of expensive distantly produced goods.

These complex interactions have an enormous bearing not just on specific resource-use management and territorial outcomes, but also on the nature and viability of democracy itself. At the local level, communities are increasingly asserting their communal rights to natural-resource use, management, and access. Such assertions contest the liberal-individualist ideas of citizenship common to conventional Western democracy, implying that true democracy must pay greater attention to the embeddedness of individuals in their communities and local ecosystems.

Anti-globalization politics are, in this sense, struggles to define membership in communities and to assert claims to local resources. Yet, the conflagration of competing discourses, fanned by multinational corporations and environmental activists alike, exacerbated conflict in Bella Coola and ultimately reduced the ability of the local communities to withstand economic and environmental threats from Industria. If we view environmental citizenship as a collective, as well as an individual, phenomenon, such polarizations undermine the basis of ecological democracy.

Arguably, grounds exist for agreement among local groups regarding an economy based on sustainable livelihoods. Both Nuxalkmc and Bella Coolans might agree that past unsustainable logging practices, which have provided a steady supply of tree pulp and lumber to a voracious Industrian economy and tremendous profits for industrial forestry corporations, have simultaneously caused the impoverishment of the region and severely curtailed potential for future sustainable forestry. The people of the Bella Coola region, Nuxalkmc and Bella Coolan alike, have taken important first steps toward (re)asserting local control over resource utilization. At the base of these efforts is an abiding love of place and an implicit respect for the non-human members of the local community. The Nuxalkmc have asserted their sovereignty over their lands through direct actions and site reoccupations, whereas in 1990 the Bella Coolans embarked on the important initiative represented by the LRUP (though regrettably without Nuxalk involvement).

There may be grounds for reconciliation within the Nuxalk Nation. Interviews conducted with NNBC and HoS supporters affirmed that the views and objectives of these factions were not nearly as opposed as had been claimed. The commonalities among the Nuxalkmc could serve as the basis for a Nuxalk national reconciliation. Certainly, this is a precondition for the development of solidarity with the Bella Coolans and meaningful resistance to external penetration and control.

The difficulties of imagining and implementing an equitable form of local control – in the face of immense external pressure – are formidable. These first steps, however tentative, may well point out the direction for a greater journey toward political autonomy. In undertaking such a journey, the Nuxalkmc and Bella Coolans may choose to enlist the aid of outside organizations, or, in the fashion so typical of the people of this rugged bioregion, they may choose to go it alone. These two peoples may also choose to walk some distance from one another, but in sharing the same path, they must eventually walk together. By strengthening local relational networks – both intra-human and between humans and non-humans – that acknowledge interdependencies and mutual needs, the people of Bella Coola can set up barriers to outside penetration and control that Industria will find very difficult to overcome.

This is so despite the grossly unequal resources available to the corporations, the environmentalists, the local community, and the Nuxalk respectively to influence the outcomes of the conflict. Such inequalities significantly undermine the effective practice of democracy in many rural communities. However, the constitutional position of indigenous peoples in Canada and the jurisprudence of indigenous title may be mobilized as a political resource – an ace up the sleeve, as it were – to counteract state power. In Paul Patton's (2000, see especially Chapter 6) parlance, these factors mean that indigenous title can be seen as a form of "metamorphosis machine" that destabilizes and transforms the practice of state and corporate power. In order to actualize this potential, the Nuxalk must first reconcile their internal differences and proceed with the legal claim to their traditional territory. In such negotiations, corporations could be effectively pressured to dramatically increase local employment levels, to make logging practices more socio-ecologically sustainable, and to return a share of royalties to communities in the region.[5]

For the Bella Coolans, the most effective strategy would be to ally themselves with the Nuxalk, promising not to intervene as "third parties" in any land claim, in exchange for a reciprocal promise that they would share equitably in any employment or royalty benefits won in negotiations with corporations. I have argued elsewhere that a "Bioregional Charter" could serve as the vehicle for such an intra-local alliance (Hipwell 2004b; see also discussion in Aberley 1999). In light of the BC government's recent gutting of the LRUP process – LRUP areas have been arbitrarily amalgamated into much larger Land and Resource Management Plan (LRMP) zones, effectively thwarting the possibility of local input – a unilateral approach to local control seems to be the only viable option.

Environmental groups could employ the same strategy, offering to lend their support to the Nuxalk claim and to use their international networks to promote the region as an environmentally friendly tourism destination in exchange for a seat at the table in negotiations over resource-use management practices.

I do not suggest that any of these options will be straightforward – a long history of racial and political conflict must first be confronted and resolved – but any efforts toward actualizing them will increase the potential of local empowerment. The solution to Castree's (2003, 2004) conundrum, in which he struggles with the paradox created by attempting to reconcile a parochial indigenism with the "needs" or "wants" of the larger Industrial society, lies in the creation of multi-ethnic *intra-local* localisms. Such intra-local hybrid identities would need to be based upon both recognition and affirmation of differences, and an agreement on common interests.

Fortunately, in Bella Coola, as in many other parts of the global resource periphery, such common interests would not be difficult to identify. Nor are conflicts insurmountable. One hereditary chief stressed that, even if the Nuxalk succeeded in actualizing their territorial claims, they would not expect non-Nuxalk to leave and would respect "third-party interests" (Nuxalk hereditary chief 2, HoS supporter, interview June 1996). Like the people of Clayoquot Sound, the residents of Bella Coola have what Catriona Sandilands (2003) has aptly dubbed "rich attachments" to the region. In every interview conducted – with Nuxalkmc and Bella Coolans alike – the importance of the local environment, the flora and fauna, and the quality of air and water came up in some way. Similarly, archival research at the office of the *Hagensborg Coast Mountain News* revealed that, without exception, every issue examined published stories about hunting, fishing, camping, environmental quality, human-bear encounters, and/or the overriding argument that these things set Bella Coola and its people apart from urban areas. In the words of a Bella Coolan, "Here it is different. People here have stakes in this community ... This is a valley that we all love ... The common ground we have in this valley is the valley itself. To a person that *lives* here, the valley, the environment, the richness, the generosity of the *place* is something that people want to maintain" (Bella Coola resident 6, local naturalist, interview June 1996, emphasis in original). Leveraging this collective sense of attachment will be the key to creating a hybrid local identity in Bella Coola that could, perhaps, usher in socio-ecological sustainability and end the unwelcome meddling of those who "have no stake in where they're at."

Notes

I wish to thank members of the Nuxalk nation, the Bella Coolan community and the Forest Action Network for hospitality and assistance. Field research would not have been possible without the support of Dan Hara and Angela Keller-Herzog, for which I am deeply grateful. Simon Dalby offered important, constructive criticism on my preliminary research results. Special thanks go to Laurie Adkin and three anonymous reviewers, whose insightful comments and criticisms improved this chapter in many important ways.

1 "Bella Coolan" is the term used throughout this chapter to refer to non-Aboriginal, or, in local parlance, "white" residents of the Bella Coola region. Although "Bella Coolan" was once used to denote Nuxalk people, they themselves have eschewed it in favour of Nuxalk or Nuxalkmc. The differentiation is necessary for analytical clarity.

2 "Industrian," or "of Industria," refers specifically to entities and forces characteristic of, or originating from, the hierarchical network system of globalized industrial capital (Hipwell 2004a, 2007). On the significance of legal codes, see especially M. Hardt and A. Negri's (2000) discussion of the evolution of "Empire."

3 The community decided by a large majority not to allow the burial.

4 Although the regional or even Canadian state economies are often considered as separate entities, ultimately all economies interpenetrate and are interdependent. Given this, there is only one – albeit fluid and dynamic – "Industrian" economy.

5 A similar royalty-payment arrangement was made between Georgia-Pacific and the Mi'kmaq Nation of Unama'ki (Cape Breton Island) in the Melford Gypsum Mine Agreement (Georgia-Pacific Canada and the Unama'ki Mi'kmaq Communities 1998).

10

Privatization, Deregulation, and Environmental Protection: The Case of Provincial Parks in Newfoundland and Labrador

James Overton

Public parks, as one form of state regulation, have come under attack in recent years by governments pursuing neo-liberal policies. Viewed as part of what Jennifer Sumner (2005) and John McMurtry (1998) call the "civil commons," and David Marquand (2004) the "public domain," public parks exist to further and protect a range of democratic rights that are both eco-logical and social. It is these rights that are threatened by park privatization and deregulation. In the 1990s, many of Newfoundland's parks were priva-tized – handed over to business or transferred to community groups – and others were closed completely, leaving only a small core system of provincial parks (Overton 1996, 205-39; 2001).[1] Newfoundland's park privatization and deregulation is "significant because of the magnitude of the privatiza-tion" (Hanson 2000, 8) and because it is presented as a "success" by those who urge us to let "the market save our parks" (LeRoy 2004, 11). It provides the focus for this chapter's examination of the relationship between neo-liberalism, Newfoundland provincial parks policy, and citizenship and democracy.

After briefly outlining the park privatization in Newfoundland, the chapter discusses public parks as a form of state regulation, emphasizing the import-ance of understanding the politics of this aspect of state policy. An outline of neo-liberalism is then provided, focusing particularly on policies of fiscal conservatism, deregulation, and privatization, and their impacts on parks. The implications of parks policy changes for citizenship and democracy are then broached. In particular, I argue for the importance of improving our appreciation of how parks serve the public interest and for the need to protect them from privatization and deregulation.

Privatizing Newfoundland and Labrador's Provincial Parks

A Private Sector Initiative (PSI) program was introduced in the late 1980s by the Progressive Conservatives and continued by the Liberal government of

Clyde Wells, elected in 1989. PSIs were initially limited to allowing concession stands and boat rentals in parks. However, in 1991 a Cabinet decision allowed a large section of land from Frenchman's Cove Provincial Park on the Burin Peninsula to be handed over to the Grande Meadows Golf Association for golf course development. In 1995 the first major wave of park privatization was announced. Overnight, some twenty-eight provincial parks and scenic attractions were removed from the public system. The decision was supposedly informed by the findings of a task force on the parks, but even after several Freedom of Information Act requests, the Liberals refused to release the task force report. In 1996 the government initiated a program review exercise, which released a detailed report on the province's parks and natural areas (Newfoundland and Labrador 1996). This document examined the impact of the 1995 privatization, which, incidentally, it suggested had not been recommended by the never-released earlier task force report. The review showed that, of the twenty-eight parks and scenic areas removed from the system, only eleven had operated as private parks in the summer of 1996. What this meant was that seventeen parks and scenic areas had either ceased to be parks or failed to find operators. Some land that had formerly been protected by the Provincial Parks Act (1970) was left with no more protection than other Crown land in the province. The negative economic impact of park closures on rural areas and the weakening of protection for the environment were major themes of the review report, which strongly argued for keeping those parks that remained.

This recommendation notwithstanding, a second wave of privatization occurred in 1997, involving twenty-one parks and seven natural and scenic areas. Hospitality Newfoundland and Labrador, representing private investors in the tourism-hospitality industry in the province, had lobbied the provincial government for further park privatization in early 1997, arguing that the money saved could be used to expand the province's tourism advertising budget (Hospitality Newfoundland and Labrador 1997). However, the reason given by the government for this further privatization was the need to prioritize spending on essential services, the minister responsible claiming that the choice was "between hospital beds and campgrounds; classrooms or scenic attractions" (quoted in Kelly 1997, 6).

Both the 1995 and 1997 privatization measures were opposed, but such efforts failed to reverse the decision (Overton 2001). Ultimately, however, the campaign to save the parks did have an effect. In late 1997, the Liberal government of Brian Tobin, which had undertaken the second wave of privatization, revamped the Provincial Parks Act to give a greater level of protection to the remaining fourteen provincial parks, as well as other protected areas. The government also agreed to desist from further privatization, a commitment repeated in 1999 as part of an assurance that the wholesale privatization of land, rivers, and wildlife was not on the political agenda

(Newfoundland and Labrador 1999). A Declaration of Rights to the Use of the Outdoors was also issued, which indicated that natural resources were being held in trust for all citizens to enjoy. These moves came largely in response to a significant movement of opposition to privatizing the outdoors, including rivers, which emerged in the 1990s (McGrath 1998), and to the efforts of the Protected Areas Association of Newfoundland and Labrador (PAA), which continued to press for improved protection for wilderness and "ecologically important" areas. Originally named the Protected Areas Association of Newfoundland, the PAA was formed in 1989 as part of the Endangered Spaces Program of the World Wide Fund for Nature, Canada (WWF). The stated aim of this St. John's–based non-governmental organization funded by the WWF was to press for the formation of provincial and national parks. The PAA did not oppose the provincial park privatizations of 1995 and 1997. Pressure to privatize and deregulate the remaining public parks continues. However, this has mainly involved piecemeal efforts by commercial interests, which, thus far, have been resisted.

Parks, Politics, and Regulation

Public parks are one of the products of struggles to "wrest a share of control over the countryside from its firmly entrenched rulers" and give "the public a say in the destiny of the landscape" (Shoard 1987, 549). But what has been gained can also be lost. The creation, operation, dissolution, and privatization of public parks are political in the broadest sense of the word (MacEachern 2001; Shultis 2005). Park creation and operation has often been a contentious and conflict-ridden process, parks being opposed by those who feel that their creation will negatively affect their use of the land and their interests (Jocoby 2001; Neumann 1998; Overton 1996, 172-90).

Park policies also reflect shifting ideas about landscape aesthetics, ecology, and the value of animal and plant species, as well as the human use of nature (MacEachern 2001). Park protection is, by definition, selective, protecting those places, environments, and species that are considered important for regional development, recreational, ecological, and national-symbolic reasons. It follows that parks often serve quite different and perhaps what are seen by some as conflicting purposes, such as protecting the environment versus providing for recreation.

Economic Liberalism, Privatization, and Deregulation

By the 1980s, state regulation was under attack as programs of market liberalization, including privatization and deregulation, were implemented by neo-liberal governments in various parts of the world. Neo-liberal ideology is, as William Graf (1995, 141) describes it, "in its core profoundly and restlessly anti-state," viewing state regulation as a mere obstacle to the efficient functioning of markets, entrepreneurism, and economic growth. Canada proved no exception to the neo-liberal trend at both the federal and provincial levels (Teeple 1995; Shields and Evans 1998). The privatization of the Newfoundland and Labrador parks was driven by such ideology and involved both deregulation and privatization.

Deregulation means removing, weakening, or failing to enforce some of the regulations that have been developed to protect what is defined as the public interest. Legislative changes and funding cuts to those agencies responsible for regulation weaken their capacity to effectively perform their designated functions. Deregulation is justified by governments as a cost-cutting measure in the context of national debt or budgetary deficit, as well as a way to aid business profitability (by reducing the costs of meeting regulatory requirements). Often, deregulation is justified using "jobs blackmail" arguments – that is, by claiming that regulation will lead to the failure or relocation of business activities (for a discussion of this in regard to environmental regulation, see Adkin 1992, 1998b). Such arguments are particularly effective in areas where unemployment and poverty are serious problems, as is the case in Newfoundland.

Privatization aims also to roll back state capacity to influence market actors and may take the form of the transfer of the provision of services that can be commodified from the public to the private sector (see Leys 2001). Privatization may also involve handing over responsibility for public services to what is called the "third sector," that is, to churches, unions, or other non-state organizations to be run on a not-for-profit or volunteer basis. The term is also used to describe the introduction of user-pay or other "conditions which typify the private sector" into the public sector (Swann 1993, 124). Although privatization is justified as a cost-cutting measure, it facilitates private-sector capital accumulation by transferring public assets to private hands (Sumner 2005, 40).

One of the main sources of pressure for states to pursue neo-liberal policies has been the corporate activism that emerged as a powerful political force in the 1970s (Harvey 2005). The neo-liberal agenda, including deregulation and privatization, has been pursued actively and with considerable success in weakening environmental legislation, by corporations themselves, business lobby groups, and a range of general and specialist conservative think-tanks (Rowell 1996; Beder 2002, 2006).

Privatization, Deregulation, and Parks

Neo-liberal fiscal policies have had a major impact on public parks. John Shultis (2005) looks at their effect on "park science, management and administration," drawing on examples from Canada (especially Ontario and British Columbia), the United States, and New Zealand. By the 1990s, public parks systems had long been underfunded and were unable to maintain their infrastructure or to deal with increasing recreation and tourism demands. When the major impact of neo-liberalism was felt in the 1990s, "park agencies were extremely vulnerable to the additional budget and staff cuts" (ibid., 56)

Shultis' observations are echoed by Kerry Van Sickle and Paul Eagles (1998) in their work on funding cuts to Canada's national and provincial parks. National parks saw significant cuts in the 1980s and early 1990s in real dollars and more drastic cuts in the mid-to-late 1990s (ibid., 226). The damaging effects of federal funding cuts on Environment Canada, including the Canadian Parks Service, in the period after 1993 have been documented by Glen Toner (1996-97, 99) and James Fulton (1996). The impact on national parks is documented by Rick Searle (2000) and the two-volume report *"Unimpaired for Future Generations"? Conserving Ecological Integrity with Canada's National Parks* (Canada Parks Agency 2000). The Harper government is currently using its expenditure review program to cut funding to Canada Parks (Galloway 2007).

Ontario park funding was cut in the period 1980-93, leading to the elimination of all waterfront life guarding and rescue services, cuts to management planning and to research, and the loss of trained managers. Eight provincial parks were closed in 1994 and a further fifteen in 1996 following a 20 percent cut in the operating budget (Van Sickle and Eagles 1998, 226-27). Seriously inadequate funding resulted in major staff cuts in all Canadian park jurisdictions in the 1990s, leading to lower levels of maintenance and research (ibid., 229). Parks and camping areas were closed and/or privatized across the country; services were eliminated; user fees were introduced or raised; and free access to parks for day-use or special groups such as the disabled (British Columbia) or seniors (Newfoundland) was discontinued. Private concessions proliferated in parks, as did the use of volunteers in park operations, as park agencies sought to generate revenue and cut costs (ibid., 226). It is true that land with protected area status is expanding in Canada (Dearden and Dempsey 2004, 226); however, given inadequate funding, this puts park agencies under even greater financial strain and may create pressure down the road for cost-cutting and revenue-generating measures, including privatization (Shultis 2005, 53).

Similar trends are in evidence outside Canada. In Britain, serious budget shortfalls in 2006 caused by below-inflation grants and spending freezes from the Department for Environment, Food and Rural Affairs led to the announcement of major cuts in the national parks, including staff layoffs and the suspension of environmental projects (Lewis 2006). Gregory Barton (2002, 166) reports "the almost complete decay of the parks system in Africa and the destruction of forests in the Indian parks that once set the example for environmental resource management for the world." In Africa, it is suggested that "the underfunding of national park agencies is now probably the single most important threat to the conservation of the areas under their control," and because of this, it is argued that "commercialization is essential" (de la Harpe et al. 2004, 190, 198).

The crisis in public parks systems provides ammunition for those promoting privatization and deregulation. Various "free market" think-tanks have played a major political role in both promoting the fiscal conservatism that is an essential feature of neo-liberalism and pushing policies of privatization and deregulation in relation to public parks; these include the Fraser Institute in Canada, whose Sylvia LeRoy (2004) has used the deteriorating state of Canada's public parks as a springboard for promoting "the market" as a way to "save our parks." These efforts draw on the "free market environmentalism" promoted by organizations such as the Political Economic Research Center at Bozeman, Montana (Anderson and James 2001).[2] Their self-supporting parks model, endorsed by the Fraser Institute, forces parks managers to cut staff and wages, to contract out, to raise revenue through user fees, and even to promote "environmentally safe resource extraction" (Green and LeRoy 2005). Non-self-sufficient parks are closed or privatized.

Pressure to deregulate and privatize parks also comes from those who aim to use parks to provide new, or to improve existing, opportunities for business. In Newfoundland, private campground and even motel operators have long complained that state-subsidized camping in parks undermines their business (Overton 1996, 205-39). Business interests see privatized public parks as profit-making opportunities, especially where these assets can be acquired at fire-sale prices. Others see public parks, developed and maintained at public expense, as prime locations for business activities, including the high-end private-for-profit resorts that are currently being developed in British Columbia (SourceWatch 2006). Plans to sell former park land for cottage lots are currently being considered in Newfoundland and Labrador. It is not surprising that private investors are increasingly interested in public parks. These protected areas have great amenity value and are attractive to a growing eco-tourist industry. It may well be that such moves are part of a gentrification process that also discourages poorer users of parks by extracting user fees.

Some deregulation and privatization may be accepted by park agencies and environmental lobby groups as necessary to fund fiscally hard-pressed parks systems (Shultis 2005, 54). They may also be seen as a quid pro quo for expanding the area of wilderness that is given park protection. More generally, Sharon Beder (2006, 90) argues that the use of "market-based instruments" has come to dominate "the whole public discussion of sustainable development." She suggests that many environmentalists have accepted "free market environmentalism," through a process of institutionalization, co-option, and marginalization, "in order to be taken seriously in the policy arena and to have a voice in the public debate" (ibid.).

Parks, Citizenship, Democracy

Jennifer Sumner (2005, 96) asks, "What do the following items have in common?" and goes on to list parks, public education, universal health care, old-age pensions, and the Charter of Rights and Freedoms. Her answer is that they are "all examples of the civil commons at work." The "civil commons" idea originated with John McMurtry (1998, 376), who defines it as "the organized, unified, and community-funded capacity of universally accessible resources of society to protect and to enable the lives of its members as an end in itself."

The institutions that protect and enhance life by providing access to goods such as clean air and water, that prevent their pollution, and that regulate against the marketing of unsafe products or exposure to health hazards in the workplace perform a set of "collective functions which have been won over many decades of social struggle" (Sumner 2005, 97; McMurtry 1998, 216). In this view, public parks – along with public education, libraries, public services, building regulations, sewage systems, and human rights codes – are life-enhancing systems that increase human security and well-being.

As Robert Fine (1999) suggests, a precondition for most of these rights is "membership of a nation-state," which is largely responsible for protecting citizenship rights. Rights are unevenly distributed, globally, and they are limited, conditional, and fragile. It is widely recognized that the neo-liberal counter-revolution involves an attack on many of these rights. Deregulation and privatization undermine democracy in the relatively simple sense that they exclude the public from participation in the setting of social priorities or the criteria that will guide the distribution of goods and services, as well as from decision making regarding the use of resources, the design of urban environments, and so on. Governments may establish advisory committees and consult with those whom they define as "stakeholders," but this is a way of limiting – not promoting – meaningful public participation.

Governments may also deny access to the information needed to evaluate their policy decisions. Newfoundland's park privatization provides perhaps an extreme case of what many of the chapters in this collection reveal is a common problem: information provision and consultation tend to occur only when this is required by legislation, and the weakness of existing legislation no doubt reflects the lack of power of those campaigning to improve environmental regulation. Van Sickle and Eagles (1998, 230) identify lack of public knowledge regarding the serious impacts of 1990s funding cuts on Canadian parks as a major factor limiting effective political action. However, in this case, they suggest, not only park agencies but also outdoor recreation and environmental groups failed to monitor the situation and to provide information to the public about what was happening.

Citizenship and democracy entail more than public participation in decisions: they are also about the creation and protection of other rights (Rowell 1996, 69; Beder 2002). Public parks serve the public interest by providing a degree of protection for the environment in general: air, water, animal and plant species, ecosystems and habitats – what many, including Susan George (1998, x), term the "global commons." Parks offer protection to cultural and natural features and whole landscapes that are valued for aesthetic, recreational, national-symbolic, and other reasons. But a frequently ignored aspect of public parks is that they form part of the state-provided network of recreational facilities that have a social policy function. Parks are, in part, a result of efforts to establish recreation and leisure as a basic social right, in T.H. Marshall's (1968, 71-134) sense, by providing all citizens – particularly the less well-off – with health-giving and educational recreational opportunities (Alexander Wilson 1992).

The social functions of parks were recognized by M.B. Williams (1936), when he argued that the establishment and use of parks were exercises in democracy. The basic idea here was that – in the words of the Government of Canada's 1944 Advisory Committee on Reconstruction (Canada 1944, 25) – "recreation should be made available to all classes of Canadians without regard to their financial status." This included the experience of nature, which should be accessible to all citizens. Arguments about the importance of leisure activity for reproducing labour power were also important, and "room to wander" in forests, moorlands, and other open areas – in both urban and rural settings – became an objective of workers' struggles (Donnelly 1993).

With the emergence of the welfare state, the importance of recreation received further recognition. To this end, it was suggested that subsidizing recreation by, for example, the "provision of low-price accommodation such as hostels in the national and provincial parks" should be studied (Canada 1944, 26). Governments became heavily involved in outdoor recreation in the period after the Second World War (Alexander Wilson 1992, 44). In Newfoundland, the provincial parks were expanded rapidly from the late 1950s until the second half of the 1970s (Overton 1996, 209). Just over ten years before the privatization of the 1990s, the government recommended that "all areas" be provided with "the services of a park" (Newfoundland 1982, 40), so that "the right of equal recreational opportunity for all citizens of our province" could be enjoyed. Article 22 of the United Nations Declaration of Human Rights was drawn upon to validate the treatment of access to nature and to recreation as human rights (ibid., 10-12). Recognizing the right to recreation, the 1982 *Green Paper on Recreation* (ibid., 40) argued that, "prior to any further consideration being given to transferring provincial parks to private enterprise, a committee, external to government be established to present the implications of same upon resident park users and the level of service they receive."

Park privatization and deregulation undermine the role of parks in providing a right to recreation. They cause a shift in "the distribution of benefits across social classes," as the old, the disabled, the poor, the working classes generally, and those in disadvantaged regions are disproportionately negatively affected by user fees, privatization, and park closures (More 2002, 57, 63-64). As they deepen social inequalities, deregulation and privatization also lead to what Marquand (2004) calls "the hollowing out of citizenship" as basic civil and social rights are eroded. In the process, decisions about key aspects of our lives are removed from the public sphere, where actions must be justified as being in the public interest, where decisions may be scrutinized, debated, and challenged in relation to criteria of fairness, justice, and equality. But, deregulation and privatization also give private interests the control of the resources and environment on which all humans depend. Park privatization and deregulation contribute to "the escalating depletion of the global environmental commons (land, air, water) and proliferating habitat degradations [that] have ... resulted from the wholesale commodification of nature in all its forms" (Harvey 2003, 148).

The Politics of Parks in Newfoundland

Newfoundland's partial dismantling of the provincial parks system was a top-down political initiative. Privatization and deregulation were key components of the economic liberalization program spearheaded by the Economic Recovery Commission (ERC) created in 1989 (Newfoundland and Labrador 1992; Overton 1999). The aim of the ERC was to circumvent the power of the "Old Guard" in the civil service and give business interests a major role in shaping Newfoundland's economy (House 1999). Privatization and deregulation were justified as cost-cutting measures and as means of providing business investment opportunities. Assisted by the London-based Adam Smith Institute's Madsen Pirie, the head of the ERC and Premier Clyde Wells became active boosters of privatization, the latter stating that he was "prepared to consider any reasonable proposal for privatization" of any activity that "can be more effectively carried out by the private sector" (Wells 1995, 3).

By 1994 Newfoundland's provincial parks system was on the ERC's privatization hit list (Newfoundland and Labrador 1994b). This was despite the fact that the government's strategic economic plan had noted the need for more effective regulations to "protect the public, the environment and the wilderness resource" (Newfoundland and Labrador 1992, 50), and a 1994 government tourism study had argued that the province's "long-term comparative advantage in adventure tourism depends almost entirely on the preservation of its natural environment, which is rapidly being eroded as a result of resident pressures and industrial expansion" (Newfoundland and Labrador 1994c, 23). The weakness of the Provincial Parks Act (1970) was also noted by the Protected Areas Association (PAA 1993, 9): "The Minister, with the approval of Cabinet, may expand or decrease the boundaries of any provincial park and even repeal the park status completely, although reversing a park designation is rarely done. There is no legal or written policy requirement for public involvement in the establishment of parks under the Provincial Parks Act." It may have been this weakness that led the government-appointed Newfoundland and Labrador Round Table on the Environment to suggest that the Provincial Parks Act be "strengthened" and that "public hearings well in advance of any changes to Park boundaries or use" be "required" (Newfoundland and Labrador 1994a, 7).

Expressed concern for the environment notwithstanding, this was a period in which government was seeking to weaken its already weak body of environmental assessment legislation (Newfoundland and Labrador 1995, 17). Existing legislation gave the minister responsible the power to waive environmental assessment when no harmful or negative impacts were anticipated. The privatization and deregulation of provincial parks occurred without the public being informed or consulted and without environmental assessment. However, we also need to explain why privatization was not stopped by public opposition once the policy had been announced.

From the late 1970s, inadequate funding had undermined Newfoundland's provincial parks system, making the parks less attractive to users (Overton 1996, 205-39).[3] This meant that, during the deficit-cutting hysteria of the 1990s, privatization and deregulation could be presented as the only ways of dealing with the crisis in the parks system by a government arguing that the choice was between hospital beds and parks. Many people accepted the argument that the private sector should be given a chance to operate parks and invest money in upgrading them.

The state's reluctance to spend money on parks and the decision to privatize cannot be viewed as financially driven in a simple sense. While public parks were being starved of funds, closed, and privatized, significant sums of public money were being spent on private-sector tourism-related developments, including parks, hotels, ski facilities, and golf courses (ibid., 287-88; Overton 2001, 82-86). This suggests that parks privatization was less about lack of funds than it was about limiting state involvement in parks while using state expenditures and public assets to improve private-investment opportunities.

Only limited protest followed the 1995 privatization. However, in 1997 the second privatization generated a movement to save the parks. Important in this mobilization was the Newfoundland Association of Public Employees

(NAPE), which sought to protect both the jobs of park employees and parks as a public good. NAPE and various labour activists were already heavily involved in fighting job and wage cuts as well as privatization generally, and their efforts were thinly spread. Others joined what became a significant movement in defence of the public sector, including publicly owned parks. Public meetings, letters of protest, and petitions failed to reverse the government's decision, although they did have some effect – forcing government to continue protecting some ecologically important areas within parks. However, protest notwithstanding, by late 1997 over 80 percent of former park land had been deprived of the protection of the Provincial Parks Act.

Park privatization also had supporters. Hospitality Newfoundland and Labrador, a group representing businesses, had apparently been consulted before the 1995 privatization. In early 1997, it advocated further privatization as a way to expand profitable business opportunities for its members. Both the province's Protected Areas Association (PAA) and the World Wildlife Fund (WWF) had allegedly been consulted by the pre-1995 task force. Shortly after the privatization announcement of 1995, the PAA and the WWF issued a report that criticized the Newfoundland government for not creating new parks or reserves but that failed even to mention the park privatization (Stacey 1995). Moreover, although expressing some concern about privatization during the anti-privatization protests of 1997, the PAA (1997, 1) stated that it was "cautiously supportive of a degree of privatization." This was an oddly irrelevant (and politically unhelpful) position to take, given that most of the province's parks system had already been dismantled.

The PAA argued that "provincial parks as a general rule do not play the major role in conserving the province's biodiversity that national parks, wilderness and ecological reserves do," but it added that "full protection under the Parks Act must be maintained in areas important for *conservation purposes*" (emphasis added). The PAA was willing to see park areas privatized that, in its view, were significant only for recreation (camping, picnicking), but not for "conservation." It seems also that the PAA saw park privatization as an opportunity to become involved in operating parks – an example of what David Harvey (2005, 177) calls "privatization by NGO [non-governmental organization]." According to its former president, Jon Lien (2005), the PAA made a proposal to government to take over the parks that were being offloaded in the 1990s.

The ambivalent resistance offered by some environmentalists to parks privatization, grounded as it was in their adoption of market-based approaches to conservation, also reflected their constructions of ecologically significant places as "pure and pristine nature" exclusive of the human presence (Alexander Wilson 1992, 41; Soper 1995, 272-74; Procter and Pincell 1996, 685). In this view, "one particular kind of countryside" – known as "wilderness" – is especially valued and worth protecting (Shoard 1987, 106-7).[4] Seeing the recreational use of parks as unimportant in environmental terms, or seeing the presence of humans as a threat (or as somehow disqualifying areas from being considered worthy of protection), limits the grounds upon which environmentalists may take a stand against the erosion of public goods and connect ecological concerns to citizenship rights. Allowing one bit of the commons to be annexed by private interests opens the door, as Marquand (2004, 132) argues, to further annexations.

There was some opposition to park privatization and deregulation on the part of people living close to affected parks, especially where parks were being closed. In rural areas, many people are interested in the benefits to be derived from having a park. However, relatively weak opposition to privatization, overall, may be explained by the preceding erosion of the benefits derived from the parks due to staff layoffs, loss of amenities for local users, and their declining value for tourism. Some people saw privatization as a business opportunity; others supported it because they hoped it would provide the investment needed to revitalize the parks and create greater economic benefits.

The case of Windmill Bight Park provides an example, however, of some success in resisting privatization. The park had been partially privatized in 1997, but 2001 saw efforts to privatize it further and turn it into a golf course, a move spearheaded mainly by a local growth coalition armed with government support and funding (Tracy Barron 2001, 2002). The coalition's failure to obtain land within the park was due to the mobilization of an effective PAA-led campaign to save it (Quesnel 2005). The PAA became involved because it considered the land in question to be "ecologically significant." Interestingly, the conflict was portrayed as one between environmentalists and developers; the former were represented as valuing "birds and bogs" more than people's need for livelihoods, whereas the latter were seen as offering economic opportunities to a community in need of employment. Hence, environmentalists were accused of "ruining the economic development we need as a species to survive" and of wanting to "turn this whole place into one great big park" (Murphy 2002, 18). Seeming disregard on the part of some environmental campaigners for those who feel that regulation or conservation harm their interests can, as Raymond Williams (1989, 220-21) observed, fuel organized opposition to parks. Such conflict is intensified by Newfoundland's high unemployment, which was worsened by the closure of fisheries in the early 1990s, displacing about thirty-five thousand people, and by cuts in income support and other state programs (Overton 1999). Anti-conservation arguments have been widespread in Newfoundland since the early 1990s, manifested, for example, in popular opposition to declaring the Atlantic cod an endangered species. In this case, it was argued that "special designation [of cod] would destroy rural life" (Rowsell 2004, A6).

Protecting Parks and the Public Interest

Public parks are a necessary – if insufficient – venue through which to regulate destructive human activity and protect and further human rights. They are, however, currently under pressure from neo-liberal policies. How can we defend the park systems that exist and advance more progressive management and policy programs for parks in the long term? Thomas More (2002, 61) suggests that "the long-term health and well-being of natural areas requires an interested population with the political will to solve problems, protect endangered species, and protect and enhance the areas themselves." This is a lesson confirmed by events in Newfoundland where efforts to mobilize a population of sufficient size and force to oppose a single-minded government committed to privatization and deregulation proved largely unsuccessful. Moreover, More (ibid.) suggests that "indifference ... is the real long-term enemy of natural environments. Policies that discourage participation among broad segments of the population ultimately promote indifference [and] actually reduce public support for management agencies." Changes in policy have made public parks less attractive and less effective, and have weakened support for them. One of the keys to preserving parks, long-term, is to increase the number of people who recognize that they have an interest in a well-funded and adequately protected parks system. As the conservationist M.B. Williams (1936, 128, 135) argued in the 1930s, the "best defense" for public parks is "to make them 'people's parks' in the broadest sense of the term."

One of the reasons that the Newfoundland government was able to dismantle the park system was that there was limited understanding and discussion of "what public recreation resources are *for*" (More 2002, 75, emphasis in original). For example, the role of the parks in providing citizens with "their right to recreation" was only touched on in the debates around privatization (Greg Wood 1997). Discussion of the social functions of public parks (and "*public* campgrounds, hiking trails, play grounds, and recreation facilities") is urgently needed (More 2002, 73, emphasis in original). These functions may certainly be defined to encompass more than recreation – ecosystem services, habitat protection for biodiversity, and carbon sinks, in addition to both physical and spiritual well-being for humans all come to mind. If parks are not linked to public goods and to the rights and responsibilities of citizenship, they will remain vulnerable to the constant pressures to deregulate and privatize. Phil Macnaghten and John Urry (1998, 35) observe that environmental concerns have often entered the public consciousness via attempts to preserve the integrity of parks in the face of development pressures. Through the same struggles, we can raise questions about democracy and citizenship.

Conclusion: Politicizing the Commons

Public parks are a limited and in some ways problematic and inadequate response to the needs to protect the environment and cultural landscapes as well as to meet human material and cultural needs, but they do represent a major, if fragile, achievement. The analysis offered here suggests that those who look toward state regulation (and parks) as a way of limiting the tide of environmental destruction have reason to be pessimistic. It is the pursuit of neo-liberal policies by governments (instigating funding limits or cuts, deregulation, and privatization) that have created major problems for public parks systems in recent years.

As Donna Harrison argues in Chapter 4 of this volume, we are experiencing a new phase of enclosure of the commons, one driven by ever-expanding capitalist accumulation (Harvey 2003; Roosa 1997). The privatization of parks is a form of accumulation through dispossession – in this case, of citizens' entitlements to benefit from the many services provided by public parks. It was the recognition, in an earlier era, of the function of parks as a public good that led to their establishment. What Sumner and McMurtry call the "civil commons" and Marquand terms "the public domain" is, for all its limitations, "literally a priceless gift," one won over a long period of time at the cost of great sacrifice by many people, which is "fundamental to a civilized society" (Marquand 2004, 32). In the neo-liberal era, these lands are subject to pressures for recommodification; leisure has become big business, and governments are not supposed to interfere with market forces.

The battle being fought is to prevent the erosion of democracy in the form of those civil, political, and social rights earlier won. This means preventing the weakening of public involvement in decision making and stopping deregulation and the privatization of public goods and services. Colin Leys (2002), too, asserts that the protection of public goods is a condition for civilization and democracy. He states that public services need secure funding and a clear philosophy, that their interests should be promoted and defended in public policy, and that the boundaries between the public and the private should be effectively policed. Although Leys' research pertains specifically to health care and public broadcasting, his argument is equally relevant to other public goods and services, including public parks. The story of parks privatization in Newfoundland should serve as a warning of what can occur when the political linkage of public goods to citizenship rights and responsibilities is let slip.

Notes

1 In 1974, when I was briefly employed as a parks planner in Newfoundland, the issue of park privatization was being discussed. In the 1980s and 1990s, I participated in debates about parks policy and was involved in opposing the two waves of park privatization.

2 More information on this body, now called the Property and Environment Research Center (PERC), is available at http://www.perc.org.

3 Lack of maintenance on bridges and trestles in T'Railway Provincial Park created a public hazard, leading to the temporary closure of the park in February 2008.

4 For various reasons, some within environmental circles express considerable ambivalence about parks, fearing that preserving a few "showcases of nature" detracts attention from "the problem of bad management of natural resources" and that "the protected area approach gives people a false sense of security" (Day 2002, A12). In this vein, Michael M'Gonigle and Jessica Dempsey (2003, 112) see parks as examples of "the traditional protectionist approach" that "fails to counter the powerful centralizing forces that make protected areas necessary in the first place." However, this kind of criticism of the way in which bits of nature are split off for preservation while the rest is left open for degradation was not a significant theme of environmental discourse in the conflict surrounding Newfoundland parks privatization.

11
Managing Conflict in Alberta: The Case of Forest Certification and Citizen Committees
John R. Parkins

A recent trend in natural-resource management is a transition from "command-and-control" management systems, with levers of power held primarily by centralized government agencies, toward a more diffuse network of management structures that involve an extensive field of actors from government, industry, and civil society. This transition to a more network-oriented mode of management holds important implications for the role of citizens. In observing this trend, analysts might suggest that the political landscape for citizen involvement is more open now than at any time in recent history. Yet, there is a sense of uneasiness about this emergent political terrain. Although more opportunities for public dialogue now exist, observers wonder about the quality of these experiences. To what extent does this opening of the discursive landscape translate into opportunities for critical reflection and debate? Do the acute asymmetries of power within seemingly decentralized modes of policy making negate opportunities for meaningful dialogue or democratic outcomes?

This chapter addresses these questions by exploring the case of forest certification and public advisory committees in the forest sector. Advisory committees, an increasingly dominant mode of public participation in the forest sector, represent a unique opportunity to examine the changing context under which citizens are invited inside the spheres of forestry regulation and monitoring. Arguably, one of the strongest catalysts for the growth of public advisory committees is forest certification. The evolution of certification systems involves the construction of a post-bureaucratic political terrain that brings together state, local non-state, and industry actors to establish standards to which commercial operators must adhere in order to apply "green" labelling to their products. Taking forest certification as a specific mode of environmental governance, this chapter uses a case study in Alberta to focus attention on the public processes within these systems, the ways in which lay knowledge is valued, and the opportunities for citizens to play a meaningful role in the development of environmental performance standards. In this way, the linkages are made between environmental governance, forest certification, advisory committees, and citizenship.

Although certification has become inextricably linked to public dialogue within small-group settings, these ubiquitous group processes have only recently gained the attention of the research community. They have been subject to the common gamut of evaluation techniques, such as checks on the adequacy of local representation and levels of satisfaction concerning them, but they nonetheless require a much more thorough critical evaluation. The present analysis draws heavily from deliberative democratic theory, which understands dialogue, debate, and conflict to be critically important components of contemporary politics. Rather than viewing conflict as a corrosive element, it perceives it as an integral component of the public sphere, a sphere that forges a critical bulkhead against colonizing forces such as elitist representation and capitalist expansion. In the spirit of deliberative democracy, this chapter attempts to understand the ways in which public advisory committees in the forest sector contribute to critical public inquiry and therefore lead to opportunities for resistance and social change.

Tiny Publics in Network Governance

In order to frame the central questions of this chapter, it is important to develop a sense of the potential role and function of small groups, or what some scholars refer to as "tiny publics" (Fine and Harrington 2004), within a theory of capitalist society and social change. According to B. Jessop (1997), this relationship between capitalism and the social spheres in which it is embedded is best understood as a "structural coupling" whereby both "systems" (economic and social) co-evolve and are influenced by each other in a way that creates a kind of "blind co-evolution." Jessop (ibid., 563) states that, "within this overall blind coevolution (which always and necessarily escapes any general or global control), there is some, albeit limited, scope to try to coordinate capitalist economic development with the operations of other systems and to anchor it more firmly in the lifeworld. These attempts can take the form of top-down imperative coordination (centralized planning) or more decentralized forms of governance. It is the latter ... that are now being emphasized in economic and political discourses and practices."

The decentralized forms of governance that are so prevalent in environmental management systems correspond with several key developments within Keynesian welfare states over the last several decades – two of which are particularly important for understanding the case of certification systems in the Canadian forest sector. The first general trend involves a process that scholars have come to understand as the "hollowing out" of the nation-state (ibid., Jessop 1999). This involves a kind of denationalization whereby the state experiences a loss of autonomy – through the internationalization of power and control, and through the emergence of subnational levels of political authority. The acutely international nature of modern issues such as climate change and international security intensifies pressure from above. Conversely, pressure from below is rooted in the chronic inability of nation-states to deliver social and economic benefits within certain subnational zones, resulting in the growth of regional political and economic structures.

A second general trend involves the de-statization of politics (Jessop 1997, 1999). This trend is commonly understood as a shift from a top-down command-and-control state government to a more network-oriented style of governance. Although an explicit diminishing of government decision-making authority occurs within this process, several authors have observed that government retains a role as visionary and facilitator (Davidson and MacKendrick 2004). Under this system, government continues to project state power by steering a host of agencies within the private and public sectors. The expansion of network styles of governance at the super-national and subnational levels places an increasing emphasis on identifying effective mechanisms and strategies for guiding such ubiquitous governance networks.

Within the field of environmental governance, several scholars have documented similar trends. For instance, tracing the history of public participation and environmental politics in Canada, A. Dorcey and T. McDaniels (2001) identify several stages of what they call citizen engagement in environmental management. They identify a movement toward greater lay involvement in planning and policy-making initiatives, one that began in the late 1960s. The managerialist orientation gave way to pluralist approaches to political participation in which civil society actors take a more active role – the prototype in Canada being the Mackenzie Valley Pipeline Inquiry (Berger 1977). In the European context, J.P.M. Van Tatenhove and P. Leroy (2003) identify a transition through three stages: early modernization (characterized by enthusiasm for instrumental rationality and a strong state-driven agenda), anti-modernization (growing skepticism regarding scientific approaches to problem solving and the opening of a philosophical wedge toward more enhanced forms of lay involvement), and late modernization (typified by decentralized politics and an integration of state, market, and civil society).

These general trends in the co-evolution of capitalism and its environing institutions have allowed the entry of new actors into governance networks – from both above and below the nation-state. This openness provides what Jessop (1997) identifies as the interface between the developmental logic of capitalism, on the one hand, and the communicative rationalities of its environing institutions, on the other – an interface that offers the possibility for struggle and resistance. Many of these ideas are also articulated within the environmental governance literature, where it is noted that decentralized and coordinated decision-making structures are playing an increasingly significant role.

As these processes unfold, the importance of local-level politics becomes strikingly obvious, and the outcomes of such decentred and small-group deliberations are no longer understood as merely trivial exercises in public relations but as potentially important locales for citizen involvement. Crucial battles between dominant and alternative discourses are won and lost within these spheres, and if local actors are to have any opportunity to influence the system, they must cultivate and sustain the possibility for struggle and resistance; that is, they must cultivate a political terrain for environmental conflict. What is striking about much of this literature, however, is the extent to which the steering tactics of dominant interests and the power asymmetries within a decentralized governance context are discussed in one of two polarized ways: as a highly problematic realm of public life or as a relatively benign and constructive aspect of public life. Very little work examines the dynamics and outcomes of actual cases of "decentralized" governance structures. It is in the local political arenas that citizen and resource manager, expert and layperson engage each other, where the meaning of citizenship is defined, and the opportunities for conflict and resistance are forged.

Deliberative Democratic Theory

Deliberative democratic theory offers important insights into how democratic processes within society can function outside of formal parliamentary systems and how the role of citizen may find full expression. The deliberative democratic ideas associated with inclusiveness, argumentation, and opinion formation match closely with the ideals associated with governance-oriented approaches to environmental politics.

The deliberative turn in democratic theory is based on a critique of episodic forms of democratic involvement, where citizens participate through voting and where public deliberation is severely limited to what could be described as issue sound bites and popularity contests. The literature is concerned with "debate and discussion aimed at producing reasonable, well-informed opinion in which participants are willing to revise preferences in light of discussion, new information, and claims made by fellow participants" (Chambers 2003, 309). Since the early 1990s, deliberative democratic theory has grown into an influential body of work (see ibid. for a recent review of the literature). This work views particular settings, as well as more diffuse networks that extend throughout civil society, as forums in which individuals can inform each other about matters of public interest and develop public-spirited opinions on a range of social issues. Whereas the public sphere was understood historically as a geographically specific and unique locale associated with openness, inclusiveness, and free-ranging debate, contemporary theory sees it as embedded within the fluid dynamics of civil society (Fine and Harrington 2004). A leading theorist in this field, Jürgen Habermas (1996, 307), describes the public sphere as "an open and inclusive network of overlapping, subcultural publics having fluid temporal, social, and substantive boundaries. The currents of public communication are channeled by mass media and flow through different publics that develop informally inside associations. Taken together, they form a 'wild' complex that resists organization as a whole."

These notions of the public sphere coincide with ideas by Jessop and others who are interested in the potential for civil society to provide some coordination for capitalist economic systems. Distinct from instrumental procedures for consensus building or decision making, deliberative democratic theory presents a strong normative claim for the activities that should be taking place within a deliberative context. Critical reflection and opinion formation represent the foundational elements of this sphere and are brought about by encounters with claims and counter-claims, arguments and counter-arguments, as well as the rhetorical and personal reflections of those who observe issues from perspectives other than one's own. At the end of these discursive engagements, personal opinions are likely to be better informed, better tested by other arguments and ideas, and articulated to some conception of general (as opposed to particular) interest.

The Citizen in Public Deliberation

Although deliberative democratic theory generally agrees that the citizen is a key player in the more diffuse systems of governance that mark contemporary approaches to environmental management, some strong differences remain concerning who should be at the table, what citizens should be doing, and how they should behave. Scholars focus on a variety of normative criteria for public deliberation, such as fairness and competence, as discussed by Darren Bardati in Chapter 7 of this volume. Others focus on issues of representation and voice as important considerations for public discourse within an environmental management context (Parkins and Davidson 2008). In this chapter, however, I focus on forms of knowledge that are embraced or repelled within these venues as a primary concern. Differential approaches to citizen knowledge provide a starting point in understanding the divergent behaviour and outcome of citizen actions in small-group settings. The comparative case study in this chapter provides two distinctly different ways in which citizenship is construed, with important implications for conflict, debate, and resistance. These distinctions find their origins in two contrasting arguments about the role of active citizenship in public life. For some scholars, the central challenge in public life involves bringing a small group of citizens up to a level of technical competence sufficient to effectively

Table 11.1

Approaches to citizen involvement and environmental conflict

	Minipopulus	Post-normal science
Theoretical paradigm	Citizens as learners of science	Citizens as contributors to science
Primary contribution	Expressions of scientifically informed personal values	Knowledge that is outside the realm of normal science
Approach to conflict	Conflict as dysfunctional and a procedural failure	Conflict as an essential component of learning and social change

discuss and debate scientific issues. R. Dahl's (1989) notion of the "mini-populus" aims to reduce the cognitive disparities between lay people and experts, and to broaden public debate about the benefits and costs associated with complex systems. Focused specifically on the quality of scientific discourse, the minipopulus serves as a kind of scientific training ground for lay people that is related to complex systems of concern. S. Funtowicz and J.R. Ravetz (1985, 1994), on the other hand, propose a more radical form of citizen engagement. Widening its scope beyond traditional scientific peer review, which is extended to a select group of citizens in Dahl's formula, post-normal science draws on epistemological insights from physics and the social sciences, including the awareness that science is conducted by social actors influenced by power and social relations. According to these authors, a turn to post-normal science is required precisely when the contradictions and continuous novelty of complex systems are involved. Under such conditions of scientific uncertainty, Funtowicz and Ravetz propose a necessary paradigm shift in the way that science is conducted and decisions are made concerning complex systems (see Table 11.1).

Two key issues are identified here. First, the extension of peer communities does not mean the displacement of traditional scientific peer review, but rather the inclusion of actors who might be more intimately acquainted with the landscape, such as local fishermen in the Atlantic fisheries, cattle ranchers in the oil and gas fields of Alberta, or trappers in the boreal forest. Such peers would be central to the construction of knowledge around environmental risks that are both highly complex and subject to continuous change. Second, along with more inclusive communities of peers would come extended facts in the form of local knowledge, anecdotes, beliefs, and even the perceptions of local people regarding the environmental impact of various industrial activities. These features have the effect of democratizing science by broadening the range of knowers. Unlike that of Dahl, which seeks to raise the general public's cognitive abilities to a "scientific level," this approach validates and integrates particular and unique forms of local knowledge that extend beyond the scope of normal scientific inquiry.

This example of a difference in the way that citizenship is perceived at the local level provides a basis for understanding the role of citizen within the context of environmental management and how conflict and debate are perceived. Under Dahl's formula of citizenship, debate and conflict are perhaps less related to competing sources of technical knowledge and scientific understandings and more related to the values that citizens hold, what specific values and benefits will be managed, and how costs or trade-offs will be handled. Conversely, within a post-normal science context, debate and conflict may be more likely to focus on sources of technical knowledge and fundamental understandings of the issues at hand than on the "trade-offs" between benefits and costs. The case of public advisory committees and forest certification provides an opportunity to observe these distinctions in the way that citizenship is understood and how these distinctions either accentuate or attenuate possibilities for conflict and resistance.

A Case Study of Forest Certification

According to some analysts, forest certification since the early 1990s has been the leading example of non-state market-driven governance systems (Cashore, Auld, and Newsom 2004). Other sectors, such as the petrochemical industry, have developed third-party certification systems, and the International Standards Organization has developed a more generic environmental management standard through ISO 14001. The forest sector, however, represents a unique case primarily because of the intense competition between several high-profile certification systems. Emerging in the early 1990s, the Forest Stewardship Council (FSC 2008) developed an international standard that was supported by several large international environmental organizations. This system provoked businesses in Canada and the United States to develop competing certification systems that were more "business friendly." One of the key twists in the business-friendly approach was to de-emphasize predefined environmental performance standards and instead generate procedures and processes for standards development. In Canada, this system, created by the Canadian Standards Association (CSA), is known as the Sustainable Forest Management Standard (CSA 2002); in the United States, it is called the Sustainable Forestry Initiative (SFI) (SFI 2002). These governance systems, which are based in the private sector, draw on market incentives (such as informed consumers) and third-party monitoring for compliance (Cashore 2002; Cashore, Auld, and Newsom 2004; Eisner 2004).

B. Cashore, G. Auld, and D. Newsom (2004) discuss three reasons for the generation of non-state market-driven systems during the mid-1990s. Two of these relate to developments on the international scene at that time. First, the late 1980s saw an increased focus on the pervasive destruction of tropical

Table 11.2

Forest area in Canada by type of certification, January 2008

Certification system	Forest area (million hectares)
CSA certified	77.8
FSC certified	24.3
SFI certified	39.1
Total forest area in Canada	401.5

Source: Canadian Sustainable Forestry Certification Coalition (2008).

forests. Boycotts of several large-scale forest retailers in Europe and Canada drew attention to the tropical source of wood products and wood fibre, and the extent to which forestry practices were harming the environment. Second, the Earth Summit in 1992 failed to sign a global forest convention, causing international environmental organizations to search for market-based or other regulatory mechanisms to improve environmental and social outcomes in the forest sector.

The third – and perhaps most significant – factor is the turn by governments to market-oriented policy instruments to address matters such as forest conservation or carbon sequestration that extend beyond the regulatory environment of any single province, nation-state, or group of states. The success of market-based boycotts and the preference of corporations for "voluntary compliance agreements" rather than state regulation typify this trend. Also, B. Cashore and I. Vertinsky (2000) point to the dual challenge for policy makers in the US and Canada – to deal with decreasing state resources, on the one hand, and increasing pressure from civil society for more stringent environmental standards, on the other. Under these conditions, some argue, business-friendly third-party performance standards such as the CSA or the SFI fill a growing desire to maintain oversight of industrial activity while limiting the need for command-and-control regulation and state-based monitoring.

In Canada, the clear leader among the three competing certification systems is the CSA standard (see Table 11.2). According to the Canadian Sustainable Forestry Certification Coalition (2008), almost 78 million hectares were under CSA certification in 2008. If we consider forest area alone, the influence of CSA and other certification systems on the management of forestlands in Canada is not trivial. Furthermore, several provincial governments, including those of Alberta and Ontario, have taken steps to harmonize provincial forest planning regulations with the CSA standard, making it easier for forest companies to meet provincial planning and reporting requirements while also aligning with CSA requirements and reducing the need for duplicate reporting systems. The successful application of this governance system across a wide area of the Canadian forest landscape represents an important framework for environmental management – one with clear implications for citizen involvement. This particular aspect of CSA certification is discussed in detail below.

Certification and Public Participation

Although a forest certification system is comprised of many components of forestry practice and management, our interests in this chapter pertain primarily to the ways in which forest certification, as an increasingly dominant mode of forest management in Canada, contributes to our understanding of how citizens are involved within this sector of environmental governance and how this institutional environment amplifies or attenuates the opportunities for critical reflection and debate.

With respect to these democratic principles, CSA lays claim to very progressive mechanisms for public participation. Unlike the FSC and SFI systems, which afford less direct opportunity for local citizens to participate in the standards development process, the CSA standard puts this relationship between citizen and forest company at the centre of the certification system itself. The introductory remarks of the CSA standard (CSA 2002) state that "The public participation requirement of this Standard is one of the most rigorous of its kind in certification standards in the world today. Because Canadian forests are primarily publicly owned, it is vital that a Canadian forest certification standard involve the public extensively in the forest management planning process. Forest management that meets the SFM requirements of this Standard involves a positive relationship between the organization and the local community."

The CSA performance standard is more process oriented than the FSC system, with its prescriptive elements, and involves a dialogue between the applicant (a forest company) and a broad cross-section of local community representatives who constitute the public advisory committee. Within this system, the principal role of the advisory committee is to work with the applicant to develop values, objectives, indicators, and targets for a defined forest area. Applicants must adhere to a general evaluation framework that involves six key criteria (such as conservation of biological diversity and maintenance and enhancement of forest ecosystem condition and productivity), along with some specific elements under each of these criteria, but the more specific indicators and performance targets are established entirely locally, on a case-by-case basis in consultation with the public advisory committee. These performance measures then form the basis for third-party audits and CSA certification. Once the applicant achieves certification, the public advisory committee continues to play a role in auditing and recertification.

The first company-sponsored advisory committee in Alberta was formed in 1989, but most of these committees have existed since the mid-1990s and are now involved in CSA certification initiatives. In most instances, the company seeks broad-based representation from the local community to commence activities, and then committee members develop "terms of reference" for subsequent changes to group membership. Members typically include representatives from the local labour unions, educational institutions, local elected officials, outdoor recreation clubs, the business community, and Aboriginal and environmental organizations. Members are not paid for their time, but the company often provides meals and covers out-of-pocket expenses. Those who do participate typically have the time, the intellectual curiosity or concern for forest management activities in their region, and the support of a constituent group to act as its representative. (See Parkins 2002 for a more detailed description of committee structure.)

Although other certification systems require some form of public participation, none of them undertake such a highly decentralized process whereby, together, the applicant and the host community identify a set of environmental performance standards specific to the local forest area. This kind of local control over standards development is anathema to many critics of the system because of the potential for manipulation by dominant interests at the local level. However, in opting for the CSA system, many forest companies in Canada are choosing to take their chances with local-level processes, anticipating that the environmental standards developed within the CSA system and the perceptions of these standards within the marketplace will continue to prop open doors to product markets throughout Europe and North America.

A Closer Look

By examining forest certification systems such as that of the CSA, one can draw some general conclusions about the role of public participation in forest management. One can also examine the results of such processes by reviewing the performance standards that are developed. Martin von Mirbach (2004), for example, conducted a review and analysis of the CSA standards of several forest companies and, as a sharp critic of this system, laments the inconsistencies among them. His analysis suggests that company officials have a great deal of influence over the outcome of the public advisory processes. He (ibid., section 2) also states that "local public advisory groups may lack adequate independent expertise with knowledge of the latest advances in science or in best management practice. The advisory groups may instead rely on the selective information given to them by the applicant organization, and thereby miss out on the opportunities to apply relevant progressive models from elsewhere in the country."

This analysis provides some insight into the functioning of advisory committees, but there remains a lack of documentation of the group processes that result in a specific performance standard in one locale and a very different performance standard in another. Can we conclude that citizens always lack adequate independent knowledge, or is citizen involvement somehow constrained in other ways? Do outcomes result from lack of knowledge or from lack of opportunity for meaningful debate and influence over the process? A comparative study of two advisory committees in Alberta provides additional insight into such questions. I studied both committees through direct observation of committee meetings for a period of ten months between September 2001 and June 2002. In addition, document analysis and semi-structured interviews with committee members provided the basis for a critical ethnographic analysis of committee activities. Assurances of confidentiality in research prevent further descriptions of the names of companies involved and their whereabouts in Alberta.

During my observations of one Alberta committee, members were involved in discussions with the forest company over the specific values, indicators, and targets for their CSA performance standard. This process led them into a protracted discussion of the need for a set-aside (a protected forest area functioning as a benchmark against which the impacts of industrial activity in another area can be measured). The vignette presented below offers an example of persistence on the part of committee members in achieving a certain performance standard and the eventual denial of this standard.

A local citizen with extensive training and experience as a biologist introduced the topic of designating a benchmark area. At a committee meeting, this citizen suggested that "in any kind of management scheme you are going to need a control area, to determine whether or not your management is effective, particularly if you are talking about maintaining soil and water. To me a value would be to set aside a portion of land as a control to test your management activities against." Initially, the company responded to this comment by holding it in a "virtual parking lot" until a later meeting where soil and water quality issues would be discussed at length. At the following meeting, another local citizen with extensive forest management training tried to resurrect the subject: "I like the idea of having a buffer or a cushion of forest that may be sitting in the bank for some other reason." Repeating the claim that discussion of a set-aside would be premature, the company official suggested that the committee should "hold that thought until we get to that specific goal." Five months elapsed between that point and the presentation to the committee of the draft version of objectives, indicators, and targets. This document, which represented extensive discussion with company officials, members of the public advisory committee, and numerous special guests with expertise in forest science, contained a summary of values and their associated measurements for the local forest areas. It included an indicator titled "undisturbed site from forestry" and an objective "to establish and monitor an undisturbed site from forestry used to determine success for achieving goals." Although company officials had put off addressing this value for several meetings, it did find its way into the first draft of the performance standard. But the matter did not end there.

In a subsequent meeting, the forest company came back to the advisory committee to recommend that the indicator and the objective be dropped from the standard on the grounds that they "may be against provincial policy." This policy required that, in forest areas, active measures must be taken to prevent fires and to deal with invasive pests. Thus, establishing a set-aside, which could function as a benchmark only if it remained undisturbed by human activity, would contravene provincial forest policy. The advisory committee replied that "all advisory committee members (excluding [local small-scale loggers]) would like to see the indicator regarding an undisturbed site be left in the [performance standard]." This strong recommendation generated another series of discussions about the size of the undisturbed area and the extent of its protection from industrial activity. Committee members proposed several potential sites, including one that was registered by the provincial government as having some existing protection from industrial activity. In spite of these efforts, the forest company successfully resisted the inclusion of a large protected area in the performance standards, stating that it was looking for "a large area of the defined forest area that is not planned for harvest in the next 20 to 30 years. If such an area is not available, we will not agree to set a benchmark aside."

This vignette from an advisory committee in the throes of developing a performance standard offers some insight into the tenacious and persistent efforts of some committee members. Although a knowledgeable group of local citizens advanced reasonable and convincing arguments for a local benchmark, their efforts did not translate into an element of the final performance standard. Ultimately, their wishes were blocked by existing forest legislation and the veto power of company officials. In other words, although advisory committees can make general recommendations through consensus or majority rule procedures, company officials reserve the right to reject them and to take action on the basis of internal company policy, provincial regulations, or other rationale. These decision-making rights are afforded to forest companies by way of their status as forest management licensees of public forest lands.

A second vignette from another committee in Alberta provides an important counterpoint to the relatively dynamic relationship between citizens and company officials in the first account. This second committee was also involved in the CSA performance standards development process. But in this case, no discussion or debate occurred between local citizens and company officials. Instead, the second committee was characterized by a disturbing lack of attention to detail and a seemingly depoliticized discursive landscape. This committee had reached a later stage than the first one in the development of performance standards, which had already been certified by the CSA. The certification process required renewal, however, and it was during the periodic review and renewal of the performance standard that a company official brought forward the performance standard known as the Sustainable Forest Management (SFM) Plan. During a meeting, he presented the SFM plan to committee members and invited them to submit feedback to him prior to the following meeting. At the next meeting, the company official commented, "At the last meeting we talked about our SFM objectives and indicators and passed out where we were at in that process and requested feedback from the members. Since that time, I have not received anything from any of the members, so we are continuing on with our process and coordinating our SFM plan." For this committee, the lack of critical dialogue was a central tendency during specific performance target reviews. Environmentalists on the committee did push for higher standards but were easily defeated by internal voting procedures because other committee members did not support them. As a result, those who held alternative points of view felt highly disenfranchised and disengaged. Expressing frustration, the activist members of the committee resigned themselves to a more passive role in deliberations. As one activist, interviewed by the author in August 2002, put it, "From our perspective, now I think we're really going there to see what direction the company is going in and, if there's a way that we can influence through other bodies, then we'll use that information to influence." Furthermore, most members of this particular committee appeared to entertain high levels of trust in the forest company. Rather than seeing themselves as catalysts for change and as citizens with special knowledge to inform the decision-making process, they were much more content with the passive role as receivers of information. Discussing the issue of trust, one committee member suggested that, "if there were a graph to be drawn of it, it would probably be in the first two years on the [committee], because that sense of trust develops very rapidly. And then it's just sort of a booster shot once in awhile."

In this vignette, we observe a highly attenuated discursive environment where debate and conflict, if it occurs at all, rarely generates a coherent set of alternative discourses or a strong mode of resistance to the status quo. Rather, through long years of shared experience, where much of the discussion is oriented around scientific understandings and forest management practices from the perspective of a single forest company, a group-think sets in that is difficult to challenge.

Despite the democratic shortcomings observed in these cases, such processes continue to pass for meaningful public participation. In the next section, the implications of these findings are discussed in relation to environmental conflict within the forest sector.

Deconstructing the Governance Discourse

Some social coordination of capitalist economic development may occur by creating venues where ruptures may be opened in the hegemonic discourses of capitalist expansion. This kind of hegemony is perhaps best described by Chantal Mouffe and Ernesto Laclau (1985), following Antonio Gramsci, as a kind of common sense in perpetual motion. Notions of network governance or decentralization, while appearing to be counter-hegemonic in their ideal forms, may serve the primary interests of social and economic elites. This research provides support for these general claims by revealing what decentralized decision making means in the specific context of forest management in Canada. In this case study, at least two specific meanings become clear. These meanings hold implications for ongoing struggles within the public sphere and the potential for such struggles to affect the coordination of capitalist economic development.

First, decentralization means *political isolation*. In the late 1980s and early 1990s, several large-scale land-use planning processes were underway in Alberta. For instance, the Alberta Forest Conservation Strategy (AFCS) provided opportunities for many groups within civil society to work together on a common vision for the province. The AFCS spanned three years of dedicated effort and involved input from over eight hundred Albertans. More than sixty groups participated, "representing essentially every organization in the province with an interest in forests" (Schneider 2002, 9). Although the official government response to the AFCS was deeply disappointing to many participants, the process represented an opportunity for large groups of people to work together on a common vision – and with the hope of providing some strong counterpoint to full-scale industrial development throughout the forest regions of the province. In contrast, the more recent modes of decentralized decision making provide none of the same kinds of synergy that can be created within large-level processes. This synergy is especially important for the not-for-profit sector, which is often strapped for resources and in need of close collaboration to sustain successful counter-hegemonic discourses. In this context, decentralization means disconnection and isolation for many citizen-based groups who would otherwise share a common vision and work toward common goals.

Second, decentralization means a *truncation of public participation in favour of local and industry-supported publics*. Although approximately 80 percent of Albertans live in urban areas, the system of decentralized public advisory committees involves a high proportion of members who are from local resource-dependent communities. Furthermore, these members tend to be male and are wealthier and more educated than average (McFarlane and Boxall 2000). They are commonly in positions to benefit directly from the local extractive industries, and their values are more oriented toward human uses of the forest than are those of the general public. Due to this, decentralization tends to create a public dialogue and debate that is disconnected from the values and interests of many, especially those who live in urban areas. Within this context, the friendly terrain of a host community offers a comparative advantage in garnering the public support necessary for ongoing industrial activity.

This process of decentralization, which is embedded within contemporary notions of environmental governance, constructs a "public sphere" that is greatly disconnected from the larger realm of public discourse. This ghettoizing effect results in minimal impact on civic discourses and public opinion formation. The average citizen rarely hears these debates, never has an opportunity to understand the issues, and infrequently reshapes and revises his or her personal opinions. Advisory committees become subterranean zones of local contest and debate with little influence on public knowledge and understanding. According to Habermas (1996, 360), effective streams of communication within the public sphere must at regular intervals "coalesce into bundles of topically specified public opinion." If it lacks a mechanism for upward movement to legislative and administrative spheres, this type of bundling and coalescing of knowledge is even less likely to occur within a decentralized system of public dialogue. These findings suggest that decentralized governance in the forest sector – in the form of the CSA forest certification system's public advisory committees – serves legitimation ends more than ecological ones. These processes do not provide the procedural guarantees necessary for meaningful deliberation. Moreover, due to their isolation, "tiny publics" may be easier to dominate, or manage, than those involved in broad public processes.

One could, however, derive more optimistic conclusions. If advisory committees become effective as sources of knowledge exchange and political interaction, forcing experts to question conventional wisdom and lay people to question expert assumptions, they may play a key role in reinvigorating the public sphere. The first case studied provided some evidence that citizens attempted to take this role but were thwarted by undemocratic procedures. The complexity of ecological decisions and the uncertainty of scientific knowledge may lead decision makers to attribute greater value to local knowledge of places and to appreciate the importance of various kinds of knowledge. According to G. Auld and G.Q. Bull (2003), recognition of science's uncertain or imperfect knowledge of forest ecosystems may lead to inclusion of local perspectives in the creation of performance standards. Yet, if they are to realize some of these knowledge gains, small groups must be, at the very least, coordinated with broad-scale public processes where dissenting voices and forms of civic science can find a more even footing within the public sphere.

Citizenship and Environmental Conflict

This chapter examined the relationship between citizenship, conflict, and environmental management through a study of two public advisory committees in the forest sector. The political opportunity for citizen involvement in environmental management is arguably greater today than at any time in recent history, but the possibilities for constructive dialogue and debate are probably as constrained as ever. The functioning of public advisory committees in ways that promote the deliberative democratic ideas of debate, critical reflection, and informed opinion formation is likely to remain well contained by the dominant interests that govern our social and economic lives. This claim is consistent with those of other authors in this volume, such as Peter Andrée and Lucy Sharratt, who state in Chapter 2 that "the risk evaluations of farmers, consumers, and other lay people are often dis credited, even though these groups have valuable experience and insights to offer the process" (30). Although absolute exclusion of these voices remains a problem in some venues, these chapters demonstrate that exclusion is also about whose knowledge gains standing. In other words, exclusion involves the deployment of consensus-building strategies within a decentralized mode of environmental governance. To break from this exclusion, the possibility for a more radical role for public advisory committees comes in part from a more radical conception of citizen involvement in environmental management. It also involves a restructuring of environmental governance away from tiny and isolated publics, and a repositioning of environmental politics at the centre of political life and the public sphere.

For those who start from the position of Dahl's minipopulus, the opportunities for constructive conflict are few. Within this paradigm, the work of citizenship involves an effort to come up to speed on resource management issues, to think like a scientist, and to begin to understand the scientific complexities of the forest ecosystem and then to contribute to the scientific debate through the expression of scientifically informed values and preferences. This notion of citizenship entails an economistic interest in the formation of personal values and the explicit expression of those values as a contribution to the management process. Conflict is thought to be a dysfunctional aspect of this formulation – an expression of failure within modern management systems to cope with system complexity and the achievement of optimal solutions.

For those who start from the position of post-normal science, the opportunities for constructive conflict are enhanced. Within this paradigm, the work of citizenship involves an effort to understand the contested nature of knowledge and knowledge productions, the vested interests that support the scientific enterprise, and the valuable contributions that citizens can make to our understanding of complex systems. This knowledge involves the exploration of alternative explanations and local understandings of issues. It requires that peer communities be actively expanded to include citizens who possess special kinds of knowledge that are not easily accessible by Western scientific methods and that facts, ideas, anecdotes, and other articles of evidence be incorporated into the decision-making process. This notion of citizenship treats lay people as equal partners in the exploration of satisfactory solutions to modern environmental problems. It goes well beyond an appreciation for citizens as bearers of informed or uninformed values and creates political space for the citizen as bearer of ideas. Conflict is thought to be a crucial component of this formulation because discussion, debate, and argumentation stimulate critical reflection and the democratic basis upon which better ideas become a force for social change.

12
Beyond the Reach of Democracy? The University and Institutional Citizenship
Jason Found and R. Michael M'Gonigle

> Students learn that it is sufficient only to learn about injustice and ecological deterioration without having to do much about them, which is to say, the lesson of hypocrisy. They hear that the vital signs of the planet are in decline without learning to question the de facto energy, food, materials, and waste policies of the very institution that presumes to induct them into responsible adulthood. Four years of consciousness-raising proceeds without connection to those remedies close at hand. Hypocrisy undermines the capacity for constructive action and so contributes to demoralization and despair (Orr 1992, 104).

The University of Victoria (UVic) is a growing institution on a finite property within a suburbanized landscape. In 2001, under the pressures of increasing enrolment and new building proposals, the university completed a first draft revision to its founding 1961 campus plan. The draft, however, continued to reflect the modernist assumptions long embodied within the institution; its release provoked extensive criticism and sparked a movement to change UVic's development practices. The authors' research group, the POLIS Project on Ecological Governance, has been a central part of that movement from its inception and continues to participate in it. Over the years, we have collectively produced numerous research reports, published dozens of handouts and flyers, sponsored meetings and coalitions, drafted petitions and a protest letter (signed by over half the faculty), and demonstrated against new building construction. The campaign even led to a several-month-long "tree-sit" (see Figure 12.1).[1] This chapter is a report on some of the lessons learned from that movement.[2]

This activity led the university to adopt a much improved campus plan in 2003 (University of Victoria 2003). From an administration that had not even heard of "smart growth" a couple of years earlier, the new campus plan

Figure 12.1   This University of Victoria tree-sit attempted to protect trees slated to be cut down to make way for construction, January 2003. Photo John McKay, *Victoria Times-Colonist,* 14 January 2003, D1.

had smart-growth principles embedded throughout. In the years since, the university's land development practices have evolved incrementally to embrace "green building" practices and a diversity of efficient technologies. Nevertheless, the document was explicitly cast as advisory and without "regulatory" effect, and a large gap remains between potential and practice. Many constituencies affected by development remain without a forum for meaningful dialogue with the decision makers; administrative accountability is minimal. At a time when sustainability issues are of critical national and international concern, as evidenced by the growing anxieties regarding the catastrophic effects of climate change, leadership from the university sector is lacking across the country. Given the largely unremarked but nonetheless massive significance of the "higher education industry," this failure demands major remedial action.

Conflict on Campus

As George Fallis (2004, 21), former dean of arts at Toronto's York University, noted in a 2004 submission made to an Ontario government commission on the future of higher education, the university "is one of the chief innovative forces of society, one of the chief determinants of social opportunity and social stratification ... The emergence of post-industrial society moved the university's mission to centre stage." By mid-decade, a campus sustainability movement had taken root across North America. In 2006 the Association for the Advancement of Sustainability in Higher Education (AASHE) was founded in Portland, Oregon, to promote the sharing of best practices among groups and universities. At UVic, the campus planning movement began with the belief that it could reach agreement with the senior administrators to seize the opportunity for the university to become a model of regional innovation.

For campus activists, the presupposition has been that universities are sites where conventional ideas and attitudes can, and should, be challenged for the benefit of society in general. Outside its walls, new ideas are subject to the laws of the market and to the bureaucratic barriers imposed by multiple layers of government. In contrast, with the protections offered by the relatively autonomous (self-governed) institutions for higher education, it is envisioned that novel ideas might be nurtured and developed, and a supportive context provided for their implementation in order to bring comprehensive sustainable solutions into greater social prominence. At the same time, should the larger world move dramatically toward sustainability, one cannot imagine that the university would fail to follow suit.

From the beginning of the debate over the new campus plan, POLIS identified the structures of decision making as the critical variable for such innovation (MacNair and MacDonald 2001). There is certainly no shortage of opportunities for innovative practices. Numerous physical sites exist in which new ideas for practical change could be played out: for example, certain parking lots and lawns could be redeveloped; existing and new buildings could promote state-of-the-art principles in new urbanism, smart growth, and green building; disturbed landscapes could be subject to ecological restoration techniques; and the whole campus could shift from the sprawl-style growth that colonized greenspace to urban densification that could transform a suburban landscape.[3] The benefits would be tangible for cities and regions beset by urban sprawl, automobile congestion, food insecurity, and the loss of agricultural land. However, these ideas also go beyond a particular set of sectoral approaches by recognizing the prerequisite need for more innovative thinking, more collaborative processes, and new forms of internal organization that would benefit both the institution and the communities that comprise and surround it. The challenge, then, is to consider how university governance might be restructured to permit change that is both visionary and practicable.

University Governance

At UVic, the president and four vice-presidents (Finance and Operations, External Relations, Academic, and Research) play central roles in development, but campus planning effectively falls under the mandate of Finance and Operations. As is common at most universities, primary authority for sustainable practices is allocated to Facilities Management, a division of Operations. Almost every university has a campus planner; in recent years, some have acquired a "sustainability coordinator." Both offices, however, exist as line positions within Facilities Management. Given this alignment, and regardless of how committed they might be, the occupants of these positions have limited, even delicate, power. Their actions have to be tip-toe strategic; many actions are simply proscribed by higher-level authorities. Meanwhile, despite their formal authority, the Board of Governors (BOG) and Senate exercise minimal leadership, choosing instead to defer politely to the day-to-day managerial prerogative of senior administration.

At UVic, two committees control planning: the newly formed Campus Planning Committee (CPC) and the Facilities Development and Sustainability Sub-Committee (FDSS). The committees are composed of administrators and line staff, with representation from faculty and students. Most members are appointed; a few are elected. These committees act as advisors without decision-making power. The process is driven by the administration, the gatekeeper that appoints about half of the committees' participants, sets the agenda for meetings, and chairs deliberations.

Potentially the most powerful group shaping university actions is the faculty, but it has historically focused on non-campus academic issues and been largely uninterested in campus planning. Professors are generally absorbed in their own research and even in their personal interests – including convenient parking and comfortable facilities. Sessional instructors are typically too pressed, and lack the job security, to involve themselves in university governance. To the extent that they are interested in the larger collective interests of the university, they show this through existing channels such as Senate committees that are not geared to social movement concerns, let alone campus-based operational innovation. Their pervasive apathy reflects a university structure of incentives that rewards bringing in research dollars and peer recognition, and that discriminates against any "scholarship of engagement" that might cut into more traditional forms of teaching and research achievement. In addition, some cash-pressed Western (particularly European) governments have implemented academic performance indicators for universities while reducing public financial supports. On the other side, community-based research has gained a foothold by offering productive new participatory methodologies, but these are largely directed to the outside community rather than the physical operations of the university community.

For most students, concern centres on tuition fees, a pocketbook issue with an important social justice component. Student groups are largely unaware of the style of development being implemented at the university.

Much like the professoriate, most remain preoccupied with their own studies, a situation that has been exacerbated by the debts they rack up to pay for escalating tuition fees. One new student group, the University of Victoria Sustainability Project (UVSP), has begun to advocate more accountable and sustainable planning measures. Since 2007, a new student-led group, Common Energy, has worked to focus the energies of regional groups onto changes at the university. Meanwhile, staff are constrained from taking an active role, given their rigid hours, the bureaucratic hierarchy within which they work, and the limited interest (and time) of union staff in non-traditional issues.

Local community groups have a tumultuous relationship with the university. The institution is a major source of revenues for the region, yet also imposes significant costs on the municipalities, such as traffic congestion on local roads and pressures on the local water and sewer infrastructure. The UVic campus is divided between two municipalities with different bylaws and must gain acceptance by both municipalities for development permits and variances. Community groups are represented inside the university through a community liaison committee that meets with a junior university officer and has no official authority. We have observed that this committee is essentially a vehicle for information exchange; many participants see it as an unresponsive "black hole" where their concerns are "managed" by the administration rather than involved in a constructive way. Our group, the POLIS Project (which involves faculty, independent researchers, students, and community members), has advocated administrative reform to bring these groups more directly into planning. Given its presence on campus and its academic and research expertise, POLIS has played a unique role in attempting to take the issues outside the university's frame so as to advance alternatives. Its role is, however, limited by financial constraints and by the administration's extensive discretion in decision making.

Due to this institutional structure, dialogue among these actors is weak, ad hoc, and unfocused. Groups in positions of power (senior administration) or potential power (faculty) remain unsupportive or uninterested in a more open planning process, whereas those community groups who feel strongly about planning processes and goals struggle without a forum to generate dialogue. Progressive change demands political mobilization, but such mobilization confronts the need for more collaborative and democratic planning processes if new voices are to be heard. This circular situation tends to generate two contrasting outcomes – co-option or conflict.

Essential to reforming university governance is a renewed reflection on the purposes of universities. This task is especially timely, given the recent corporatization of the "higher education industry" that has subjected progressive constituencies to new forms of disciplining that stifle both critical reflection and social innovation.[4] The planning conflicts at UVic demonstrate how this institution's governance mechanisms contain dissent and create consent but also stimulate active resistance from groups committed to the university's larger social potential.

History of the Modern University

In Canada, the modern university is a corporation established by government charter, with a president and an administrative bureaucracy modelled on late nineteenth-century principles of public administration. The formal governing structure includes a two-tier system. At UVic, the Board of Governors (with an elected minority and a government-appointed majority) has final responsibility for all matters of financial management and development. The Senate – composed of professors, students, and staff – is responsible for all academic matters. In practice, a third tier of administrators plays a major role in the operation of the university through authority delegated by the president. The president, who is responsible for leadership, management, and governance, wields broad executive power on campus (McLaughlin 2004).

The modern university is a uniquely hybrid institution – a proto-public entity. It is not entirely public in that its Board of Governors is partly elected, partly appointed. And, although it relies on public funds for its operations (and increasingly on private funds, especially tuition and donations), no clear lines of accountability exist between the institution and either its tax-paying funders or its students. It is not entirely private in that its financial success is independent from "productivity," although this is changing somewhat in relation to corporate interests. Overall, its "products" are diverse – from academic publications to research patents to thousands of job-seeking graduates. The value of these products – their quality and impact – is difficult to judge, but some measures include pass/fail rates, the number of graduates who find work in various fields, the amount of research dollars and awards attracted, the number of citations of its scholars' publications, and the media presence of its experts. These are all loosely deemed to create forms of social value. Other unique aspects of universities include their often large land base (the campus itself provides a sense of place unparalleled by most other institutions), which is itself of value beyond that of simple real estate.[5] And, of course, there is the energy of young students at a formative time in their lives and the cluster of elders (academics) that is unmatched in its range and complexity by any other social institution.

The Mission of the University

Although many aspects of academic life trace their roots back to Plato's Academe, the institutional origins of the modern university date back only nine hundred years. A recent study suggests three periods of university history exemplified by three universities: the University of Paris in France, the

University of Halle in Germany, and the University of Phoenix in the United States, representing what Marcus Ford (2002) calls Christianism, nationalism, and economism respectively.

The University of Paris, which was founded in the twelfth century, offered a liberal arts education through the lens of Christian theology – the "queen of the sciences." Its mission was to serve and advance Christian civilization (ibid., 22-26). Five centuries later, higher education changed significantly when the University of Halle was founded to serve a secular German state. Studies were focused on matters that would benefit society through the development of the nation-state (ibid., 26-31). Influenced by the powerful philosophy of Immanuel Kant, the university's approach was to seek knowledge through scientific inquiry governed by reason alone. Because it was enmeshed in the entire Enlightenment process, this meant the compartmentalization of knowledge, particularly through the rise of disciplines. As knowledge became more specialized, Marcus Ford (ibid., 44) argues, it also produced a "loss of intellectual coherence" to the point that universities such as Halle lacked any integrative worldview. In this way, German universities became distinctly modern, and their new mission spread throughout the world.

The modern university is now situated in a third transformation period, shifting from nationalism to economism. To Ford, the exemplar is the University of Phoenix, which embodies the shift from basic to applied research and the increasing demand for commercialization (Fallis 2004, 52).[6] Critiques of the corporatization of the university are widespread (see, for example, Blomley 1994; Bourdieu 1988).[7] Some suggest that commercialization leaves no room for radical disagreement in higher education (Castree and Sparke 2000). In contrast, even business theorists argue that conflict and dissent are essential to continuing innovation and thus inherent to successful business practice. More broadly, dissent and debate are viewed by political theorists as essential to good government and to democracy (Dryzek 2000; Gutmann and Thompson 2004; Mouffe 2000). For many commentators, for-profit thinking and interests – from monopoly contracts over food services to direct funding and control of scientific and applied research (Bok 2003) – pose a serious threat to the university.

Throughout these historical transformations, people have asked what the role of universities should be. To cultivate political citizenship? To provide credentials for prospective employees? To supply research for corporate development, or new policy ideas for central governments? Fallis (2004, 6-7) sees an almost irreconcilable conflict between universities as providers of knowledge for its own sake or for the service of particular organizations. As university mandates change, the social contract between society and this proto-public institution must be continuously re-examined. But who should determine the mission of the university? Since, in Canada, public funds are at the core of the public university, it is presumed that its elected representatives are accountable for the institution's priorities and performance. And yet, universities behave autonomously; academics are relatively free to determine their own research. As a result, an inherent tension exists between accountability to society, on the one hand, and freedom from political direction, on the other. Insofar as it is a public institution, how should the university demonstrate its public value? In neo-liberal regimes, universities are pressured to prove their worth by reducing their demands on government revenue and relying on private or market sources of income. Turning away from public service, however, risks eroding the broader social legitimacy and accountability of universities. How can universities today shape identities and mandates that escape such contradictions? And in the process, how might they respond to the growing social challenge of sustainability? Our answers point in a common direction.

Education for Social Change

> That "sustainability" is a messy, ill-defined concept gives universities the opportunity to grapple with the concept and develop new ways of thinking about the concept. Sustainability provides colleges and universities an opportunity to confront their core values, their practices, their entrenched pedagogies, the way they program student learning, the way they think about resources and allocate these resources and their relationships with the broader community (Wals and Jickling 2002, 230).

As the Halifax and Talloires declarations signed in the 1990s suggest, universities should be leading the social revolution to address "the unprecedented scale and speed of environmental pollution and degradation, and the depletion of natural resources."[8] Such hortatory documents do not, of course, translate easily into coordinated action. Few universities have embraced this role. Some specific innovations have been applied in an incremental way to the university campus itself, even while comprehensive approaches to regional reform are eschewed (Shriberg 2002). To understand why these institutions are unable to fulfill a larger social role here, one must look at their internal structures and how they might be changed. Again, the UVic case is illustrative. Experiencing a period of rapid growth, UVic is the recipient of grants from the provincial government for much-needed new building on the condition that it meet ministerial timelines – often determined by the timing of provincial elections. Before the last election, the BC government announced nineteen hundred new full-time spaces for UVic by 2010, which would bring the campus population to over twenty thousand. With an expanding population comes the expansion of facilities.

Contestation has thus revolved around unsustainable development practices. Researchers at POLIS (POLIS Project on Ecological Governance 2003) have argued that a traditional building-by-building approach and incremental planning policies simply exacerbate pressures on the built and natural infrastructures. Needed instead are new governance and planning mechanisms that can generate more comprehensive and integrated approaches that achieve synergies among projects. In a recent report (Found and M'Gonigle 2005), POLIS presented specific proposals for on-the-ground innovation. Two sites were suggested – the Urban Village and the UVic Farmlands – that offer opportunities for growth while reducing resource use, for improving community character and sustainability, and for expanding the educational value of the campus. The report focuses on what might be called the "place" potential of the university, as compared to what is presently a mere "space" in which to learn (Orr 1992). These proposed sites would allow UVic to develop sustainable growth patterns that maintain greenspace, to become a model for innovative, equitable urban development, and to build a "complete community" that enhances campus life (Found and M'Gonigle 2005).

As it is, UVic's 2003 campus plan acknowledges progressive notions such as smart growth, but its actions on the ground are often contradictory and incoherent. Development moratoria are placed on forests, but new buildings are nevertheless sprawled throughout the rest of the campus. The university's commitment to build on parking lots and thus reduce car travel is followed by its proposal, made shortly after it created the new position of co-coordinator for "transport demand management," to build a five-hundred-stall parkade. Sustainable building design is now part of the set of development considerations, as innovative programs have resulted in significant cost savings for many universities. But progress everywhere is incremental and certainly not transformative. For example, the proposed Urban Village at UVic offered smart-growth benefits for the university and the municipality but was treated as a marginal notion, with no discussion in any formal setting. Without either a regulatory document containing meaningful standards or a forum for real participatory planning, would-be "citizens" anxious to move the university into the forefront of ecological transformation are blocked, unable to hold the administration accountable for its actions, let alone get it enthusiastic about what could be done.

The proposed UVic Farmlands – an area of 30.7 acres at the southeast edge of the campus – offers another unique opportunity for the university to address sustainability. Historically, universities have promoted the dominant and unsustainable global food system characterized by the infiltration of market-based profit-oriented processes on a grand scale into networks of food provision (Goodman, Sorj, and Wilkinson 1987; Kloppenberg 1988; Watts and Goodman 1994). The evolving system compromises food security and environmental stability (Friedmann 1993; Whatmore 1991) as it undermines historic knowledges, cultural values, and skills. In response, the organic and food security movements propose diversified small-scale community-based approaches to agriculture as a way to step back from the brink of crisis (Allen et al. 2003; McMichael 1994). Vancouver Island is host to an active organic farming community with work under way to institute a Food Charter. UVic remains an indifferent community member with no agricultural or comprehensive food studies program and no demonstrated plans to do anything about it.

Long fallow, the Farmlands present an opportunity for the university to develop a unique profile in the area of sustainable organic agriculture. The site could address a variety of contemporary needs and research questions, from wastewater recycling, energy efficiency, and sustainable food systems to youth education, aging and gardening, and urban health. A hands-on outdoor learning environment would establish UVic as a model of teaching and research in these areas. But, surprisingly, this public land is less accessible to progressive social uses than it could be because of the institutional structure in which it is situated. Facilities Management has turned down one request to use this "vacant land" for agricultural purposes and has pushed back its master plan for the site by another year.

Democratic Deficit

It is simply unrealistic to expect that the governance mechanisms developed decades or even centuries ago can serve well either the contemporary university or our society more broadly. It seems clear that the university of the twenty-first century will require new models of governance and leadership capable of responding to the changing needs and emerging challenges of our society and its educational institutions (Duderstadt 2000, 257).

A disconnect exists between what is taught at, and what is practised by, the university. Courses educate citizens about the values of democracy (Fallis 2004, 40-45), yet the daily operation of the university itself is often controlled in a top-down fashion. One scholar describes the university as an "administocracy" (Roberts 2000). Thus, universities are key examples of non-state institutions that promote democratic ideals in society while rendering them illegitimate in their own operations. Given their ability to establish the "truths" of society and to suppress alternative knowledge systems (Foucault 2003), universities are perhaps the most important sites for addressing the nature and implications of this social contradiction.

When a serious and broad challenge is posed to the university, its democratic deficit becomes evident. Supporting piecemeal approaches (Newman

2005; Shriberg 2002), but not willing to critically examine its own workspace (Silvey 2004), it fails to take advantage of opportunities that are possible even within the constraints of government funding. Too often, efforts are limited but milked for their public relations benefits. Conflict continues to exist "between the revolutionary changes called for by leaders in the sustainability in higher education movement and the incremental changes which typically occur in colleges and universities" (Shriberg 2002, 2).

The situation on the ground reflects the situation in the classroom where a discipline-based learning structure creates a body of knowledge marked by the intellectual fragmentation of complex realities whose systems are actually interconnected. Disciplines do not speak the same language, let alone engage in a collective conversation about the social context that they all share (spatially at least). At UVic, for example, a letter on campus planning submitted to the president and signed by 306 of the 650 faculty members included only 43 signatures from the natural and applied sciences, a disproportionately low number. Separated from each other, faculties are also removed from the operational side of the university, so that Operations does not learn from the academic side of the university, and vice versa. Were this not the case, university pension funds (to take a controversial example), presently invested in massively unsustainable industries at a far remove from Victoria, might become an object of research and action in the interests of shifting some of these funds to more sustainable industries that reflect the insights of community-based research. Ultimately, the result is a "shadow curriculum" of university-in-practice that teaches the opposite of the formal curriculum of university-as-educator.

Nowhere is the gap between promise and potential greater than in the institutions of governance. The state of the art in institutional reform is the lowly sustainability coordinator in a small office; eschewed are the larger changes called for in much business literature, such as P.M. Senge's (1990) well-known demands for a business to become a "learning organization." For example, at UVic, the administration responded to demands for improved governance with an extensive and very promising third-party review process, which was chaired by the former president of the University of Winnipeg, Dr. Marsha Hanen. The resulting Hanen Report (Hanen, Austin, and Higgs 2003), released in late 2003, featured innovative proposals for change including an "office of integrated planning," new planning bodies with elected members, and dozens of procedural improvements. However, the implementation of this report was turned over to the very agency that was to be reformed (Operations), a fact that prompted more student demonstrations of dissatisfaction at UVic. In the end, only a semblance of implementation occurred, one that involved no substantive change and that left decision making to function in an ad hoc fashion controlled from the top (M'Gonigle and Starke 2006b). Ironically, though the university has already gleaned benefits from changes made as a result of public pressures, the administration opposes a shift in the locus of control that would occur were decision making formally opened to authoritative participation from the universities' constituencies.

Citizenship and Good Governance

> The deliberation of democracy will surely draw upon existing knowledge and call for new knowledge; it will require the adjudication of competing knowledge claims. It will require the involvement of public intellectuals and engaged, informed citizens. All these are the stuff and substance of the university. The deliberation of democracy is not just in the political process, it must be throughout society (Fallis 2004, 51).

If the opportunity to criticize, to initiate ideas and actions, to vote on specific proposals, and to ensure accountability when action is taken does not exist within an institution, as is the case at UVic, meaningful dialogue cannot be maintained there. We have already seen how this situation has adversely affected transportation demand management and sustainable investment at UVic; yet, the university has an important opportunity to explore and promote a further crucial innovation: deliberative democracy. Indeed, the academic literature is now replete with examples of "democratic experimentalism" (Dorf and Sabel 1998; Fung and Wright 2003) that have been applied to regulatory agencies and even to corporations (to help prevent internal abuses and even to enhance their competitiveness through improved internal communications). Instead, the "vulgar" corporatization of the university in recent years further distances the individuals who work and study there from their potential as citizens of a self-governing community. When the traditional corporate model is followed, citizens are reduced to "employees" and "customers," and physical space is neither public nor open for displays of dissent. Privatized space is highly controlled, even though (like a shopping centre) it may maintain the appearance of being public (Klein 2000). Such a transformation of the university has been occurring with relatively little disruption. As one critical business theorist comments, academics are too busy interpreting the writing on the wall to realize that the wall has been sold (Parker 2002, 132). Take, for example, the attempt to conceive of a university as a planned community: Simon Fraser's "Univer-City." This massive development provides some excellent social elements, such as low-income housing and a commitment to the local economy by rejecting all retail chains in favour of independent businesses. Yet, UniverCity is a city in name only; it has no elected council and mayor, let alone democratic decisions on development.

In the context of opposing university corporatization, it is not easy to consider what being a "citizen" might mean without also considering that universities have obligations beyond serving their own needs as conceived by those who now control them. Good, or "effective," governance has been variously defined by both its process and outcomes. Some have defined it as the process that leads to "quality" decisions (Kezar 2004, 35). Others define good governance in relation to an institutional culture in which the campus citizens must decide for themselves how decisions are made (Birnbaum 1991). For example, if good governance results from processes of negotiation to seek consensus, then specific processes would seek such consensus (Kezar 2004, 36). Overall, we might define good governance as involving processes that are responsive to citizens – that help to produce consensus and that implement consensus-based decisions.

In the literature on university governance, two major streams have addressed the issue of how to create change. One has argued that structural reform (such as elected positions controlling development) should allow for more effective decision making (Benjamin and Carroll 1998). In contrast, the second approach asserts that trust relationships are more central than formal structures to university reforms (Kezar 2004; Tierney and Minor 2004). A recent survey of thousands of institutions even concluded that "structures of governance do not appear to account, in a significant way, for variance in outcomes among institutions of higher education" (Kaplan 2004, 31). This stream suggests that bureaucracies are political, relational, even anarchical such that formal structure does not generally account for behaviour (ibid., 39). In this latter approach, process, interpersonal relations, and group membership are key issues to improve campus governance (Kezar 2004; Tierney and Minor 2004). One study in particular pointed to leadership as having the greatest impact on the efficacy of governance (Schuster et al. 1994). Thus, effective governance is not defined primarily by the structure but "pertains more to the understanding and management of meaning such that the core values of the faculty and of the institution are not merely preserved, but advanced" (Tierney and Minor 2004, 94).

Yet, these relationships themselves take place within, and interact with, formal structures. The director of Facilities Management would have a different relationship with the campus planner were the planner independent of that director's control – which is why there is no such independent position. Structure has a role. John Dryzek's (2000) theory of discursive democracy emphasizes the need for meaningful dialogue and the importance of *political* contestation, something that is structurally influenced. In addition, all organizations have experience with weak leaders who act as innovation blockers if only to protect their personal positions. More open and democratic dialogue would pose a challenge to this pervasive bureaucratic phenomenon. Amy Gutmann and Dennis Thompson's (2004, 4-6) analysis of deliberative

democracy outlines the following four preconditions for democratic deliberation: all affected members must be present, the process must be accessible, deliberation should be binding, and the process must be recognized
as dynamic. Within most universities, the formal structures in place do not
allow for these preconditions to be met. Community members, both on
campus and off, are shut out of closed sessions, with key deliberations occurring within a small cadre of upper echelon administration. The resulting
decisions are then forwarded to planning bodies whose role is focused
largely on details, not the decision itself. Everywhere, campus growth is
seen as good and desirable, with limited discussion occurring on either this
overarching goal or even what sorts of growth might – or might not – be
desirable.

To foster egalitarian decision making requires a two-pronged approach
that corresponds to two streams of democratic theory. On the one hand are
what might be called liberal or modern rationalists, such as John Rawls
(1971) and Jürgen Habermas (1984), who seek to overcome political conflict
and achieve consensus through institutional designs that promote communicative agreement. On the other hand, theorists such as Chantal Mouffe
(2000) and John Dryzek (2000) argue that conflict is ineradicable and, indeed,
is a necessary condition for democracy. Mouffe advocates a conception of
agonistic democracy in which antagonisms remain within the sphere of
democratic politics. In this regard, the Hanen Report was a modernist attempt at structural reform from within, its goal being to produce a planning
structure to achieve Habermasian "ideal speech situations" where all voices
would be heard without intimidation. The report was initiated in response
to "Mouffian" insurgent protests such as the tree-sit pictured in Figure 12.1
– a situation in which uncontainable conflict was clearly necessary if reforms
were to be obtained from an administration with an entrenched set of beliefs
about the university's core interests and its own role in advancing them.
From the point of view of the university administration, such conflictual
dialogue was something to be avoided.

The administration's success at "managing" the tree-sit out of the forest
and its parallel failure to implement the recommendations of the Hanen
Report attest to the need for universities to embrace both understandings
of democracy: deliberative democratic procedures, along with openness to
continued challenge from "outside" interests. In other words, the two streams
of theory reinforce, rather than contradict, each other. Top-down change
can make it difficult to gather support from below, whereas bottom-up
change suffers from being slow and cumbersome (Rowley and Sherman
2001). Thus, shared governance between top and bottom is ideal because it
can be efficient and productive while also minimizing resistance and increasing legitimacy in an egalitarian decision-making environment. Reaching
this happy state will be a challenge as those who hold established power are

always prone to adopt models that are risk-averse, controlling, and without the complexities of time-consuming participation.

If there is a benefit to the corporatization of the university, it is the wealth of new literature on corporate innovation that may now find traction. Many business theorists make the case for a change to more adaptive and innovative management structures. José Fonseca (2002) argues for processes of "communicative interaction" that incorporate critical discussion in an open dialogue. Forums in which conflicting positions can be equitably aired are essential to successful process (Rowley and Sherman 2001). Rather than merely creating a forum for the solidification of conflicting positions, interactive processes can lead to solutions that no one party would have brought to the table (Dryzek 2000). This is where true innovation can occur.

Reinventing the University

> The University has a sort of de facto – and de jure – monopoly, which means that any knowledge that is not born or shaped within this sort of institutional field ... is automatically, and from the outset, if not actually excluded, disqualified a priori (Foucault 2003, 183).

For any reader who takes the crisis of sustainability seriously, this argument will lead to a simple, but powerful, conclusion: we need to reinvent the university. The recent trend toward its "vulgar" corporatization increases the urgency of this task. The model of development that virtually every Western university promotes – from their departments of economics to those of purchasing and grounds maintenance – reflects the very mindset and practices that underlie this crisis. Universities are, in other words, part of the problem. In contrast, an ecologically focused institution would have huge intellectual and social impact, and must be fostered.[9] This is no small task – one demanding that we localize our gaze while at the same time thinking critically about what is at stake beyond the university. Today's globalized university is deeply disconnected from place. Students can study the effects of climate change even while cars idle in sprawling parking lots, and cafeteria food is imported from thousands of kilometres away. Researchers and academics see themselves as part of a global community of experts in their respective fields, but they often fail to examine the chronic problems outside their office windows.

In order to address this in the context of planning at UVic, POLIS has proposed a strategy of comprehensive local innovation (CLI) (M'Gonigle and Starke 2006a, 2006b). CLI is a place-based strategy that draws an imagined (porous) boundary around the institution, and within this place it engages in a comprehensive and integrated re-examination of all practices and processes – from governance to recycling, pension funds to parking lots, and curricula to research – with an aim to achieving greater economic, social, and ecological sustainability. This complicated proposal involves many institutional reforms (see M'Gonigle and Starke 2006b for a detailed discussion). If such a strategy were integrated throughout the institutional culture, planning would take on greater importance for academics and students across many disciplines, engendering an awareness of local knowledges that are not only ignored, but often disqualified by the university. CLI would facilitate what Michel Foucault (2003) calls the "insurrection of subjugated knowledges." This insurrection allows knowledges that have been disqualified as naive, hierarchically inferior, or below a required level of scientificity to return in the form of local critiques of more accepted knowledge systems. Restoring a connection to place, and developing local knowledges, would provide a common focus for interdisciplinary action. By acknowledging the importance of local relationships, a community comes into being. Citizens can recognize their self-interest in the group interest. And, when these local concerns are linked to the vast, far-reaching array of networks on which universities are already founded, the possibility emerges of a diverse and locally driven but nonetheless global movement for change.

As Foucault (1980, 133) also wrote, "the problem is not 'changing people's consciousness' ... but the political, economic, institutional regime of the production of truth." Universities exercise tremendous power in determining the accepted truths of society, legitimating certain knowledges while delegitimizing others (Foucault 2003; Ford 2002). Their adoption of a modern worldview based on Cartesian and Kantian metaphysics has had significant consequences for environmental sustainability and social justice. As Marcus Ford argued, in the history of the Western university, the organicist, coherent worldview of the medieval era gave way to a disciplinary thinking that fragmented the world (in theory and practice) into separate units operating in isolation so that the findings of one are not contradicted by the truths of others. Ford (2002, 49) gives the example that economics deems endless growth to be a possible and desirable outcome, whereas the natural sciences argue that the earth's systems cannot sustain the present rate of consumption, let alone infinite growth.

Values and knowledge are held separately in this system. Gregory Cajete, the Native American philosopher, writes, "What underlies the crisis of American education is the crisis of modern man's identity and his cosmological disconnection from the natural world ... The psychological result is usually alienation, loss of community, and a deep sense of incompleteness" (quoted in ibid., 45). This lack of relationality in knowledge creates a serious difficulty in trying to understand the prospect of sustainability. As Fallis (2004, 3) comments, "Informed reflection and a vigorous civic conversation are required as we redesign our system of postsecondary education." Ford (2002,

46) suggests that a need exists to integrate the "truths" of science with those of ethical traditions in order to foster more sustainable patterns of thought and practices of experience.

Proposing a possible alternative, Ford (ibid., 8) advances a notion of the postmodern university, one that "rejects disciplines, philosophical materialism, and economism, and that seeks to address the major environmental issues that confront us today." Ford bases the transformation on the adoption of renowned British philosopher Alfred North Whitehead's "philosophy of organism," which breaks with the Western tradition of duality where the real world is separate from human consciousness. This tradition stems from the Cartesian ontological and the Kantian epistemological division between thought and matter. Ford thus advocates a reordering of the university to allow for more holistic approaches, building relationships across what have become isolated abstract disciplines. This is at once a structural and a relationship-based change as it seeks to connect the fragmented dialogues among academics. Under this model, economic analysis would be required to take into account biologic constraints and ethics. Indeed, this approach leads one to reconceptualize the university in countless ways – particularly how it could refashion itself in the practice of place, that is, in the practice of the very physical relationships that are of concern to Whitehead.

Conclusion

Although the vision of a "constructivist" postmodern university is percolating within the academic community, practical changes must be made in the institution's governance structures to ensure that democratic dialogue is allowed and encouraged. Incorporating dissenting positions in an open process creates the possibility of innovative outcomes. By holding on to the status quo, or following political dictates, universities undermine their own role in society and, indeed, their capacity for excellence and relevance. Conflict has been a necessary element in advancing this debate. When no inclusive forum exists for dialogue, groups are forced to turn to insurgent tactics, in the process disrupting existing trust relationships. The challenge of democracy at the university is to create open and transparent processes in which conflicting views are freely debated, consensus is developed without coercion, and standards of legitimacy and accountability reflect substantive rather than promotional criteria. The rewards for achieving this balance are potentially great: universities could become true models of societal innovation, promoting advances in sustainability at the local, regional, and global levels.

Notes

1 For a discussion of the tree-sit, see R. Michael M'Gonigle and Justine Starke (2006b, ch. 1).
2 The POLIS Project on Ecological Governance is a research group that formed in 2000 out of the Eco-Research Chair at UVic, following the end of support from (and dissolution of) the Eco-Research Secretariat in Ottawa. POLIS takes its name from the Greek term that, among other aspects, situates social life in collective self-regulation. This chapter emanates from a larger project on university sustainability and governance conducted by researchers at the POLIS Project. For more detailed analyses of the ideas discussed in this chapter, see M'Gonigle and Starke (2006a, 2006b). For more information on the POLIS Project, consult http://www.polisproject.org and http://www.planetaryuniversity.org.
3 New urbanism is a broad development philosophy that, for one thing, values the role of public spaces in daily life. These spaces have been neglected and demolished by modernist planning models that have brought suburban development, with its sprawling malls, highways, and big box stores (see Chapter 16, this volume, and Leonie Sandercock 1998). In the Eco-Research publication *Smart Growth: A Primer,* Deborah Curran and M. Leung (2000, 5) explain that "Smart Growth refers to land use and development practices that enhance the quality of life in communities, preserve the natural environment, and save money over the long term. The goal is to limit urban sprawl and save taxpayers money: developments that conserve resources cost less and increase property values when compared with conventional sprawl development."
4 For an excellent recent discussion of this situation, see Fallis (2007).
5 Some successful for-profit corporations have attempted to replicate the model as a way to entice employees, a case in point being the Microsoft campus in Washington State.
6 Victoria itself has not proved immune to this transformation. The city is home to BC's first private for-profit university, University Canada West, which is focused predominantly on business degrees and funded by private donors. For information on University Canada West, consult its website at http://www.universitycanadawest.ca.
7 We must carefully examine the way in which the private/public debate is framed. On the one hand, privatization is seen in the limited terms of the ownership of property and a distinct freedom from the state. Under this definition, UVic is certainly a public institution, whereas University Canada West is not. Yet, "public" can also be defined according to how institutions derive legitimacy and have accountability to the broad community that is developed through cooperative dialogue. In this sense, it is unclear that UVic is indeed a public institution, whereas private universities could claim a greater public role through a market justification of "the customer comes first."
8 The Halifax Declaration was signed by sixteen Canadian universities in 1991, who thereby declared that universities hold a major responsibility "to help societies shape their present and future development policies and actions into the sustainable and equitable forms necessary for an environmentally secure and civilized world" (full text available at http://www.iisd.org/educate/declarat/halifax.htm). The subsequent Talloires Declaration included 270 signatories from around the world. Talloires included a series of ten actions for universities to address "the unprecedented scale and speed of environmental pollution and degradation, and the depletion of natural resources" (full text available at http://www.iisd.org/educate/declarat/talloire.htm). UVic is a signatory to both declarations.
9 See, for example, the current attempt by the New University Cooperative to found a university in Quebec based on ecological principles. For details, consult http://www.newuniversity.ca.

13

The Myth of Citizen Participation: Waste Management in the Fundy Region of New Brunswick

Susan W. Lee

"It's All a Farce" – Situating a Landfill in the Democratic Process

Sitting at her kitchen table, Marlene, a homemaker in her fifties, revisits her feelings about the River Road Action Team's failure to prevent the siting of a municipal landfill at Crane Mountain, in the northwestern part of Saint John, New Brunswick – and weeps. When I ask her about public involvement in the siting decisions, she responds simply and directly: "It's all a farce." Roberta, a retired university instructor in her sixties, says that she initially believed that citizens' groups could succeed if they meant well, had strong scientific backing, and went through all the proper channels. After her experiences opposing the landfill siting, she has a totally different view of the so-called democratic process and the way power works. Fran, a project manager in her forties, views her involvement in environmental politics as a "matter of duty," a "question of justice." But she says that she and other women involved in similar struggles are often seen as loose cannons and are socially ostracized.

I spoke with these women during my research into the decision to locate a municipal landfill at Crane Mountain. Their stories and the stories of other citizens involved in the struggle against the siting of a landfill on the aquifer for their wells are the foundation for this chapter. The Crane Mountain Landfill case raises three key questions for thinking about environmental conflicts and democracy: How did the language used by industry, various levels of government, and the judicial system perpetuate the myth of a democratic, participatory process while simultaneously undermining such a process? What does this case tell us about the extent to which citizens' groups can exert influence in such an inhospitable political context? Finally, how do the experiences and perspectives of these particular citizens inform us about the possibilities for developing organizing strategies that do not legitimate "farcical" democratic processes, but advance the struggle for a more just society, socially and environmentally?

In the mid-1980s, the New Brunswick government began the process of closing small local dumps and developing solid-waste facilities in thirteen regions across the province. In 1989 the Fundy Solid Waste Action Team (Fundy SWAT), which consisted of representatives from the city of Saint John and twelve surrounding townships and unincorporated districts, was officially empowered to undertake "the challenge, by means of a democratic process, to resolve how to manage the Fundy region's solid waste" (FRSWC 2003). Citizens who tried to participate in this process found it to be an insurmountable challenge; many said that, in retrospect, it seemed more like a "done deal" than a democratic process.

My interest in the siting of the landfill began a few years after my parents sold our family home and moved into a semi-rural area on the outskirts of Saint John. They chose the area because it was a beautiful stretch of land on the Saint John River, just outside of what Saint Johners call the "fog belt." Each time I visited them, our dinner-table conversations centred, with increasing concern, on the proposed landfill and how it could affect their new community's health. As I listened to them, I began to notice the discourse surrounding this situation and to recognize its social effects. I was drawn further into the case by examining the kinds of texts that were used to placate area citizens who were asking pointed questions about the wisdom of siting a landfill in fractured rock, on an aquifer that formed the recharge area for nearly a thousand down-gradient domestic wells.

What struck me about many of these texts was the variety of ways in which they perpetuated myths of democratic citizen participation. I realized that, when citizens become politically engaged, they confront head-on a dissonance between official lip-service to these myths and the reality of attempting any genuine participation. Marlene, for example, directly articulates this dissonance in calling this political drama "all a farce." The situation in her community, like others in which citizens struggle to participate in the political realm, exposes massive inadequacies in the decision-making processes pertaining to environmental issues in Canada. The dilemma of these particular citizens sheds light on larger struggles and on the possibilities and limits of political engagement in our current political context. As David Harvey (1999, 159) states, "All environmental-ecological arguments are arguments about society and, therefore, complex refractions of all sorts of struggles being waged in other realms."

I investigated the struggles involved in the Crane Mountain Landfill case by examining texts, conducting informal interviews, listening to group discussions, and engaging in e-mail dialogues with members of the citizens' groups. The texts on which I have focused include responses to the actions of citizens who, between 1994 and 1997, opposed the landfill siting as well as critiques of the landfill since it began operation. I have concentrated on those discursive practices that spin myths about "a democratic process,"

Figure 13.1 Protesting the Crane Mountain Landfill, July 1997.
Photo David Nickerson, *Saint John Telegraph Journal*, 7 July 1997, A7.

"local solutions," "community," and "participation," and on the ways in which officials used these practices to subvert the efforts of the citizens' groups. I gathered examples from websites, correspondence with various levels of government, media reports, court documents, and pamphlets.

Using recent work on environmental discourse, I examine how environmental arguments are constructed and how, in this case, various players uphold or resist the political ideologies underpinning these constructions. My analysis is also informed by links made between environmental and social justice: the connection between "the pressures to exploit the Earth" and those "that lead to social injustice" (ibid., 159). I have also chosen to weave in the experiences and stories of three women – Marlene, Roberta, and Fran – who took part in resisting and monitoring the landfill. Each played a crucial role at various stages in the process of opposing it. Their stories give us insight into the marginalization and exclusion that people face when they engage in environmental struggles. Moreover, these women believe that their gender partially explains their level of engagement, their approaches, and the way in which they were treated. I view their stories as part of a larger history of women playing important roles in shaping the environmental justice movement.[1] Finally, I believe that health issues for women make their involvement all the more urgent. The carcinogens present in groundwater contaminated by a landfill are particularly harmful for women and children.[2] Above all, the women involved in this struggle saw themselves as defending the health of their communities.

Citizens like Marlene, Roberta, and Fran are concerned because significant evidence indicates that leachate (the liquid that seeps from garbage in landfills) contains toxins that, when released into groundwater, can migrate to domestic wells and cause cancer, birth defects, and genetic damage to those who use their water (Coates 2003, 24). As Roberta points out, the lack of public awareness about leachate is accompanied by a dangerously lingering misconception that leachate-contaminated well water can be made usable simply by boiling it: "What they don't understand is that if there is leachate in our well water, boiling it will not help because it will contain toxic chemicals, not bacteria. You can't drink it, you can't shower in it, you can't wash your clothes in it, you can't do anything with it." It is this message that the citizens have worked so hard to convey to the public.

I gathered the stories of Marlene, Roberta, and Fran via written exchanges and informal interviews with each woman. We discussed how their motivations and notions of "participation" changed throughout their struggle, the impacts these changes had on their strategies, and how they felt gender issues may or may not have played out in the process. Their voices expose the many obstacles to political involvement and highlight a (false) tension between reason and emotion in issues of social justice. For those of us who are involved in local environmental issues, their stories also hint at ways to reimagine ourselves in the struggle, to experiment with approaches that disrupt various points of power, and to seek alternative strategies once our faith in the system is lost. All of these stories offer diverse, passionate, and insightful perspectives on the challenges of locating and resisting power.

"We Never Could Find the Smoking Gun": The Myth of a Democratic Process

In 1993 Fundy SWAT publicly announced that it was in the process of selecting a site for a regional landfill to replace smaller dumps in the area that were to be closed. As part of this process, it advertised and held public meetings in several communities. When Crane Mountain was mentioned as a potential site for the landfill, residents living along the Saint John River on the western margins of the city were alarmed because the landfill would sit atop the aquifer for their domestic wells. This moment became a significant turning point in their lives, because for many it marked their first involvement in politics generally and environmental politics specifically. As the people in this part of the city, commonly known as the River Road area, became increasingly concerned, they revitalized an existing community group and instigated another.

The existing group, the River Road Concerned Citizens (RRCC), had originally been formed to oppose the location of a rock quarry in the area and had won its fight. Now, RRCC members began attending Fundy SWAT's public meetings and asking questions. Meanwhile, Marlene formed another citizens' group, the River Road Action Team (RRAT). From 1993 until the end of 1997, both groups took on the task of researching the site, educating themselves about landfills and groundwater, and opposing the siting of the landfill at Crane Mountain in any way they could.

Both groups were informal, with no set membership or organizational structure, but people generally identified themselves as "members" of either the RRCC or the RRAT, sometimes because of where they lived on River Road. However, they shared a bank account and held joint meetings, sometimes at Marlene's house, sometimes at that of David, the middle school teacher who chaired the RRCC. Together, they organized community meetings, circulated a petition that was signed by over a thousand people, raised funds, lobbied three levels of government, obtained scientific evidence that the landfill was being situated in "the worst possible place" (Gale 1997, 11), filed two civil suits, accessed documents through the Right to Information Act, and garnered extensive media coverage. Anywhere between twenty and thirty people attended these ongoing meetings (generally equal numbers of men and women). However, 250 to 300 people, and even up to 500, participated in special rallies, demonstrations, and marches. Despite all these efforts and actions, the Crane Mountain Landfill officially opened in late 1997 (for a timeline, see Appendix 13.1). This story raises serious concerns about the legitimacy of community participatory processes, reinforcing similar concerns of activists throughout Canada.

Because Fundy SWAT, later called the Fundy Region Solid Waste Commission (FRSWC), was tasked with implementing the democratic process, many citizens' accounts of their thwarted attempts to participate focus on their dealings with it. Reflecting on RRCC and RRAT interactions with FRSWC, Roberta remarked, "We never could find the smoking gun." By this she meant the impossibility of discovering what forces lay behind siting the landfill in a location so clearly prohibited by all environmental guidelines, as well as the difficulty of ascertaining to whom the FRSWC answered. Although the commission was created by the New Brunswick government, an early version of its website touted it as a separate entity: a statutory corporation under the Clean Environment Act. When the River Road citizens' groups tried to solicit information or support from either the municipal or provincial governments, they were redirected to the commission. What became clear to them was that this nebulous appointed body (its members were not elected) had somehow co-opted responsibility for ensuring "a democratic process," although it was never clear to what body it was accountable.[3]

On its website history of the Crane Mountain Landfill siting, the FRSWC spins its own narrative about how this democratic process played out. In November 2003, the website described the process as follows: "The goal was to gather residents' opinions regarding proper requirements for waste handling" (FRSWC 2003).[4] The commission gathered "valid concerns" and used them to establish "guiding principles." Based on these guiding principles, the decision was narrowed down to two possible sites: Crane Mountain and neighbouring Lorneville, where a municipal dump had been located for several years. According to its website, the commission rejected the latter because the Lorneville community "put forth emotional suggestions." The website does not indicate what these might have been or how they affected decision making.

The FRSWC (2005) also describes what, for it, was a low moment in the process – when both Crane Mountain and Lorneville, its two preferred sites, became unavailable to it. According to the FRSWC, Crane Mountain was eliminated only because efforts to purchase land there had failed. It makes no mention of how local resistance affected decision making at this point or of what politics lay behind the fact that "a local solution" to the problem of waste disposal "could not be generated" (ibid.). Spurred by a proposal to ship the area's garbage to Moncton, the FRSWC played its trump card: it authored "a factual and sobering financial report" (ibid.) in which it revealed that the twenty-five-year cost of exporting the region's waste would exceed $200 million. The report "provided the incentive for the Cabinet to approve the Crane Mountain site," and the FRSWC duly embarked upon an "aggressive schedule to open a new containment landfill" there (ibid.). By portraying its difficulties (and their solution) solely in financial terms, the FRSWC erased all other problematic aspects of the Crane Mountain issue. And by packaging the site selection as "a democratic process" in which "everyone was welcome to speak at the meetings as either an individual or a representative of a group" (ibid.), it effectively concealed the truth – that the Crane Mountain process constituted a death of democracy.

The story told by Marlene and other citizens is vastly different from that of the FRSWC. She recounts the bitter disputes that erupted between neighbours and within families: those who wanted to profit from selling their land disagreed with those who refused to sell because they strongly opposed the rezoning. Marlene also presents quite a different picture of the public meetings. "Those running the meetings were very disrespectful to us," she says, recalling a moment when she stood up at an FRSWC-organized public meeting and presented evidence from an independent scientific study to support her arguments. "I've never been a public speaker before, but I put myself forward. When I stood to speak and tell facts, they told me I was grandstanding." In addition, from the residents' perspective, the "opinions" of citizens were gathered but were not connected to decision making. Ultimately, the "emotional" environmental and health concerns of local citizens seemed to be disregarded in favour of "factual" financial considerations. Thus, in both its explicit structuring of the process and its ongoing reactions to citizens who tried to participate, the commission asserted its power to make decisions and discouraged any meaningful participation – all the while presenting an illusion of a democratic process.

"A Local Solution to Our Own Problems": Political Initiatives and Responses

The River Road citizens also solicited support from the municipal, provincial, and federal governments. Responses to their appeals reveal much about the ways in which governments suppress local citizen input by redirecting responsibility to other bodies and by invoking "individual responsibility," "economic realities," and "environmental necessities."

According to Marlene, Fundy SWAT cleverly used its newsletters to confuse and pacify local citizens. She pointed out that, for several years, Fundy SWAT published newsletters titled *SWAT Talk*, which focused on protecting the environment by creating a maximum recycling facility (MaxRF). Local citizens generally supported the creation of a MaxRF as a sound environmental decision, wherever one was located. Marlene believed that the intent of these newsletter articles was to lull the community into a false sense of security. She noted that, when the final environmental impact assessment (EIA) report came out, it made little mention of a MaxRF. Another alternative supported by citizens was shipping Saint John area garbage by truck or train to two already existing landfills (one, in southern New Brunswick, actually needed more garbage and is now importing waste from Maine).

In Saint John, the Board of Trade supported the Crane Mountain site. An article in the *Saint John Board of Trade Business Today* (1996, 1) reported that the board was concerned about "the potential impact on the community at large, on taxpayers, and on the business community, if we are forced to transport our waste to some distant location." Initial responses to RRCC and RRAT arguments also came from members of the Saint John Common Council, one of whom contended that citizens must deal with their own garbage: "We need to deal with our own garbage in our own community. We created it and we should be responsible for looking after it" (Councillor Sterling Brown, quoted in *Saint John Times Globe* 1995, A1). The mayor later reiterated this argument: "It's a local solution to our own problems. We'll be looking after our own waste" (quoted in Anderson 1996, A1). In this way, the mayor claimed to be localizing the problem even while she disregarded the particular location where citizens would be most directly affected. Thus, River Road citizens were cast as selfish and irresponsible for challenging the dump.

At the provincial level, Minister of the Environment Vaughn Blaney offered employment opportunities as another argument for situating the landfill at Crane Mountain, announcing to the media on 12 August 1996 that "the creation of thirty to forty jobs in the area could be considered an economic incentive for the [Crane Mountain] site" (St. John Board of Trade *Business Today* 1996, 1). The significance of this remark in a province with one of the highest unemployment rates in Canada cannot be underestimated. This argument allowed politicians to cast themselves as concerned for the economic well-being of the people of New Brunswick. The local activists, focusing on the health of those who would live in the community in years to come, were effectively branded as obstructing jobs for good people who needed to feed their families in the here and now. Roberta commented that she particularly resented being portrayed as not caring about the needs of her friends and neighbours.

In a 28 November 1996 letter, Premier Frank McKenna, responding to a 5 October 1996 letter from the RRCC, argued that a "new and modern waste management system" was required to replace the "environmentally problematic dumps in the region." He added that the site would be monitored by the commission, the Department of the Environment, and "a committee of area representatives, who will be aware of any potential risks and who can oversee whatever corrective measures may be warranted." Thus, the premier suggested that the environmentalists were in fact acting *against* the best environmental interests of the province.

In a letter of 12 December 1997, Joan Kingston, who succeeded Vaughn Blaney as environment minister, responded to a 22 October 1997 letter from both River Road groups, assuring them that if the groundwater did become contaminated "as a direct result from landfill operations," local citizens would be offered a "safe, uninterrupted and adequate water supply." The recipients of this letter were particularly struck by the absurdity of the paternalistic assurance that the province would take care of people only *after* their water supply had become contaminated. They also realized that the phrase "as a direct result from" implied an onus to provide causal proof and might entail costly litigation. They understood that the costs of cleanups and of piping fresh water to individuals or to an entire community would far exceed the financial and human costs of finding a more suitable location for the landfill. Fran also asserted that this kind of statement violated "the very basic principles" of the New Brunswick Clean Water Act, which requires that contamination be prevented from occurring in the first place.[5]

When the citizens appealed to Ottawa, citing the Canadian Environmental Protection Act (CEPA), an official from Environment Canada's Environmental Protection Branch responded in a 9 April 1997 letter that, although there were no "specific federal regulations which apply to municipal landfills," Environment Canada had reviewed the project and had determined that, "overall," it was "satisfied."

Fran believes that citizens have a duty to keep elected officials honest. She says that exercising one's democratic right involves using the approach employed by Lois Gibbs in the Love Canal case: confronting individual politicians and making them accept personal responsibility for the outcomes of their decisions (or non-decisions). As a result, she has been "labelled as a Norma Rae." The challenge of addressing environmental issues, Fran argues, is that many politicians see no personal political gain in concerning themselves with an issue that may become a problem only after they themselves have left office. Furthermore, "the invisibility of the issue," as Fran puts it, is also a stumbling block – "To top it off, it's not like anyone can go out there with a camera and say 'Look what they're doing,' because it's underground. It's not like you can show horrific pictures or point your finger and say 'These people did it.'"

"If We Did Enough Research": Experts and the Myth of Democratic Citizen Participation

"Risk assessment" has become a highly contentious issue in engineering circles and among environmental groups. Leo Marx (1999, 336) argues that, "in the dominant culture," engineers tend to be honoured and respected when they "play the central role assigned to them in the historical narrative about progress." In the Crane Mountain Landfill case, Gemtec, the engineering company hired to perform the EIA of the site, not only fulfilled this role but also secured the contract to plan and construct landfill operations, and to act as an ongoing consultant for them.

In *Groundwater*, a hydrogeological engineering journal, Allan Freeze and John Cherry (1989, 462) wrote an editorial in which they raised concerns about the increasing ethical blurriness in the engineering world: "The bending of hydrogeological interpretations to meet the legal rather than technical needs of a client is unethical and more of it is creeping into our business than most of us would like to see." Has Gemtec bent to client needs in the siting of the Crane Mountain Landfill and as a consultant to the Fundy Region Solid Waste Commission? In fact, one engineer who took part in the EIA acted as the commission's public relations officer when he spoke to local reporters on 15 December 1995: "Nobody wants a landfill in their neighbourhood ... Very few people see any advantage to it, and many people do not understand the details of engineering or do not, unfortunately, trust the technical people" (quoted in Rising 1995, A1-A2). Using the NIMBY (not in my backyard) label to discredit citizen involvement in the process, he implied that decisions should be left to the "real" experts.

By collecting money door-to-door and by various fundraising projects, the citizens were able to finance two independent reviews of the proposed landfill. One was undertaken by Harvard-educated environmental engineer Dr. G. Fred Lee, president of G. Fred Lee and Associates, El Macero, California, in collaboration with Brian Gallaugher of Gallaugher Associates, Toronto. The other was performed by Dr. John E. Gale, president of Fracflow Consultants, St. John's, Newfoundland. Dr. Gale, also professor of earth science at Memorial University, had been recommended by Dr. John Cherry of Waterloo University, who considered Gale to be the foremost North American expert on the transport of contaminants in fractured rock. These reviews detailed major problems with both the landfill siting and the EIA. For example, Gale (1997, ii) reported that "the proposed Crane Mountain landfill has been sited in the worst possible location within the drainage basin, namely in the main recharge area" and that "it is underlain by intensely fractured rocks whose hydraulic conductivities have been poorly characterized." He concluded, "It is our assumption that contaminants will be released from the proposed Crane Mountain landfill into the underlying fractured bedrock either through ruptures in the liner or by leakage from the leachate collection system" (ibid.). Gallaugher and Lee (1997, 15) reached a similar conclusion: "Experience has shown that this type of landfill will produce greater quantities of leachate than predicted and that it will inevitably leak leachate, thereby polluting surrounding groundwater. This groundwater, once polluted, will never again be useful for human domestic purposes."

River Road citizens also obtained memos and letters from the provincial Departments of Health and the Environment through the Right to Information Act. These documents expressed concerns about the site because of its importance as a known aquifer and recharge area, underlain by fractured rock. For example, a hydrogeologist with the New Brunswick Department of the Environment wrote in a 1994 internal memo that "The groundwater beneath the site, including the most mobile leachate components, could move to a position approximately two kilometers downgradient [to wells] after 4 years. On the basis of the site situation and of the specific hydrogeological information presented in the report, the Department's hydrogeologists have unanimously expressed concern regarding the potential danger to groundwater supplies" (quoted in David Young 1997, A1).

The citizens' groups assumed that these documents offered enough evidence to prevent the siting of the landfill at Crane Mountain.

Here we see expert knowledge operating in competing ways within environmental struggles: on the one hand, it ushers concerned citizens out of political discussions and strongly implies that such discussions should be left to the experts; on the other hand, it arms citizens with the hard evidence they need to influence the decisions of power holders. The danger to citizens' groups in relying on professional consultants is that it can result in the duelling-expert scenario. But facts, solid or not, may not resolve the conflict or even address the health concerns of citizens. Further, reliance on knowledge and expertise means that citizens' struggles become less about the creation of a fair democratic process – about changing the rules of the game (including its privileging of economic interests and its dismissal of the precautionary principle). Roberta later reflected on the process: "We naively felt that if we did enough research and had enough factual scientific material we would win."

"As They Were Writing Them, They Were Breaking Them": Guidelines, Ministerial Discretion, and the Lack of Judicial Recourse

Throughout the battle for Crane Mountain, the River Road citizens perceived litigation as their final resort. By 1997 they had exhausted every other means of opposing the landfill siting and so launched a lawsuit against the Province of New Brunswick for failing to follow its own guidelines as laid out in a 28 May 1993 document titled *Site Selection Guidelines for Municipal and Industrial Sanitary Landfills* (New Brunswick. Department of the Environment 1993).

The citizens focused on the facts that the province had not followed hydrogeological guidelines related to groundwater and that the minister of the environment had failed to ensure that the site selection process and the EIA met these guidelines before approving the site. In presenting this case before Judge David Russell of the New Brunswick Court of Queen's Bench, the groups' lawyer, Juli Abouchar, emphasized the responsibility of the provincial government to adhere to its own guidelines, arguing that "there is reasonable expectation that the guidelines be followed" (Court of Queen's Bench of New Brunswick 1997). Although this expectation seemed reasonable to River Road community members, the judge's ruling presents another picture, one in which citizens could not be assured of a fair process: "Given the number of sites that were explored, the studies that were done and the input the applicants had with knowledge that the general guidelines were being altered, I cannot conclude the applicants, by the time a decision was reached, could be said to have a reasonable expectation that the 1993 guidelines were going to be followed" (ibid.).

Judge Russell's ruling communicated two conflicting messages: that the groups' "heavy involvement in the process" should have informed them that rules were not being followed and that, even while participating in the public process, they should have taken legal action sooner to improve their chances: "The evidence before me is that the applicants had all the information in hand some two years ago to make this application in that they had the final guidelines and the EIA report. That they delayed in doing so until after the public hearings and after the Order-in-Council is detrimental to their position. I conclude they are stopped from now raising the argument that the EIA report did not comply with the guidelines" (ibid.).

This encounter with the judicial system exposes a process in which guidelines are little more than hollow reminders of a system that does not exist in law. Canadian environmental lawyer David R. Boyd (2003, 22), discussing Canada's lack of water quality regulations, highlights the main problem in relying on "environmental guidelines" – namely, that they have come to mean nothing in the legal system: "Having soft guidelines instead of hard standards means citizens have limited remedies when drinking water is not safe. Guidelines are interpreted as goals to be aspired toward, whereas standards provide certainty because they must be met." Boyd suggests that the political function of preferring guidelines to standards is "to limit public concerns." He goes on to note the absurdity of not having standards for something so vital as clean drinking water: "Canadians would never tolerate guidelines to protect them from crime, or guidelines to ensure airplanes are properly maintained." This absurdity is reflected in the Crane Mountain situation where written guidelines served no purpose in regulating the process. Marlene summed up this situation: "As they were writing them [the guidelines], they were breaking them."

In the fall of 1997, one River Road citizen filed an application for an interlocutory injunction on behalf of a class of River Road area residents. Judge John Turnbull denied the application on the basis that there was no evidence that irreparable harm could occur to the wells of the residents before the case came to trial. Turnbull also said informally that the residents had "dawdled" in filing this application and that any harm that might eventually come to them and their wells could simply be remedied by financial compensation. After Turnbull's decision, River Road residents conceded defeat.

In contemplating the challenges that arose once the groups entered litigation, Marlene recalls, "I would cry and I would worry because we needed these lawyers, and how could we raise the money for them?" "At the same time," Roberta adds, "the powers that we were fighting were public institutions. They had all the clout. When we did finally go to court, we had one lawyer against their four or five. They had all the taxpayers' money, going against a bunch of citizens who raised money door-to-door. We're not an affluent area." Thus, these court appearances constituted a huge financial burden and loss for the River Road citizens. Perhaps more devastating, however, was that their legal defeat shattered their illusion that a system was in place for citizens to ensure a legal adherence to a fair process. The judges' rulings simply continued the systematically unjust treatment that these group members had experienced all along.

Manufacturing Grassroots: Grassroots Front Groups and the Construction of Democratic Citizen Participation

In *Global Spin: The Corporate Assault on Environmentalism,* Sharon Beder (2002, 26) describes a decade-old practice that she refers to as "fronting for industry" or "manufacturing grassroots" and adds that this ubiquitous practice allows corporate interests to participate in politics "behind a cover of community concern." Penina Glazer and Myron Glazer (1998, 100-1) point to the example of Cerrell Associates, a consulting firm in California hired to analyze strategies for dealing with local opposition to landfills. Using demographic and sociological variables, Cerrell's studies indicated that the best communities to target for a landfill were "rural, low-income neighborhoods, whose residents had limited education, and tended to be older in age." More importantly, Cerrell Associates also recommended "an experienced public affairs director and a credible advisory committee who together could manage public reaction."

The River Road community does not entirely fit the Cerrell model, but it *is* marginalized by its location and by the fact that it is primarily composed of working people rather than those with wealth and power. Further, the Fundy Region Solid Waste Commission borrowed Cerrell's public relations tactics to set up its own "grassroots" group called the Fundy Future Environment and Benefits Council (FFEBC). On the cover of an attractive pamphlet on recycled paper designed by the commission, it labelled itself "an *independent* [emphasis added] community-based advisory council that will monitor all aspects of the Crane Mountain Solid Waste Park." It was also empowered to distribute $80,000 (from landfill tipping fees) as benefits to community groups. The first FFEBC chair was a member of the Saint John Board of Trade (which had approved of the Crane Mountain location), and most of its members, who had also supported the site, were residents of a neighbouring municipality whose wells would not be affected by the landfill. Not surprisingly, this municipality has received the majority of the $80,000 benefits each year. The former chair of Fundy SWAT, a leading advocate for the landfill site, recently served as chair of FFEBC (now called Crane Mountain Enhancement, or CMEI). As Beder (2002, 43) points out, membership in such "manufactured grassroots groups" tends to be stacked by the companies or their consultants to exclude voices of local community members. Through such structures, corporate interests enter the discourse disguised as citizen interests, dividing and discrediting actual citizen perspectives.

Roberta joined CMEI in hopes that her resulting insider status would succeed where outsider status had not. She acknowledges that the CMEI strategy in accepting her was to legitimate itself and make it seem to serve the community it so deeply affects. However, she feels that she has had *some* influence on a landfill monitoring committee. "I'm on this committee with a bunch of men, most of whom don't have any stake in our community, and

I'm willing to do the grunt work," she remarks. She says that her willingness to do the work meant that, in 2005, CMEI actually spent $55,000 for an independent external review of the landfill by ADI, a respected Maritime engineering company.

After the River Road citizens conceded defeat and the landfill construction began, several group members formed one incorporated organization, the River Road Community Alliance (RRCA), to represent the interests of the community and to serve as a genuine watchdog of the landfill.[6] Following a few years of rest from the struggle, they began making presentations before the FRSWC and applying for funds from the $80,000 allotted to FFEBC for projects by local community groups. However, for four consecutive years, their applications for grants for community well-testing projects and independent assessments of various landfill operations have been rejected.

"Heartsick": Disengaging from the Myth of a Democratic Process

These ongoing struggles have exhausted and disenfranchised some members of the group. Marlene attended the River Road Community Alliance's first meeting only to apologize for being too burnt out to work on anything related to the landfill. Marlene says she still feels "heartsick" about the landfill site and that recovering from what she perceived to be a devastating loss took a long time. Her view, which was reinforced by some community members, was that the efforts to stop the landfill had been futile: "Some in the neighbourhood said it was a done deal – 'They're gonna bleed you dry,' they said ... and they did!"

According to Barbara Epstein (1999, 180), local environmental groups led by local individuals like Marlene have great potential but are also limited by their lack of connections to a larger social movement. Epstein believes that, because those involved in such groups tend to be ordinary people concerned about health and property values, their arguments have more local credibility and mainstream accessibility than the more radical positions taken by the larger environmental groups. However, Roberta, Marlene, and Fran all spoke of the emotional toll of the struggle, for both themselves and others in the community.

In her studies of homemaker environmental activists, Harriet Rosenberg (2004, 453) refers to this tension as the "internal contradictions that constrain mobilization." She and Epstein have found that some members of such groups became uncomfortable with the "radical" labels they were assigned. Indeed, most members of the River Road citizens' groups did not identify themselves as "activists," or even "environmentalists," although they were cast as such. As Roberta says, "I don't like being stereotyped as an environmental activist because I think that is kind of a putting you in a category. The implication is that you've always got a cause there. And you're always on the outside." Roberta added that some community members were reluctant to maintain their involvement with the River Road Community Alliance because they did not want to be viewed as radicals and outsiders. In their minds, they were not challenging the system, just asking for prudent and reasonable safety procedures. As Marlene put it, the citizens in this struggle simply believed that their demands were a matter of "common sense."

Conciliatory Approaches

In a *Saint John Telegraph Journal* article titled "Trash Talk: Lifespan of Landfill in Dispute" (Ducharme 2006, B2), the River Road Community Alliance is depicted as "hovering like a landfill hawk." The chair of CMEI, in a letter to the local newspaper *River Valley News* (Harrigan 2006, 2), wrote, "It has been troubling to witness the shabby tactics, by what I believe to be a small number of residents, who in my opinion continue to exhibit a limited and misguided vision." Perhaps these citizens can take a small measure of reassurance, from hostile allusions to their presence, that they are having some impact. However, the language used to describe the dissidents raises questions about whether genuine dialogue can occur in this community.

Several environmental discourse theorists suggest that conciliatory language opens up possibilities for "empowerment" and discussion, and minimizes the kind of conflicts and polarization the River Road citizens experienced. Thus, Michael Spangle and David Knapp (1996, 20) offer four tactics for consideration: reduce moral position, focus on common ground, downplay dramatic approaches, and create a new model for data disputes. Connie Bullis (1996, 136) also suggests that those involved in environmental struggles reframe themselves outside of the "war metaphor," which simply "reproduces the dominant paradigm of dualistic, oppressive hierarchies in the extreme."

Conciliatory approaches may offer some opportunities for environmental gains. Roberta, who has experimented with them, explains how they have changed the way in which she positions herself in the struggle: "We have to be careful not to be self-righteous. In other words, if we view ourselves as the righteous few who care about the environment as opposed to the others – if we go at it in that way – we immediately put people off." At the same time, Roberta points out the challenge of being conciliatory in a hostile environment: "You feel like when you're nice, they don't listen to you. Sometimes you have to let them know you mean business to get their attention."

"We're Sort of the Energy"

When Fran joined the River Road Community Alliance, she brought new energy into the group. Her fearless "in your face" methods, combined with her financial expertise, have attracted the attention of some of those in authority. According to Fran, women have played and can continue to play important roles in organizing the community around the landfill. She says that, particularly in traditional communities such as hers, men may be reluctant to express the emotional and human dimensions of the issue: "I think men worry they might come across like a wuss. But a mother would just say 'How can I bath my baby and feed my baby when I don't know what that landfill is doing to my water?' That's very emotional and very real."

Fran's comment illustrates the maternalist discourse that Rosenberg (2004, 453) identifies in the environmental justice movement. Rosenberg suggests that the real power of this discourse lies in its ability to communicate that children's health is not "a negotiable category." But she adds that the typical counter-strategy is as simple, time-honoured, and patriarchal as dismissing the expression of emotion concerning environmental issues as the trait of "hysterical" women. We have already noted how Marlene was characterized as "grandstanding." Similarly, Fran was criticized as being "too pushy" and "off the wall" when she made a compelling presentation before the Saint John Common Council. When Roberta remarked at a CMEI meeting that the landfill threatened domestic wells and especially the health of women and children, a CMEI member retorted, "You'll go down to your grave believing that!"

Keeping On: A Sliding Scale of Modest Gains

The River Road citizens to whom I spoke expressed a great deal of pain regarding what they had been through and what they had learned about politics and the "democratic process." Marlene told me that she gave her whole self to the struggle and has had to disengage from it because it was negatively affecting her health. She also spoke about her continued community support "outside of politics" through her involvement in social and church activities. Roberta has managed to keep working actively with the RRCA and with CMEI, although she has often felt discouraged and been tempted to quit altogether. After several years of engagement in RRCA work, Fran has moved to a peaceful rural setting, where she cultivates her own land and participates in larger environmental groups.

In her study of participatory democracy, Rebecca Abers (2000, 4, 19) contends that citizens' concerns about "participatory forums" are valid, for citizens can be "easily manipulated" by "elite interest groups" and "corrupt politicians." At the same time, she adds, interactions between state and society always offer some possibilities for change: "Empowerment and democracy fit on a sliding scale. More is always better, but that does not mean that we should not appreciate modest gains, understanding them as windows of insight into better possibilities."

Although democracy seemed to sit at the low end of the sliding scale in this New Brunswick community, citizens there have made modest gains.

They have cultivated considerable connections in their neighbourhoods through meetings, events, yard sales, protest marches, and social events. They have formed a network that has been recognized as a force that will not go away. This recognition was expressed in a 10 March 2004 letter from the Technical Review Committee of the Department of the Environment to the environmental coordinator of the FRSWC after the latter applied to the province to modify the landfill so as to extend its life from twenty-five to forty-five years. The letter states that the application would probably be "highly controversial" because of the "various public issues associated with the landfill" and that "an open and transparent public involvement program ... will be required." Although, disturbingly, the letter focuses largely on how the FRSWC might launch a public relations campaign rather than on the wisdom of extending the landfill's life, the RRCA can view its remarks as a modest victory. And, though past experience indicates that the implementation of a truly "open and transparent public involvement program" is doubtful, the fact that the letter mentions it offers some possibility that continued citizen engagement could begin to disrupt what has been a carefully blinkered public process.

A more significant success came on 22 October 2007, when the River Road Community Alliance appeared before the Saint John Common Council to outline the short- and long-term needs of the area (the former include lighting, sidewalks, and road safety; the latter include planning for water and sewerage). The presentation concluded with these words, which exhibit the confident tone of a well-established community organization (RRCA 2007):

> The River Road Community Alliance has come before you tonight as a proactive step to ensure that a responsible design for the development of the western-most part of the city is on the radar. In planning for that development, we do not want to be part of a public relations exercise that allows us merely to react to decisions that have already been made. We want representatives from our organization to be at the table from the earliest planning stages.
>
> We are not a local service district. We are not a group of individuals with personal complaints. We are proud representatives of a forgotten and marginalized, but beautiful and promising part of the City of Saint John. We are advocating for the safety and health of our men, women and children, the protection of our environment, and the enhancement of our quality of life. We want to be recognized and respected as co-planners of our own future.

By delineating a clear community vision and identity, the RRCA successfully gained the attention of the Saint John Common Council, which responded by allocating funds in the 2008 budget to develop a comprehensive plan for rehabilitation and improvements to the area. RRCA representatives were indeed at the table throughout the planning process.

Along River Road, on the western fringe of Saint John, a stronger and more unified community has emerged in response to a potentially devastating environmental threat. It is perhaps the final irony that authorities who would have liked nothing better than merely to ignore citizens' concerns and to quietly build a suspect landfill have themselves helped to create this community.

Appendix 13.1

Chronology of the Crane Mountain Landfill Conflict, 1995-97

1995

2 February:	A public meeting begins the environmental impact assessment (EIA) process for siting the landfill at either Crane Mountain or nearby Lorneville (open public meetings are a required part of the process).
17 February:	The Fundy Solid Waste Action Team (Fundy SWAT) applies to the Saint John Planning Advisory Committee to rezone land at Crane Mountain from rural to heavy industrial. At meetings in February, April, and December, the Saint John Planning Advisory Committee rejects – on environmental grounds – Fundy SWAT's applications and reapplications for the rezoning.
19 June:	The Saint John Common Council votes against siting the landfill at either Lorneville or Crane Mountain. (Two councillors, Christopher Walschultz and Dennis Knibb, subsequently bring the issue to the table for reconsideration.)

1996

February:	The New Brunswick Cabinet rejects Lorneville as the site for a new landfill. Vaughn Blaney, minister of the environment, states that southern New Brunswick does not need another landfill and suggests that the Fundy region ship its waste to another regional landfill in the province.
March:	Despite Blaney's advice, Fundy SWAT decides to proceed with siting the landfill at Crane Mountain. It purchases remaining options on land there.
8 August:	SWAT – now named the Fundy Region Solid Waste Commission (FRSWC) – chairman Danny Harrigan recommends to Cabinet that a landfill be constructed at Crane Mountain.
9 August:	Vaughn Blaney states publicly that the Department of the Environment will study the (FRSWC) recommendation for two weeks and then take it to Cabinet.

28 August:	Members of the River Road Action Team (RRAT) go to Fredericton and meet with Blaney and three other MLAs, presenting their position on the landfill.
4 September:	RRAT sends information packages to all provincial Cabinet members.
11 September:	Cabinet officially votes to site the landfill at Crane Mountain.
12 November:	At a meeting of the Municipal Capital Borrowing Board, the FRSWC applies to borrow $17 million. Representatives from RRAT and the River Road Concerned Citizens (RRCC), accompanied by a lawyer, present their objections. The board turns down the FRSWC application.
December:	A number of individual River Road residents file Notices of Objections to the Expropriations Advisory Committee, protesting expropriation of certain parcels of land for the landfill.
19 December:	The Municipal Capital Borrowing Board reverses its stance and approves a $10 million long-term loan requested by the FRSWC.

1997

March:	RRAT and RRCC file a lawsuit against the Province of New Brunswick for failing to follow its own guidelines. Judge Russell rules against them.
19 September:	A hearing is held for a class action suit against the FRSWC for an injunction to stop work on the landfill.
6 October:	Judge Turnbull hands down a decision against the injunction.
10 November:	The landfill officially opens.

Notes

1 Lois Gibbs figures predominantly in the history of women and environmental justice. Gibbs became known as the "Mother of the Superfund" after she started the Love Canal Homeowners Association and led the group in fighting for relocation of nine hundred families who lived near a hazardous waste landfill. This – and Gibbs' organizing efforts – led to stricter environmental legislation around landfills in the United States as well as the Superfund program created by the US Congress in 1980 to locate, investigate, and clean up contaminated areas. Gibbs founded the Center for Health, Environment and Justice (http://chej.org/about.htm). For more information about the Superfund program, see http://www.epa.gov/superfund/about.htm. For work on the role of women in environmental justice movements, see Celene Krauss (1994), Jennifer Peeples and Kevin DeLuca (2006), and the Women's Environmental Network (UK) (http://www.wencal.org/).

2 Jim Tarter (2002, 223) points out that environmental justice is inextricably a feminist issue because women are more vulnerable than men to some of the most dangerous carcinogens: those that are stored in adipose tissues (fats). See also S.J. Krupp (2000).

3 River Road citizens have continued to ask who is responsible for the administration of the Clean Environment Act. When they posed this question to the manager of the Waste Management Section of the provincial Department of the Environment, he responded in a 10 November 2004 e-mail: "Where no reporting to the Minister is stipulated in the legislation, it is not the department's [the Waste Management Section's] role to ensure commissions are using or not using their powers as stipulated by the Act. This responsibility rests with

members of the commissions." He added that litigation was the only recourse for citizens who wished to question a commission decision.

4 The FRSWC website still has a page describing the history of the commission, which depicts the siting process as extensive, comprehensive, and democratic. See FRSWC (2005).

5 Section 12(1) of the New Brunswick Clean Water Act states, "No person shall directly or indirectly release a contaminant or waste or a class of contaminant or waste into or upon water if to do so would or could (a) affect the natural, physical, chemical or biological quality or constitution of water, (b) endanger the health, safety or comfort of a person or the health of animal life, (c) cause damage to property or plant life, or (d) interfere with visibility, the normal conduct of transport or business or the normal enjoyment of life or property, unless the person is acting under and in compliance with authority or permission given under an Act of the Legislature."

6 Retired engineer Horst Sauerteig has been an invaluable source of technical and scientific expertise for the RRCA. Although he does not live in the River Road community, he has volunteered hours of his time in furtherance of the group's efforts. This has brought considerable new energy to the organizing effort.

14
Neo-liberalism, Water, and First Nations
Michael Mascarenhas

This chapter relates First Nations experiences of recent drinking-water reforms in southern Ontario. Unlike the federal and Ontario governments, which display optimism in the announcement of neo-liberal reforms, First Nations view these new institutional arrangements as analogous to modern-day enclosures that limit their access to, and control over, the social and environmental resources that are critically linked to their well-being. It is for this reason that indigenous people continue to be on the front lines of contemporary environmental and social conflicts associated with neo-liberal policy reform.

Recent reforms in the provision of drinking-water services in Ontario have been particularly harmful to First Nations communities. They follow the general trends in neo-liberal governance, including governmental preferences for self-regulating markets, free trade, and privatization. In particular, the erosion of public goods and services that were once considered sacrosanct – in areas such as water, food, education, housing, and health care – has left many people and communities without the basic material conditions with which to ensure the reproduction of meaningful and complete lives. Furthermore, the environmental conditions that neo-liberalism produces on a daily basis are particularly discriminatory toward the poor, minority communities, and women because they serve to exacerbate existing social inequalities.[1] In this respect, the prosperity of neo-liberalism is predicated on specific forms of unsustainable social reproduction that disproportionately impact oppressed peoples of colour, the working poor, women, and indigenous peoples worldwide.

This research argues that contemporary neo-liberal reforms constitute an *environmental state of exception,* in which indigenous peoples fall outside the environmental protection of the state. Furthermore, because this racial formation results in ongoing environmental health crises within First Nations and Native American communities, governments are then forced to step in and provide emergency-type governance measures, thereby supplanting indigenous authority. First Nations resistance to the neo-liberal agenda in Canada and elsewhere is linked directly to the defence of land-based cultures and livelihoods (Nue and Therrien 2003). Indigenous resistance is also a struggle over the terms of Aboriginal citizenship within colonial settler states such as Canada.

The Ethnographic Approach

In analyzing the effects of neo-liberal water-provision reforms on First Nations communities, I have drawn heavily upon ethnography. Ethnography, according to Martyn Hammersley and Paul Atkinson (1995, 2), "bears a close resemblance to the routine ways in which people make sense of the world in everyday life." An ethnography is a description, interpretation, and analysis of a cultural or social group or system. Both a process and a product, an "ethnography has never been a pure or purely scientific discourse" (Grindstaff 2002, 276). As such, an ethnography is as much an art as it is a science. It is both a matter of collecting data and of collecting stories about the world we live in and the subject matter we care about (ibid.). This research blends data and theory together in a narrative that explains First Nations' struggles for safe drinking water in Canada and their resistance to private ownership and control over their land and resources.

This chapter is one of several publications (Mascarenhas 2002, 2005) that have culminated from two years of ethnographic fieldwork in First Nations communities in southern Ontario. Between 2002 and 2004, I conducted interviews and participant observation both on-reserve and off-reserve with Native and non-Native community members, politicians, administrators, lawyers, and water scientists.[2]

This chapter draws selectively upon more than eight hundred pages of transcribed interviews, as well as field notes taken in seven southern Ontario reserves. Archival data were used to supplement my data analysis. My guiding research questions focused on how the recent changes in Ontario water regulation had affected the livelihood, employment, quality of life, self-determination, governance, and general health and welfare of First Nations communities.

Ontario's Common Sense Revolution

Marxist scholar David Harvey (2003, 2005) defines neo-liberalism as a theory of political economic practices that posits that humanity's well-being can best be advanced by "liberating" individual entrepreneurial freedoms and skills. The institutional framework that best facilitates these political and economic practices is characterized by strong private-property rights, free markets, and free trade. The appropriate role of the neo-liberal state in regard to the economy is to guarantee the proper functioning of markets. Furthermore, if markets do not exist (in such areas as land, water, education, social security, health care, or environmental pollution), they must be created through state action (Harvey 2005). Erik Swyngedouw, Ben Page, and Maria Kaika (2002, 108) affirm that "the process of re-regulation in the water sector is primarily impelled by the need to make the management of water conform to market conditions." To put it differently, market relations are not simply a "natural" state that reasserts itself in the absence of regulation. Rather, market conditions are the product of a new kind of regulatory activism. Neo-liberal governments have successfully rolled backed environmental laws, worker health and safety regulations, consumer protection legislation, and other regulatory safeguards deemed to hinder the efficiency of free markets, free trade, and the expansion of private property (Harvey 2003).

Writing about the Common Sense Revolution – the title of the platform that brought the Ontario Conservative Party of Mike Harris to power – Mark Winfield and Greg Jenish (1998) highlight several important components of neo-liberal reform in Ontario during the mid-nineties. First, sweeping amendments were made to virtually every provincial statute that dealt with environmental protection or natural-resources management. As Winfield and Jenish (ibid., 130) remark, "These amendments typically weakened environmental protection requirements; expanded ministerial and cabinet discretion in decision-making; reduced or eliminated opportunities for public participation in decision-making and structures for a wide range of industries and activities having major impacts on the environment, and insulated the government from lawsuits arising out of damages resulting from the government's removal of environmental protection requirements."

Among these reforms were the repeal of land-use planning requirements intended to curb urban sprawl, the nullification of municipal bylaws intended to control environmental problems arising from "normal" farm operations, the approval of waste disposal sites without public hearings or input, and reduced monitoring and reporting requirements for industry (ibid.).

Major reductions in the budgets of provincial and local environmental and natural-resources agencies constituted the second critical consequence of neo-liberalism for environmental policy. From 1994 to 1999, the Ministry of the Environment's (MoE) operating expenditures decreased by 45 percent, while its capital expenditures decreased by 81 percent (Ontario Budget Papers, May 1998, cited in ibid.). These reckless reductions, along with reduced staffing level (one-third of MoE staff were cut during this same period) had a massive effect on the capacity of provincial and local agencies to protect the environment. Given these fiscal responses, it simply was not possible for the provincial government to adequately monitor or enforce compliance with environmental regulations.

The third major impact of neo-liberal policy reform on environmental protection, according to Winfield and Jenish (1998), has been the extensive restructuring of the roles and responsibilities of provincial and municipal governments, as well as the private sector. A major reform of the Harris government was the devolution of water management responsibilities to municipalities (Brenda Burke 2006; Prudham 2004). This encompassed responsibilities for the regulation of septic systems and the construction, operation, and maintenance of sewer and water infrastructure. In 1995 provincial laboratories responsible for the testing and monitoring of municipal drinking-water services began to close. Finally, in September 1996, the remaining government water-testing laboratories closed, forcing all municipalities to find private labs to test the quality of their public drinking water (Prudham 2004). This restructuring and rescaling of government roles and responsibilities has substantially refashioned social relations in the drinking-water sector. In general, the Harris government reforms seriously jeopardized and undermined the ability of provincial and local government agencies to manage and protect the environment. They also transferred a wide range of responsibilities and power to the private sector.[3] Jamie Peck and Adam Tickell (2002) observed that, in this form of decentralization, local institutions and actors are typically given new responsibilities without being given new powers, or resources (including revenue-raising powers).

First Nations peoples in Canada, and indigenous peoples worldwide, continue to suffer from a disproportionate share of the social and environmental injustices associated with the neo-liberal governance of water resources. For example, *The Report of the Walkerton Inquiry* (Dennis O'Connor 2002a) noted that drinking water on many First Nations reserves was among the poorest in the province (see also Polaris Institute 2008). In October 2005, persistent *E. coli* contamination in the water supply of the remote northern Ontario Cree community of Kashechewan resulted in numerous and widespread health problems. Since that time, many other communities have come forward to reveal dire circumstances regarding drinking water and general living conditions on First Nations reserves in Canada (Canadian Broadcasting Corporation 2005). Currently, as many as ninety-five First Nations reserves must boil their drinking water because it is unsafe to drink (Health Canada 2008). Furthermore, some communities have been subject to boil water advisories for almost a decade, and there are still no laws or regulatory standards similar to those in non-Native communities (Canadian Broadcasting Corporation 2005). Where local expertise exists, it is often marginalized or denigrated by state officials and drinking-water "professionals," as the statements of my interviewees attest.

What follows is a series of vignettes that detail the state of drinking-water practices on First Nations reserves in southern Ontario. These testimonies provide insights into the human and environmental conditions and consequences of neo-liberal water governance on this historically marginalized social group. In other words, what follows are examples of how neo-liberal water policy reform disproportionately affects the health and welfare of Canada's indigenous peoples.

The State of Drinking Water on First Nations Reserves

With infrastructure in dire need of repair and upgrading, ensuring a secure supply of safe drinking water has become increasingly difficult for many First Nations communities. Most of the water treatment plants I visited needed major upgrades; many simply needed to be replaced. Often the filters were too small to provide adequate filtration. Many water treatment plants were not equipped with laboratories to test water parameters, and in some cases, the laboratory doubled as a storage shed or outhouse! Several operators complained that they did not have the proper equipment and tools to make the necessary repairs and perform routine maintenance. Some had to borrow tools or bring them from home. One administrator told me that the extra repairs required to keep his plant running cost from $3,000 to $10,000 a week. At another plant I visited, the shaft to the main motor had bent, forcing the plant to shut down for several days. With no alternative supply available to them, the residents on the reserve simply drank untreated water. The feeble condition of the infrastructure and its effect on First Nations operators is clearly illustrated in the following remarks of a plant supervisor: "The other thing, when the filter gave out, was that the big paddle that keeps the water stirred, there was no other way to keep it going. We were going to run an air hose but then we didn't have a compressor in the plant that we could use. So we found some old mixing motors with L mounts, so we stuck a couple of those down inside the tank to mix the water [with the chlorine]."

In this case, we see that the operators are forced to improvise to manage repeated crises. Their job is made harder by the external pollution pressures passed on to them from upstream municipalities or industrial polluters. These circumstances have forced indigenous operators to adopt a stressful and reactive style of management rather than a proactive approach to management and control of their complex drinking water systems.

The main document to draw attention to First Nations water problems is the report produced by the Walkerton Inquiry, the judicial-scientific advisory panel charged with investigating the Walkerton *E. coli* outbreak of the summer of 2000. In Part 2 of *The Report of the Walkerton Inquiry*, Justice Dennis O'Connor (2002b) declared that two kinds of water treatment systems warranted special consideration: small water systems and those found on First Nations reserves.[4] Although this distinction may be important to highlight the high risks associated with non-Native drinking-water systems in rural areas, it is also worth mentioning that most First Nations water treatment systems are small, sometimes serving only two hundred homes on a reserve, and are predominately located in rural areas. This combination of being small, rural, and Native produces a unique set of risks that are endemic to the management of First Nations' drinking water systems. According to the Walkerton Report, small systems lack economies of scale; as a result, they may find it more expensive, on a per capita basis, to meet regulatory requirements. In addition, communities with small systems may experience difficulty attracting, retaining, and affording the expertise they need to run these highly complicated drinking-water systems (ibid.). Several water experts whom I interviewed shared this view. In fact, some experts described First Nations drinking-water systems as a "unique problem" – one that could generally be seen as the "extreme example of the rural case." It is largely for these reasons that Justice O'Connor and other drinking-water experts have concluded that the drinking water on many First Nations reserves is among the poorest in the country.

In my interview with Steven James, a quality assurance engineer for the Ontario First Nations Technical Services Corporation (OFNTSC), I began to appreciate the disparate conditions of First Nations' treatment plants in Ontario. With a master's degree in engineering and about twenty years of consulting engineering in water and wastewater treatment, James is a true expert in this field. His descriptions of First Nations plants were, to say the least, disconcerting. Citing "National Assessment of Water and Wastewater Systems in First Nations Communities" (Canada. Indian and Northern Affairs Canada 2003), a report that evaluated the conditions of First Nations water treatment plants in Ontario, James confirmed that, of the "one-hundred-plus plants scattered across Ontario, about sixty are considered high risk, and thirty or so medium risk." The report assessed 131 community water systems and classified 35 as having minimal or no problems, 35 as requiring some repairs, and 61 as putting users' health at risk. In other words, *most* of the water treatment plants on First Nations reserves in Ontario are in need of repair, upgrading, or replacement. Moreover, most operators require basic or additional training. This lack of basic drinking-water infrastructure has also led to frequent boil water advisories; as of 22 February 2008, ninety-one First Nations communities across Canada were under a drinking-water advisory (Health Canada 2008).[5] By comparison, there was only one drinking-water advisory in a large Canadian city (Vancouver) in 2007.

When I asked Larry Smith, a professional engineer with the OFNTSC, about his concerns regarding the 131 First Nations plants in Ontario, he replied,

> First of all, their treatment plants have not been built according to Ontario regulations. The regulations [were] just coming in 2002-03, and those plants were built ten, fifteen years ago. And some of the plants are just drawing water from the lake and just [adding] chlorination. No treatment. Now Ontario regulations require that any surface water has to be [treated with a] chemically cogitative [collective], sedimentation, filtration process with

appropriate disinfective agents to be supplied. So some of the First Nations are without that system. So I guess Indian and Northern Affairs Canada is trying to get those plants operated to the level of this Ontario regulation. So that will take time. I don't know how long. It depends on how quick the money flows.

Smith speaks of several important themes regarding First Nations treatment plants in Ontario. First, many of them do not meet the new Ontario water regulations. Second, many are old and in need of repairs and upgrading, and lastly, First Nations lack the resources to make these repairs. The main impetus for change, then, will come from how Indian and Northern Affairs Canada (INAC) thinks about water quality issues on Canada's First Nations reserves and how it governs itself, others (experts, consultants, and private practitioners), and First Nations communities in the provision of drinking water.

The Canadian government, through the use of "treaties, agreements, compacts, covenants and statutory obligations continues to affirm sovereignty and self-determination" of First Nations (Arquette et al. 2002, 260). Furthermore, governments continue to acknowledge the value of stakeholder participation in natural-resources decision-making processes, and research has indicated that participatory approaches tend to be viewed by citizens as the most legitimate form of governance (Mascarenhas and Scarce 2004). However, research has also shown that "Native Nations often are not respected or considered sufficiently competent to have meaningful participation in decisions that affect their Nations, lands, and resources" (Arquette et al. 2002, 260). What, then, accounts for this disjuncture? In attempting to define this question, we can look to several examples to highlight how INAC governs participation in the provision of drinking water on Canada's First Nations territories. Kevin Spencer, a First Nations' waterworks manager with more than twenty-five years of experience recalled how his water treatment plant had been upgraded in the past: "The Department of Indian Affairs [built] the first water supply for this community in 1968. [It] was a man-made dam, surface water [sourced], and we had one well. It [water] was pumped from the well to the community to our first small subdivision here, and all I did was chlorinate by hand into the well. And eventually it produced swamp water. So they [INAC] upgraded the filtration system, and that lasted until we ran into an iron bacteria problem."

Spencer's reflection that the sole Native input in the treatment of his community water supply was to dump in whatever chlorine was required suggests that INAC considers itself as a lead and legitimate authority in governing the planning and development of water treatment plants in First Nations territories. And, given that INAC is the federal department responsible for fulfilling Ottawa's obligations to indigenous peoples, this mentality regarding the conduct of water provision seems to make sense. However, the federal government has also continued to affirm indigenous sovereignty, and one would assume that this would entail the recognition of First Nations authority and expertise, not its denial. Again, we can look to INAC's past practices for further insight into government mentality regarding the provision of drinking water in First Nations territories. Spencer remarked, "First Nations, back in olden days, if they needed a water system ... they [INAC] would go in there and design the very basic, the very, very basic treatment plant. Plop it there, push a guy in there to run it, and away they go. That was it! The guy doesn't have the maintenance training, [or] operation training. So eventually it breaks down or fails and, again nobody goes out there to check on them."

Like others whom I interviewed, this operator insisted that the absence of a clear INAC drinking-water program, combined with a lack of financial support for the ongoing training that goes hand-in-hand with the introduction of new technology, has compromised the efficiency and safety of water treatment plants in First Nations territories and has undermined the competence of indigenous labour in the provision of drinking-water services (Canadian Broadcasting Corporation 2005). According to the Assembly of First Nations (2008, 2), "The federal government froze funding levels for First Nations communities over a decade ago. In 1996-97 it capped the national budget for core services at a 2 percent growth rate. It also capped for a decade the Indian Health Envelope at 3 percent which provides nursing, medical transportation, drugs and other essential health care." This inequity in funding has detrimentally affected the quality of all basic services in First Nations communities including education, housing, water, infrastructure, and health. More recently, commenting on the 2005 federal budget, Assembly of First Nations National Chief Phil Fontaine stated, "Instead of receiving more funding to finally make inroads towards improving our shameful health status and strengthening the role of First Nations governments in delivering health care, this budget actually claws back much-needed funding." "Again," the national chief concluded, "no funding has been allocated towards public health infrastructure development in First Nations communities, even though this has been repeatedly identified in national reports as the largest gap in the Canadian public health system" (quoted in Assembly of First Nations 2005, 1).

For the many communities that cannot afford to train their own water operators, this means that they must repeatedly call upon Ottawa for technical assistance. Jason Miller, an assistant manager of a medium-sized water treatment plant, provided further insight into the consequences of INAC's approach:

> We didn't have input in planning of any of these facilities that came to the
> First Nations. They are all package deals – pack-plans, I guess you would call

them. They [INAC] put a submission for tender for water and wastewater plants. And they [INAC] have three companies that bid on them.

Author: So your local knowledge is ...

Miller: No! They didn't ask us! They bring in their engineers. Their engineers tell [us] how this is, how this works, and this is how this should work. A lot of times it doesn't work like that. Or it applies to someone else but it doesn't necessarily apply here. And we call the man and say this plan here is a package deal. "Oh," he says, "there's a lot of First Nations [that] have this similar plant that we have." Because at one time the guys that were building these plants were going to all the First Nations saying, "Hey we have a plant here, your cousins down the road they have one, or the First Nations across the river here has one." And they have no problems trying to find them [First Nations that are in dire need of a new "cheap" plant].

The historical exclusion of First Nations involvement in the design, construction, maintenance, and operation of on-reserve water treatment facilities has been a consistent feature of the government's (INAC) approach to the provision of drinking water in First Nations territories. This approach suggests that INAC does not consider capacity building for self-management to be part of its mandate. One could go further, to say that the withholding of funds for operator training serves to perpetuate INAC control over a resource critical to First Nations well-being and autonomy: a secure supply of clean drinking water.

My research suggests that First Nations operators have an intimate understanding of the range of water issues confronting their communities. These practitioners comprehend the issues facing their local environs because they live with those issues on a daily basis. However, when INAC pits outside expertise against indigenous knowledge, instead of working with First Nations to include and build upon both in a mutually acceptable way, First Nations members are excluded from decision-making processes that affect their basic needs and aspirations for self-governance. Thus, in the provision of drinking-water services and infrastructure, INAC exercises a *governing of participation*, thereby undermining First Nations' sovereignty over their lands and resources while contributing to their dependency on government programs and political caprice. However, though such exclusion may, in part, be the consequence of a continuing colonial governing mentality, neo-liberal reforms – in particular, new accounting procedures – have deepened existing conflicts.

Accountability with Autonomy

"Accountability" (or the lack thereof) has become a prominent tool of neo-liberal governance, one that has significantly influenced how we think about the conduct of governments, others, and ourselves. Accountability laws, procedures, and best practices have become key features of the audit explosion that has accompanied the rise of neo-liberal governance. Accountability has become a benchmark for securing the legitimacy of a wide range of organizational practices in which the adoption of auditable standards of performance becomes synonymous with good conduct. However, as Michael Power (1997, 11) points out, auditable standards of performance are used not merely to provide for substantive internal improvements to the quality (or productivity) of particular services, but also "to make these improvements externally verifiable via acts of certification." As governments have become "increasingly and explicitly committed to an indirect supervisory role, audit and accounting practices have assumed a decisive function" (ibid.). Audit and accounting practices, then, are not simply a solution to an administrative problem: they are also a technology that makes possible ways of re-designing the practice of government.

The accountability requirements for the funding of First Nations drinking-water services provide a salient example of the tensions and conflicts that arise from this neo-liberal governing mentality. My research reveals that the lack of consistent funding and the absence of a clear drinking-water program on the part of INAC, in conjunction with provincial and federal demands for new auditable standards of performance in the provision of drinking water (see below), have provoked much criticism – even disgust – from those subjected to them. Larry Smith, a professional engineer with the Ontario First Nations Technical Services Corporation, argued, "I think better managing of the dollar [is needed]. I think channelling the dollars to First Nations [is long overdue]. If the government is pressuring them to adopt Ontario regulations [then] they should [be] flowing the monies to First Nations to adopt that regulation." Smith and others insisted that government expenditure had been neither adequate nor efficiently distributed to those who needed it most. Furthermore, though failing to provide the necessary resources, Smith said, INAC "want[s] everything [done] today," not comprehending that "it takes time for the First Nations to get operators trained, certified, and plants upgraded." However, INAC continually failed to support operator training programs, even as it insisted that all operators must be licensed by 2006. This concern has prompted Canada's commissioner of the environment and sustainable development, Johanne Gelinas, to call for stricter controls and monitoring of how INAC is trying to "fix" First Nations' water problems (Canadian Broadcasting Corporation 2006). Kevin Spencer, a waterworks manager, struggled with this contradiction: "There is still a lack of funding. My sewage plant operator, he is just on operator training [support] right now. Unless Indian Affairs [INAC] comes up with more dollars, I will have to let him go next year, come April. Like I train these guys, and bang we are stuck again, lack of funding, [and] I haven't got the staff [again] ... They [INAC] will preach to you this and that, but when you go to approach them with a proposal for dollars, that is when they clam up."

The changes in drinking-water practices alluded to above by First Nation interviewees are partially due to the new Ontario drinking-water legislation. The 2002 Safe Drinking Water Act introduced a broad array of auditable performance standards to control and regulate the provision of drinking-water systems and drinking-water testing in Ontario. This legislation not only initiated substantive internal changes to the character of drinking-water services in the province, but also redesigned the role of government from one of service provider to that of indirect supervision. INAC's comprehensive adoption of auditable standards of performance regarding the governing of water services in First Nations territories manifests neo-liberalism's equation of audit and accounting practices with good governance. Yet, most First Nation practitioners find the demands of accountability unfair and opaque.

For example, it is not unusual for INAC to take two years to respond to a First Nation request for assistance with the repair or replacement of a degraded drinking-water system. One operator told me that his community had been on INAC's "high priority" list for four years, waiting for financial support to repair its failing infrastructure. Another administrator told me that the community's water treatment plant was twenty-four years old, too small, and in desperate need of repair, and that "they had been negotiating with INAC for ten years now to build a new one." Much of this delay stems from the multiple scales that INAC has introduced to ensure that particular standards are complied with and that funds are accounted for in the upgrading of drinking-water facilities. However, many operators felt that this process was less about transparency and more about institutional control. One operator, Stan Williams, emotionally recounted his experience:

> It is just the way you are in line at the ministry. Your submissions have to show how much it is going to cost. If it costs too much you might be up there [in priority] but they'll look at it and say "That's too much fellows. You've got to revisit that, bring it down some." And that drops you down again to the bottom [of the submissions list] and you have to go through that cycle again. It has to go to [different] departments ... Their engineers got to look at it; their accountants have to look at it. It's all the same ... all INAC cares about is if ... it is feasible. That is a process in itself that we have to go through. And it is a vicious one!

Recalling the four-year struggle that occurred before his community received INAC approval to install four kilometres of watermain in its existing system, Williams noted that INAC repeatedly claimed to lack the funds for this project. Each time the community was denied, it was also asked to resubmit its application. "We had to resubmit, resubmit!" Williams remarked. And each year it did, starting from the bottom again. Finally, INAC did grant its permission, but for the same amount of money the community had requested four years earlier. Williams continued, "But in four years, the cost of everything goes up. It doesn't go down. Materials, labour, all of it goes up." Following the Walkerton Report, the federal government promised $600 million for infrastructure renewal. However, according to Kevin Spencer, this amount was utterly inadequate for the needs of the 630 First Nations across Canada.

But Kevin Spencer's frustration came not only from the insignificant amount of financial support being offered, given the desperate circumstances described above, but also from how much of it would go to meeting the government's accountability requirements:

> The first thing that they were going to take off of that $600 million dollars was to hire Ontario Clean Water Agency [OCWA] to come in and evaluate our systems, which they already did two years before that, and they want to do it again! And our water association sort of stopped that. [However,] when we caught wind of that, they had already contracted [OCWA], without posting [a competitive tender process], back in March. As soon as they [INAC] got wind of that $600 million dollars, they got these guys [OCWA] contracted to come out and do an evaluation and inspection on these plants. And it was already done. The inspections were done! The evaluations and inspections were done! OK, now give us the money to fix it up ... that is all we needed! But no, they wanted to hire somebody else to come out and do it again.

I asked him why INAC would do this. He replied, "They want to cover their own butts. That is what I have noticed with the Department of Indian Affairs; they want to shove it onto somebody, [Ontario] Clean Water Agency, or Ministry of Environment. Oh ya, they have hired Ministry of Environment to come in, too. Hey, we know how to run our plants; we just need the money to keep it fixed. But they want to spend [it on other things]. You know, to get a consultant out here, that is big bucks, that's big bucks."

According to the Assembly of First Nations (2008, 1), INAC "officials have confirmed that only about 82 percent of policy and program funds actually reach First Nations in the form of grants and contributions." The Treasury Board has estimated that 11 percent, or $600 million per year, is spent on INAC departmental overhead alone. Furthermore, it is estimated that only half (53 percent) of "Aboriginal issues" funding from other federal departments actually reaches First Nations, leaving First Nations communities to provide more programs and services to more people with less money and support each year (ibid.).

Many First Nation practitioners with whom I spoke were simply over-whelmed by the demands of accountability, which insisted that they comply with new standards and practices notwithstanding dwindling government support. They found that, though these new governing measures were driven by programmatic commitments to greater accountability, they contributed little to equity or justice for First Nations. Furthermore, though new legisla-tion has generated markets for "quality assurance" and inspection in the provision of drinking water and other public services in Ontario and else-where (Power 1997), it has done little to ensure that First Nations commun-ities can participate in these new institutional arrangements. This technique of governance, then, has shifted a further burden of accountability proced-ures to local communities. Many First Nations communities simply find it impossible to meet the new regulatory demands, whereas others refuse to make the attempt because they feel that these practices undermine local accountability measures and First Nations capacity to self-govern their own water resources.

Conclusion

The increasingly prominent role of internal control systems, rooted in explicit programmatic rationalities about water governance, has undermined in-digenous self-rule and self-determination. "It is an old struggle for the Indian people," lamented Spencer, the First Nations water plant manager. However, whereas old-style colonialism involved the direct use of state violence in forced relocation and the appropriation of traditional territories, neo-liberal governance appears in the form of programmatic commitments to greater "accountability" and "transparency." As Dean Nue and Richard Therrien (2003) point out, this form of governance presents a much more subtle incursion into the everyday lives of First Nations. In the case of drinking-water provision on reserves, many First Nations are unable to protect the health of their people because colonial governments continue to impose responsibility without providing the resources necessary for self-government. At the same time, the weakening of environmental laws and other regulatory safeguards, combined with major reductions in the budgets of provincial and local environmental and natural-resources agencies, has severely im-paired the ability of provincial ministries and local agencies to regulate and monitor environmental conditions. This is a particularly salient problem for First Nations communities, who simply do not have the capacity to provide the monitoring technologies and expertise to demonstrate scientific-ally – and debate politically – relationships between health disparities and environmental inequalities.

For First Nations, the re-regulation of water resources in Ontario has been particularly discriminatory because, while the new regulatory framework forces their compliance, it continues to limit access to resources as well as participation in relevant decision-making processes that affect their health and welfare. In these ways, neo-liberal technologies of governance adopted by Ontario and Ottawa, in conjunction with the neo-colonial relationship that continues to characterize INAC policies and practices, have led to a worsening of the forms of exclusion experienced by First Nations in Canada. We see this clearly in regard to the poor quality of drinking water on reserves and the apparent normalcy of this situation, in comparison to the standards expected for drinking water in non-Native communities. This constitutes an *environmental state of exception* in which indigenous peoples fall outside the environmental protection of the state (Agamben 2005). Creating an environmental state of exception ensures that governments and their political interests maintain their legitimacy as environmental policy makers even while they undermine other forms of knowledge and sovereignty regarding environmental governance.

Notes

1 This form of environmental inequality is particularly destructive for poor women and women of colour because they continue to subsidize their households' reproduction and well-being (Katz 2001; Merchant 1996; Shiva 1988).
2 The interview subjects quoted in this chapter are identified by pseudonyms, to protect their identity.
3 A transfer of power to the private sector accompanied the devolution of responsibilities to municipal governments. For example, self-monitoring and voluntary compliance systems were established for the forestry, aggregates, petroleum, brine, commercial fisheries, and fur industries, which were once regulated by provincial ministries (Winfield and Jenish 1998).
4 Small drinking-water systems fall into one of two categories: "The first category comprises systems covered by Ontario Regulation 459/00, which sets out water quality, treatment, monitoring, and other requirements for systems that serve more than five households or that have more than a specified capacity" (Dennis O'Connor 2002b, 15). The second category of small systems is "privately owned systems that do not come within Ontario Regulation 459/00 but that serve drinking water to the public" (ibid., 16).
5 The number of drinking-water advisories in First Nations communities across Canada fluctuates, as water quality is not static. For example, as of 30 September 2008, 106 First Nations communities in Canada were under a drinking-water advisory (Health Canada 2008).

15
Contesting Development, Democracy, and Justice in the Red Hill Valley
Jane Mulkewich and Richard Oddie

Conflicts regarding the pace and quality of urbanization have become widespread in southern Ontario, particularly during the past decade. Here, in one of the most urbanized regions in North America, large-scale development, housing, and infrastructure projects have generated land-use conflicts in numerous communities. The Red Hill Creek Expressway in the city of Hamilton, Ontario, which is the focus of one of the longest-standing disputes of this kind, presents a case study of considerable complexity and importance.[1] The struggle has involved several instances of litigation, a wide variety of environmental assessments and regulatory processes, a blockade of construction vehicles and an Aboriginal land occupation in 2003, and one of the longest tree-sits in Ontario history. The project – a four-lane highway running through the Red Hill Valley, Ontario's largest urban greenspace – was finally completed in late 2007, after decades of controversy.

In this chapter, we highlight three particularly significant aspects of the expressway conflict. The first is that, because the debate persisted for well over fifty years, it provides an important case study of the changing political-economic, cultural, and ecological forces that shape large-scale urban development projects. The various actors involved in the conflict and the arguments and strategies they used changed considerably over time, due in part to changing economic conditions, shifting political opportunities, and alterations in the use and abuse of urban nature.

The second significant aspect is the involvement of Aboriginal people in an urban area where they were dispossessed of their land over two hundred years ago. Aboriginal people from the Hamilton area and the nearby Six Nations of the Grand River territory were engaged in the Red Hill struggle (see Figure 15.1). They based their case on a three-hundred-year-old treaty guaranteeing them hunting and fishing rights; thus, the Red Hill case may constitute the first instance in which the issue of Aboriginal hunting and fishing rights has arisen in a modern Canadian city. We examine how Aboriginal

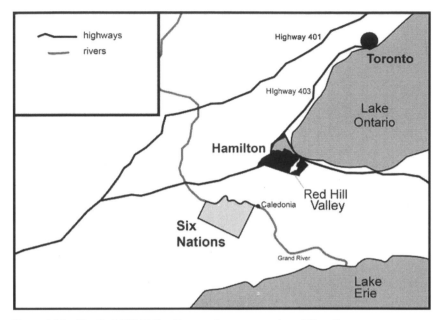

Figure 15.1   The Red Hill Valley lies on the eastern edge of Hamilton.
Map by Jane Mulkewich and Richard Oddie.

involvement in the conflict challenged the ways that non-Aboriginal actors understood and represented the debate.

Finally, the expressway conflict challenges some familiar and persistent assumptions about the nature of environmentalism in North America: these include the traditional emphasis on wilderness protection, the relative lack of attention to urban ecology and the design of sustainable cities, the neglect of the cultural framing and political economic dimensions of ecological problems, and the related neglect of social problems such as poverty, homelessness, and racism. This chapter examines how some of these assumptions were expressed and questioned through the expressway conflict, particularly with respect to the interaction between Aboriginal activists and the predominantly white middle-class environmentalist groups that spearheaded resistance to the project. We suggest that the Red Hill conflict helped pave the way for an expanded and more inclusive conception of environmentalism grounded in principles of environmental justice and radical democracy.[2]

Fifty Years of Contested Development

Hamilton, a city of approximately 500,000 people, lies at the western end of Lake Ontario. Known regionally as the "Steel City," it has concentrated its manufacturing industries along the northern waterfront, presenting a spectacular industrial skyline. Hamilton has long been associated with the negative environmental impacts of its industries, particularly air and water pollution. The deliberate siting of polluting industries in Hamilton's north end during the nineteenth century ushered in an era of environmental inequality in which land-use policies were used to legitimate and exacerbate the disproportionate exposure of working-class families to industrial pollution (Cruikshank and Bouchier 2004). By the 1920s, pollution from these factories had already begun to drastically curtail public use of the Hamilton Harbour for activities such as swimming, fishing, and ice cutting (Freeman 2001). During the post-war boom, the city established itself as a major steel-manufacturing centre and as one of the largest ports on the Great Lakes (Dear, Drake, and Reeds 1987). However, since the early 1980s, Hamilton has witnessed a gradual but dramatic decline in manufacturing, as workers have been replaced by machines and manufacturing facilities have been relocated or outsourced.

During the past thirty years or so, suburban and exurban expansion of the city has been accompanied by declining population levels in the inner city. Those who live in the southern and western sections of Hamilton tend to have the highest socio-economic status and best access to greenspace, whereas many inner-city neighbourhoods are faced with the worst concentrations of air pollution (Jerrett et al. 2001), unemployment, poverty, and homelessness (Social Planning and Research Council 2003, 2004). Racial divisions are also evident, with visible minorities and recent immigrants inhabiting the inner city, particularly its low-rent neighbourhoods. In municipal elections, divisions between suburban and inner-city residents are evident in voting patterns. These divisions have been exacerbated by the amalgamation of the various communities of the Hamilton-Wentworth Region into a single municipality in 2001. Over the past twenty years, debate regarding the city's future development has become increasingly volatile. In the effort to "revitalize" and/or redirect the city's economy, the municipal government has pursued a number of large-scale development projects that have generated a great deal of controversy, and none more so than the Red Hill Creek Expressway.

The Red Hill Creek cuts through the Niagara Escarpment in the east end of Hamilton, flowing over Albion Falls, through the forested Red Hill Valley, and eventually emptying into the Hamilton Harbour. The valley, an important site for Aboriginal communities prior to European colonization, contains numerous archaeological sites dating back as far as 11,000 BP (before the present). In the late seventeen hundreds, white settlers began farming in the valley. The City of Hamilton purchased its southern portion in 1929 and the remaining portions in 1951. During the following decades, the valley became a recreational site, with a golf course constructed during the 1960s and the addition of hiking trails. Now bounded on three sides by suburban development, the valley is the largest urban park in Ontario and the largest greenspace in the east end of the city (Peace 1998).

The first proposal for a highway through the valley was made in 1956, but it wasn't until the 1970s that provincial and municipal politicians, planners, neighbourhood associations, real-estate developers, and transportation and construction companies began collectively to advocate for the road as an answer to the challenges of traffic congestion and/or economic growth. In 1974 Hamilton City Council unanimously resolved to "permanently" protect the Red Hill Valley, while still pursuing an east-west highway connection across the escarpment. However, by 1977, under the threatened withdrawal of provincial funding for the east-west route, the majority of both the City and Regional Councils had been persuaded to support the valley route. After two years of study, which produced six alternative routes for the expressway, all running through the valley, Regional Council endorsed a route and began pursuing the necessary planning approvals from the Ontario Municipal Board (OMB) and the Ontario Environmental Assessment Board. The province soon announced that the hearings for these approvals would be combined into a consolidated joint board hearing – the first of its kind in Ontario.

This hearing began in 1984 and concluded the following year, with a two to one decision in favour of the valley route. The two favourable votes came from members of the OMB, a dispute resolution body that handles land-use conflicts between citizens, municipal governments, and developers. The dissenting vote came from a representative of the Environmental Assessment Board, who argued that the Region had failed to demonstrate the need for the highway and pointed to the negative impacts for human and non-human health. In cooperation with Save the Valley, a local citizens' group called the Conservation Authority launched an appeal of this decision, but it was defeated the following year.

Resistance to the expressway during the 1970s and '80s came mostly from conservation-oriented environmentalist groups such as Save the Valley, Clear Hamilton of Pollution, and the Hamilton Naturalists Club. These organizations largely relied upon "insider" political strategies such as letter writing, lobbying, media campaigns, and legal challenges. Bolstered by the provincial election of the social-democratic New Democratic Party in 1990 and the subsequent cancellation of approximately 50 percent of the funding for the expressway, the Friends of Red Hill Valley was formed to raise greater public awareness about the debate and the valley itself. This new group expanded its political strategies to include the organization of public events, such as cleanup days and guided nature walks in the valley, as well as ecological monitoring programs that involved citizens directly in learning about the diverse flora and fauna of the area. It also began discussing the economic impacts of the expressway, particularly the increases to local taxes and the funds that would be diverted from the maintenance of existing infrastructure and public services.

In 1995 a new Conservative provincial government was elected under the banner of Mike Harris' Common Sense Revolution (see Chapter 14). The Conservatives quickly announced their political and financial support for the expressway and accelerated plans for construction by exempting the Region from a number of provincial environmental assessments. Anticipating that the Region would soon begin construction, the Friends of Red Hill Valley mobilized citizens to lobby the federal government for a full environmental assessment (EA) of the project. The EA began in 1998, but the City of Hamilton challenged and successfully halted this assessment process in 2001, arguing that the project was exempt from federal guidelines because preliminary construction had commenced before the date specified in the federal EA legislation.

Although the growth advocates of the 1950s and 1960s had emphasized the importance of the expressway as a link between the workplaces on the industrial waterfront and the growing suburbs on the escarpment, the decline of manufacturing was met with a new focus on the expansion of those suburbs out into the rural periphery of the city. Prominent business advocacy groups such as the Hamilton Chamber of Commerce and the Hamilton-Halton Homebuilders Association argued that, because the old downtown core lacked marketable land with highway access, the expansion of commercial and industrial development on the edge of the city was required (Fothergill 2003). Many local politicians and planners were attracted to this development strategy because it promised to expand the city's tax base, and the editors of the *Hamilton Spectator,* the city's largest daily newspaper, have long provided a venue for promoting and supporting these plans. Since the late 1970s, the expressway has been presented as integral to this expansion and to the future of economic growth in the Hamilton area. Proponents have predicted that the road will reduce traffic congestion in the city, attract new investment, and provide new jobs (Hemson Consulting 2003).[3]

These plans for accelerated urban expansion sit rather uneasily with efforts to replace Hamilton's image as a polluted steel town. The City of Hamilton's website and publications celebrate the region's wealth of recreational sites and greenspaces, ongoing restoration efforts in the Hamilton Harbour, the conversion of portions of the industrial waterfront into public parklands, and new initiatives for redeveloping industrial brownfields and the downtown core.[4] Critics have made much of the apparent contradictions between the City's widely celebrated "sustainable development" plan, Vision 2020, and the pursuit of development projects that promise economic benefit but involve substantial social and ecological costs (Oddie 2003).

From Conservation to Political Ecology

Over time, opposition to the expressway expanded beyond environmentalist groups to include local neighbourhood associations, animal rights activists, and various peace and social justice groups. Although the ecological, recreational, and aesthetic value of the valley remained a prominent aspect of the arguments against the project, some opposition groups began to frame the debate as an issue of environmental justice. They noted that the highway was to be built in one of the most impoverished and polluted areas of the city, and they argued that the project would divert funds that could be spent on services such as public transit and affordable housing. Accusations of a lack of transparency, accountability, and public involvement in decision-making processes were directed at Hamilton city hall. The Friends of Red Hill Valley circulated critical analyses of City budgets and election financing contributions in an effort to demonstrate the close ties between local politicians and the various business interests that stood to profit from the construction of the highway and/or development of land on the escarpment.

The gradual expansion of the articulation of opposition could be described as a movement from conservation-oriented environmentalism to political ecology. We understand political ecology as a field of study and an approach to environmental politics that "combines the concerns of ecology and a broadly defined political economy" (Blaikie and Brookfield 1987, 17) to examine "how human practices of resource use are shaped by social relations at multiple levels over time, and the ways that these relations shape and are shaped by the physical environment" (Peter Walker 1998, 132). Analysis of the political, economic, *and* cultural context of environmental change can provide a basis for collective action that strives to challenge those relationships and forms of rationality that perpetuate social and ecological domination (Peet and Watts 1996). The recognition that "nature" and "environment" are always experienced through the lens of culture and invested with historically and spatially situated meanings and desires can also provide the basis for articulating common agendas that speak meaningfully to, and across, differences of experience and identity (Adkin 1998a, 2003).

Moving beyond a conservationist discourse, with its narrow focus on the protection of wild nature, critics of the expressway gradually began to explore the interlocking forms of exclusion and oppression that are invariably involved in conflicts regarding the use of land and ecological processes. Local activists began to better identify and understand the connections between urban governance, economic development, social well-being, and environmental change. They were forced to understand and engage with the many factors and actors, at multiple spatial scales, that perpetuate the dominant discourses and practices of urban development. Furthermore, the failure of this particular struggle challenged those opposed to critically consider the lessons learned. With the completion of the project in 2007, many on both sides of the conflict were quick to say that whatever strategies had been used to stop the road had failed. Nevertheless, we believe that it is useful to analyze the volatile events of 2003 and 2004 to identify some of the problematic assumptions and blind spots in the dominant narratives articulated by both expressway supporters and opponents. By better understanding what happened, we can understand how the Red Hill conflict set the stage for a more inclusive and effective political ecology.

Three Months in 2003: Environmentalist Blockade and Aboriginal Occupation

Following the 2001 legal decision, which halted the federal environmental assessment and allowed construction to proceed, work began again in the late summer of 2003. Local activists had been preparing to intervene by holding direct-action training workshops and a series of camp-outs in the valley so as to increase public support and involvement. This group consisted of diverse participants, ranging from experienced activists affiliated with groups such as the Friends of Red Hill Valley and the McMaster University OPIRG (Ontario Public Interest Research Group) to concerned citizens living near the valley. They dubbed themselves the "Showstoppers" in reference to comments by city councillor and mayoral candidate Larry DiIanni that construction would proceed unless there was another "showstopper." On 4 August 2003, the Showstoppers held a large rally in the valley and then began picketing at Greenhill Avenue, the area at the eastern edge of the Red Hill Valley where off-ramp construction was scheduled to begin. Using a consensus-based decision-making process and non-violent tactics, the Showstoppers prevented construction crews from entering the site for almost two weeks.

On 10 August, a short distance from the Showstoppers' Greenhill Avenue blockade, several Haudenosaunee people from the nearby Six Nations of the Grand River territory lit a sacred fire that was to be kept burning until the issues were resolved and constructed a roundhouse of tree branches where everyone was invited to discuss the issues.[5] For the next five days, there was a constant stream of visitors with a twenty-four-hour presence at both the Showstoppers' blockade and the Aboriginal occupation. But on 15 August, the City obtained a temporary three-week injunction, which stopped both picketing and construction until the matter could be settled in court. The Showstoppers agreed to stay away from the injunction zone, but many continued to supply moral and logistical support for the roundhouse. Aboriginal activists did not participate in the legal battle over the injunction and maintained that the injunction did not apply to them.

Many observers wondered why Aboriginal people had "suddenly" become involved in this conflict and implied that non-Aboriginal activists were taking advantage of, or even initiating, the occupation as way of continuing to prevent construction. However, Aboriginal activists had been voicing concerns about the valley for several years. In June 2001, Norm Jacobs, Six Nations representative for the Haudenosaunee Environmental Task Force (which addresses environmental issues on Haudenosaunee territory in Canada and the United States), posted "No trespassing" signs in the valley. The following year, he presented a petition to Hamilton City Council expressing concern that the City was ignoring Native treaty rights and that burial grounds and artifacts could be harmed by the preparations for construction. By the summer of 2003, Jacobs was very ill with cancer but issued a permit on behalf of the Haudenosaunee to members of the Showstoppers group. This permit was intended to simultaneously legitimize the presence of the Showstoppers and the rights of the Haudenosaunee with respect to the valley. Less than a month after issuing this permit from his hospital bed, Norm Jacobs passed away.

With the illness of Jacobs in 2003, other Haudenosaunee activists had initiated the occupation in the valley and sought the endorsement of the Haudenosaunee Confederacy. However, Confederacy members remained divided on the issue of direct action; in late August, the Confederacy Council agreed to create a committee to negotiate with the City of Hamilton. According to one participant, the original intention of this negotiating team was not to make a deal with the City, but rather to "buy some time" until the November 2003 municipal election, when, it was hoped, the makeup of City Council would change sufficiently to stop the expressway (*Six Nations and New Credit Tekawennake News* 2003). As the fall season approached, questions arose about how long the physical occupation could be sustained. Already, recruiting enough firekeepers to maintain twenty-four-hour coverage was proving difficult, so some non-Aboriginals were recruited and instructed to do this job. It should be noted that many of the non-Aboriginal people involved were young activists with little financial security, including students, artists, and self-styled anarchists. They dedicated large amounts of time to supporting the Aboriginal occupation and/or the later tree-sit campaign, whereas many affiliated with the more prominent environmental groups refused to break the legal injunction.

On 12 September, a Friday, the court granted a full legal injunction against interference with construction in the valley. The Haudenosaunee Confederacy met the following day and gave discretion to the firekeepers to stay at the roundhouse or to move the sacred fire to the Six Nations of the Grand River territory if they were faced with arrest. The firekeepers put out a call for help, and by Sunday afternoon hundreds of people, Aboriginal and non-Aboriginal, were working together to build a more permanent longhouse at the site. Local media reported that police would not move against the people at the encampment that week. Meanwhile, negotiations between the Confederacy and the City of Hamilton continued. Earlier, the City had signed an agreement with the Six Nations Band Council to negotiate protocols for any archaeological discoveries made during construction of the expressway. However, following the occupation of the valley, Roberta Jamieson (then elected chief of the Six Nations Band Council) issued a public statement and a letter to the *Hamilton Spectator* indicating support for the Aboriginal activists. Thereafter, the City's negotiating efforts focused on the Confederacy rather than the band council.

One motivation for Aboriginal involvement in the Red Hill struggle is the fact that the valley is a sacred burial ground. This area was used for thousands of years by Aboriginal communities and contains numerous archaeological sites (Wilson 1998). Many people involved believed that they had a sacred duty to protect nature, finding much in common with non-Aboriginal environmentalists. Some saw the Red Hill case as an opportunity to force the various levels of the Canadian state to recognize existing Aboriginal rights. For many, the assertion of Aboriginal rights is intimately connected with ecological health because, if the land and living things are degraded, the right to use that land in traditional ways will be lost, regardless of the legal existence or recognition of those rights. Finally, still others believed that they could negotiate economic benefits for the Aboriginal community while allowing the expressway to be built.

Although many non-Aboriginal activists had turned their attention and energy toward the imminent municipal election, non-violent direct action continued in late October when a number of people chained themselves to trees to prevent cutting at the south end of the valley. A week later, the Reverend Canon Paddy Doran, an Anglican priest, led a large march through the valley and proclaimed the area a sacred space that should be protected (see Figure 15.2).

This contravention of the injunction quickly attracted the attention of police. Doran was arrested but soon released without charge after fellow marchers surrounded the police car and peacefully prevented it from driving away. Days later, on 6 November 2003, police raided the longhouse, made arrests for trespassing, and allowed construction crews to enter the valley. In an ostentatious display of power, workers immediately began using machinery to tear up trees in front of the crowd of onlookers. In subsequent media coverage, police justified their actions as a response to rumours of firearms hidden in the valley and talk of the imminent arrival of armed protesters from Native territories south of the US border. Police also stated publicly that they did not consider the occupiers in the valley to be bona fide Aboriginal representatives. According to Hamilton police deputy chief Ken Leendertse, "On the morning of the arrests, the only people left in the valley were 'professional' protesters who did not represent the band council, the Six Nations Confederacy or the Friends of Red Hill Valley group. There was just those who did not have a voice at the table" (quoted in Clairmont

Figure 15.2   Protesters marching through the Red Hill Valley, led by Paddy Doran,
October 2003. Courtesy of Friends of Red Hill Valley.

2003, A5). The police raid occurred just four days before the 10 November
municipal election. After losing the physical presence in the valley, expressway opponents were dealt a further blow when City Councillor Larry DiIanni,
the chairperson of the Expressway Implementation Committee, was elected
mayor. [6]

Aboriginal Rights: Litigation versus Negotiation

On 27 November, the spirits of expressway opponents were raised by the
announcement of a $100 million lawsuit launched against the City by Larry
Green, one of the Haudenosaunee firekeepers who had lit the sacred fire in
August 2003. His case was based upon the Nanfan Treaty, signed on behalf
of the British Crown by Governor Nanfan of Albany, New York, in 1701. At

this time, the French and British were locked in a power struggle for control of North America. In exchange for a promise of peace and cooperation, the Nanfan Treaty guaranteed the right of the Haudenosaunee to hunt and fish within a massive territory that includes much of present-day New York State and southern Ontario, including the Red Hill Valley. This treaty had previously been recognized by Ontario courts to shield Haudenosaunee people from prosecution for hunting within this territory, but Green's lawsuit proposed to use the treaty as a sword, to make it theoretically and legally possible for an Aboriginal person to have unrestricted access to hunting and fishing in the area (Orkin and Klippenstein 2003). In other words, the lawsuit aimed to stop the expressway construction completely, claiming that the expressway would make it physically impossible to hunt and fish in the valley. Furthermore, if the expressway were built, $100 million would be payable to Larry Green for violation of his treaty rights. The aim of the lawsuit was not to enrich Larry Green, but to make construction of the expressway cost-prohibitive.

This lawsuit was very controversial within the Six Nations community because it was launched by an individual and without official sanction from the Haudenosaunee Confederacy or the band council. Due primarily to this lack of political support from the governmental bodies and wider public at Six Nations, the lawsuit was eventually withdrawn on 1 June 2004. Just two weeks earlier, the City of Hamilton had ratified a series of non-binding agreements with Haudenosaunee Confederacy representatives.

The discourse of Aboriginal rights is often conceptualized as one primarily concerned with land ownership and the practice of traditional economic activities such as hunting, fishing, and gathering. The Red Hill Valley case, like many others, challenges these assumptions. Green's lawsuit was not a land claim, and it applied to the assertion of traditional economic activities within an urban environment. The Nanfan Treaty granted hunting and fishing rights to the Haudenosaunee without making reference to sovereignty or land ownership. Sorting out the legalities of a land claim would have been complex in the Hamilton area, which has been home to many different Aboriginal nations. Prior to the arrival of Europeans, the land was inhabited by the Attiwandaron. They were later absorbed by the Haudenosaunee peoples that occupied this area, and by the seventeen hundreds, the land was home to an Ojibway, or Anishinabe, group called the Mississaugas. In 1784 the British Crown purchased the land of present-day Hamilton from the Mississauga Nation.[7]

Few accounts of Hamilton's history mention the colonial violence surrounding this land transaction, such as the murder of the chief of the Mississaugas in 1796 or the material deprivations brought about by land theft, sport hunting, and the damming and pollution of local creeks (Mulkewich

2003). Nor do these accounts mention that the land purchase of 1784 was a transfer of title that did not abrogate or extinguish all Aboriginal rights. Many people, including many non-Aboriginal environmentalists, assumed that the occupation of the Red Hill Valley was based on a land claim. Conversely, many assumed that, if there was no official land claim, there could be no Aboriginal rights at stake and no historical injustices to correct. In the absence of both a land claim and a clear legal definition of Aboriginal rights, many people were confused about what it might mean to assert those rights, particularly in an urban environment where hunting and fishing seemed impossible or at least incongruous.

In May 2004, members of the Haudenosaunee Confederacy signed a series of agreements with the City of Hamilton. From the beginning of the associated negotiations, it was clear that the City was unwilling to discuss the need for the road but only "the details of implementation" (City of Hamilton 2004). Similarly, as stated in the final agreements reached between the Confederacy and the City, "rather than engage in debate or dispute about the nature of their rights in the valley, the Parties resolved instead to concentrate on agreeing about the nature of their responsibilities" (ibid., 2). The legitimacy of the City's claim to own the valley was neither questioned nor debated, and the conflicts between the existence of the expressway and the exercise of hunting and fishing rights were not discussed. The agreements themselves included provisions for Haudenosaunee monitoring of archaeological work and suggested the possibility of partnerships and "economic opportunities" for the people of Six Nations to assist with ecological restoration work and the creation of a "heritage complex."

The Red Hill Valley has been cited as a successful negotiated settlement: "Even those First Nations who have previously signed treaties with the Crown may, provided there is willingness on the part of one or more of the corresponding levels of government, negotiate agreements for the protection of their off-reserve sacred sites. As with interim measures agreements, the negotiation of such agreements would provide an alternative to litigation or confrontation" (Ross 2005, 21). However, in the case of the valley, the negotiated agreement did *not* provide an alternative to litigation or confrontation. Confrontation came first, and after the police forcibly removed the Aboriginal occupation, litigation was attempted but was withdrawn when the City negotiated an outcome.

Toward an Inclusive Political Ecology
In Canada, the rights of Aboriginal people are usually tied to their "home" reserves, and they are not seen to have Aboriginal rights as urban residents. Indigenous rights have been virtually ignored in the context of Canadian cities. Aboriginal rights are conceived only in terms of land-based or territory-based rights, even though over 50 percent of Canada's Aboriginal population

lives in urban areas (Law Commission of Canada 1999). The Federation of Canadian Municipalities (2005) fosters economic development on, and coordinates services with, reserve communities (or autonomous territories created by land claims) but acknowledges that it is often unclear regarding which level of government is responsible for providing services to those who live off-reserve. In the Hamilton area, the complexities of these issues are compounded by the fact that Six Nations, the largest reserve by population in all of Canada, is only a short distance from the city.

European thought has defined "authentic" aboriginality in terms of its pre-modern nearness to nature and its spatial and temporal separation from the civilized order of Western culture, as exemplified by the city (Peters 1996). In this way, the colonial city has been represented as a civilizing force that produces order out of chaos by taming both "wilderness" and the indigenous cultures that have been identified with that wilderness (Jacobs 1996). Aboriginal peoples have been defined as external to urban culture. This is expressed through the spatial segregation of reserves and the temporal distance that is achieved by defining authentic aboriginality as an early stage of human development, beyond which Aboriginal peoples must progress if they wish to accept and integrate into modern life (Bedford and Irving 2001). This separation and erasure is evident in many popular history books of the Hamilton district, which briefly introduce the Aboriginal history of the area and the arrival of Europeans, often with little or no discussion of the brutal impacts of colonization, and then make no further mention of Aboriginal peoples within or outside of the city (see, for example, Weaver 1982; Freeman 2001).

In the Red Hill Valley, the roundhouse and sacred fire became important spaces for dialogue and cooperation, but tensions and misunderstandings were also evident. For example, many non-Aboriginal activists demonstrated a keen interest in traditional Haudenosaunee teachings and spiritual connections with nature, but some perceived claims of an *economic* Aboriginal connection to the valley as evidence of "self-interest." Some regarded such claims with suspicion and described economic interest as antithetical to the protection of nature (Showstoppers member, interviewed by Richard Oddie, 4 July 2005). This suspicion about economic motives may have been partly based on a lack of knowledge about the various Aboriginal groups involved, the political complexities of the Six Nations, and/or the poor socio-economic conditions in which many Aboriginal people live.

Arguably, these comments could also be seen as evidence of a number of problematic assumptions. The first, mentioned above, is that authentic aboriginality entails immersion in nature. The second is that economic concern necessarily entails self-interest, capitalist competition, and alienation from nature. Those who negotiated on behalf of the Confederacy did accept the narrow framing of "economic opportunities" in terms of potential employment in post-construction restoration work (such as rechannelling a portion of the creek, growing trees for reforestation, and participating in the proposed construction of "cultural features"). Nonetheless, it is important to recognize that others were interested in preserving the ecological integrity of the valley as a cultural, political, *and* economic space for the assertion of their sovereignty through the symbolic and/or literal exercise of hunting and fishing rights. In this light, it is also important to consider how easily Aboriginal struggles can be misrepresented as a form of "anti-modern preservationist politics" (Braun 2002). From this perspective, indigenous communities are seen as external to modern culture, despite the fact that they do engage in modern social, political, and economic relationships, albeit in very diverse ways and to varying degrees. Too often, those who deviate from this vision of authentic (pre- or anti-modern) indigeneity are chastised for "selling out" the preservationist principles that many environmentalists assume they share, without sufficient attention being given to the unique needs, concerns, and life experiences of Aboriginal peoples.

Related to this is the assumption that authentic environmentalism entails the altruistic protection of nature and the overcoming of human separation from nature. Such a view fails to consider that these allegedly altruistic and universally beneficial goals are often rooted in social positions of privilege. For example, those who engage in the protection of greenspaces as places of recreational or aesthetic significance are often those with privileged access to those greenspaces and the luxury of not having to worry over basic needs. As William Cronon (1995) and others have argued, the discourse of wilderness conservation and preservation that has traditionally dominated North American environmentalism is rooted in romantic longings for pristine, "untouched" nature – an ideal that tends to direct attention away from other environmental problems and often brackets out the concerns or even the existence of people who inhabit and utilize wilderness areas. Focusing on alienation from nature as *the* fundamental problematic of modern times runs the risk of obscuring the needs and experiences of those cultures and communities who, like First Nations, are struggling not to recover or re-create vital connections to natural processes and spaces, but to *preserve* them (Deloria 1983).

Mainstream environmentalism, led primarily by white middle-class activists, has paid too little attention to the ways in which our experiences and understanding of "nature," "environment," and "environmentalism" are mediated by differences of class, race, gender, sexuality, age, and ability, among others (Di Chiro 1995). Retaining a rigid conceptual division between nature and culture, it has also traditionally overlooked cities as crucial sites for environmental politics and has frequently represented the urban as the antithesis of ecological responsibility (Harvey 1996). Furthermore, it too often fails to consider the interconnections between ecological degradation, socio-economic disparity, systemic injustice, and the uneven relationships of power that operate within modern social, political, and economic institutions. In contrast, activism based on principles of environmental justice "does not treat the problem of oppression and social exploitation as separable from the rape and exploitation of the natural world. Instead, it argues that human societies and the natural environment are intricately linked, and that the health of one depends on the health of the other" (Dorceta Taylor 1993, 57).

Over the course of the expressway conflict, many participants gradually developed a better understanding of how contemporary processes of capitalist urbanization promote and support the prosperity of particular regions at the expense of others. As the Aboriginal stakes in this issue demonstrate, these processes of uneven development involve places and people both within and beyond the boundaries of the city. By exploring the intersections between structures of democratic governance, strategies of economic development, social well-being, and ecological degradation, many local environmentalists have moved beyond a narrow conservationist frame to consider the material and discursive processes through which urban development takes place. In the aftermath of the fight to save the valley, the fundamental question for many people is no longer "How can we protect greenspaces?" but rather the more wide-ranging and difficult question of "How can we create more democratic, ecologically responsible, and socially equitable cities?" Addressing this larger question of urban sustainability requires forms of activism that directly challenge the inequitable distribution of environmental amenities and hazards, that involve and engage a diversity of people and communities, and that strive to strengthen public knowledge of and control over the course of urban development.

Proponents of the expressway frequently represented the city as an undifferentiated whole, arguing that the road and associated development would bring new investment and new jobs to benefit all Hamiltonians. Critics have challenged this simplification by highlighting the socio-economic and environmental disparities between different parts of the city. A growing awareness of the connections between ecological degradation and social injustice is evident in the publications and activities of local environmentalist groups. For example, Environment Hamilton, a non-profit group founded in the late 1990s by prominent anti-expressway activists, has launched a campaign to assist citizens in collectively addressing environmental problems within the city's impoverished north end, home to numerous industrial sites and polluted brownfields.

However, much work remains to be done to foster better understanding, communication, and cooperation between environmental organizations and those struggling against poverty, homelessness, racism, and various forms of violence and discrimination. The strengthening of these connections has the potential to broaden public support and develop more inclusive and effective policy alternatives, bringing together an analysis of ecological degradation with attention to the systems of oppression (including racism and the socio-economic inequities of capitalism) that concentrate social and environmental problems in particular regions. Such alliances can diversify local environmental organizations not simply in terms of including people of colour and individuals from differing socio-economic backgrounds, but in the more substantive terms of rethinking and expanding the activities of these groups to focus on the social and ecological dimensions of issues such as housing, employment, and the provision of public services that are of central concern to marginalized communities (Agyeman, Bullard, and Evans 2003).

Beyond the Red Hill Valley

Although, at the time, the three-month Aboriginal occupation of the valley seemed lengthy, it has since been eclipsed by the Aboriginal occupation in Caledonia, a community located south of Hamilton and adjacent to the Six Nations of the Grand River territory (see Figure 15.1). In February 2006, Haudenosaunee activists occupied a partially constructed housing subdivision there, asserting their claim to this land as part of the territory granted to them by the Haldimand Proclamation of 1784. At the time of writing, the occupation is ongoing and has had a significant impact on the relationship between Aboriginal peoples and the Canadian state. The situation in Caledonia is predominantly concerned with Aboriginal rights, unlike the Red Hill case, in which the interests of Aboriginal and non-Aboriginal activists intersected. It was this very mixture of interests that, for the Hamilton police, delegitimized the Red Hill occupation and provided a rationale for its forcible termination, as discussed above. Caledonia has also involved much greater support from the people and governmental bodies at Six Nations as well as from Aboriginal communities across the country.

In Hamilton, attempts have been made to sustain the connections formed between environmentalist organizations and Aboriginal activists through local cultural events such as concerts and fundraisers. However, we found little evidence of support from environmental groups for the Hamilton Aboriginal community's ongoing efforts to work with the City on a range of issues including poverty and homelessness; nor have organized efforts been made to support land title issues in Caledonia. A number of Hamilton-based groups have been working in solidarity with Aboriginal activists in Caledonia, but they draw their strength primarily from trade unions and the labour movement rather than environmentalist groups. This is perhaps indicative of the gap that remains to be closed between social justice concerns and "environmentalist" motivations in an inclusive political ecology.

We believe that these opportunities for mutual understanding and coalition building should not be squandered. This will require a renewed commitment to forging coalitions based on recognition of the interlocking forms of exclusion and oppression that perpetuate environmental injustice and the degradation of urban environments. As Laurie Adkin (1998a, 312) writes, "A collective identity for ecology as a social movement cannot be constructed in the absence of a discourse about experiences which are spatially, temporally, and culturally located." This requires the articulation of values, goals, and political identities that resonate with the diverse social, economic, and cultural roles or positions that people embody. In this case, connections could be created or strengthened through representations of "nature," "environment," and "environmentalism" that can speak to the experiences of people who were previously unable to identify their own concerns in the dominant narratives employed by local environmentalist groups.

Coalition building requires an active engagement in anti-racism and anti-oppression work within environmental movements. Indeed, in many cases, non-white activists are greatly underrepresented in such movements, and engagement with communities of colour is often very limited – regarded as peripheral to a conception of environmentalism that excludes issues of oppression and discrimination (Gosine 2003). As Tom Keefer (2007) notes in his analysis of the non-Aboriginal left's response to the land occupation in Caledonia, white activists who have provided support for indigenous movements often fail to directly address racism within their own communities and to counter anti-Native sentiments and movements, choosing instead to "take leadership" from Aboriginal people. As the Red Hill case demonstrates, each particular situation may involve various claims to "leadership" and internal political conflicts with which non-Aboriginal activists may be unfamiliar. Furthermore, Aboriginal activists may be wary of allowing further non-Aboriginal influence upon the internal dynamics of their communities, given the historical legacy of such interference. In Keefer's (ibid., 10) words, "Building radical organizations and combating white racism within predominantly white communities, workplaces and political organizations will be particularly hard. But it remains a necessary task as a pre-condition to building meaningful solidarity with indigenous struggles."

The development of more persuasive and inclusive forms of environmental activism in Hamilton must also continue to involve critical engagement with the limitations of liberal conceptions of justice and democracy. By defining citizenship as primarily a set of individual political and civil rights to "life, liberty, and property," liberalism promotes the formal legal and political equality of individuals. Yet, it fails to acknowledge how these civil and political rights are undermined by the privileging of "market rights" to the accumulation of wealth over social rights to basic needs for food, shelter, health care, and employment (McMurtry 1998). Liberal conceptions of democracy and justice also fail to adequately consider claims to collective or group rights, particularly those that question the private ownership model of property. In contrast, a more radical and inclusive understanding of democracy seeks to extend the liberal principles of autonomy and equality. It engages in both a "politics of recognition" and a "politics of redistribution" – the former struggles against oppression that is based on cultural differences, whereas the latter struggles against socio-economic (and environmental) inequity (Nancy Fraser 1997). This notion of radical democracy provides a promising direction for an urban environmental politics that aims simultaneously to support social equity, ecological health, and democratic participation.

Although the struggle to stop the Red Hill Creek Expressway did not succeed, the groundwork has been laid for more effective and inclusive forms of environmental activism. Much remains to be done, but new connections, new understandings, and new initiatives have been established that have the potential to significantly challenge the ideology of value-free growth and to draw greater attention to the ways in which social oppression, economic marginalization, and ecological degradation work together to produce uneven and inequitable forms of urban development.

Notes

1   This road was renamed the Red Hill Parkway in 2005, but we refer to it as the Red Hill Creek Expressway as this name was associated with the project for most of its long and controversial history.
2   This chapter is based on our professional expertise in the fields of environmental studies and law, and on our direct participation in the Red Hill debate as community activists with concerns about the socio-economic and ecological impacts of the highway. We have expressed our concerns independently and in cooperation with many of the groups mentioned here, including the Friends of Red Hill Valley, the Showstoppers, and participants in the Aboriginal land occupation of 2003.
3   Many of the growth coalitions that favoured the Red Hill Creek Expressway are now supporting the proposed "Mid-Peninsula Highway" that would link the southern edge of the Greater Toronto Area with the US border crossing at Fort Erie; they are also advocating the extensive development of the rural lands surrounding Hamilton's John C. Munro International Airport (Flamborough Chamber of Commerce 2005).
4   This presentation of the City's environmental achievements and goals, which we observed in 2006 on its official website, is still evident. See, for example, the Public Works Department's page titled "Together, We're Caring for Our Environment" (City of Hamilton 2007).
5   The Haudenosaunee Confederacy (also called the Iroquois, or Six Nations, Confederacy) is a political and kinship-based union of the Cayuga, Mohawk, Oneida, Onondaga, Seneca, and Tuscarora Nations, led by chiefs and clan-mothers, which predated colonization. The Six Nations Band Council was forcibly imposed by the Canadian government in 1924 under the federal Indian Act. Those who identify as Haudenosaunee are maintaining pre-colonial traditions. The Confederacy Council is taking an increasing interest in protecting the environment throughout Haudenosaunee territory.
6   Elected mayor of Hamilton on 10 November 2003, Larry Dilanni was subsequently defeated on 13 November 2006 following a controversy surrounding illegal campaign donations from various real-estate, construction, and transportation companies, many of whom had

promoted the Red Hill Creek Expressway. On 24 August 2006, DiIanni was found guilty of violating Ontario's Municipal Act.

7 Although some individual Mississaugas were involved in the occupation of the valley, the nation itself remained silent regarding this issue. The Mississaugas are presently focusing their resources on several land claims along the north shore of Lake Ontario, including the city of Toronto.

16
Instant Gentrification: Social Inequality and Brownfields Redevelopment in Downtown Toronto
Cheryl Teelucksingh

Toronto's middle class is rediscovering the attractions of living downtown. Trendy lifestyle advertising beckons young professionals to come work, live, and play in the modern metropolis. Normally, such gentrification is associated with the upscaling of existing low-income inner-city neighbourhoods, but Toronto's current urban renewal targets underutilized and virtually abandoned industrial brownfields sites.[1] Those lying near the downtown financial centre and the coveted Lake Ontario waterfront have become hot properties. This "instant" form of gentrification (Rose 2004, 7) involves the claiming of previously undesirable industrial spaces by the middle class and the upwardly mobile. Competition for scarce resources in downtown Toronto is positioning developers, real-estate agents, and potential middle-class residents against marginalized groups and marginalized land uses. In this chapter, I examine the social and political inequalities that arise from brownfields redevelopment in downtown Toronto. Conflicts surrounding the brownfields raise important environmental justice and citizenship questions, such as who has claims to space in the city.

I explore various stakeholder perspectives regarding brownfields redevelopment in light of competing needs emerging from post-industrialization and shifts toward neo-liberal agendas. I argue that brownfields gentrification is a unique form of gentrification that has the potential to isolate and exclude low-income and ethno-racial groups, to divert attention and investment from the creation of affordable housing for Toronto's growing population, and, ultimately, to tarnish the city's image as the ideal Canadian multicultural urban centre. The brownfields conflict in downtown Toronto is resulting in social and environmental inequalities as well as new forms of resistance to these inequalities. Drawing on an analysis of planning documents, newspapers, and other public accounts, my study uncovers the social struggles over space that have arisen from brownfields redevelopment and offers examples of grassroots voices that are emerging as marginalized actors attempt to reclaim their rights to the city.[2]

Environmental Justice and Citizenship
The environmental justice movement challenges the systematic imbalances of power and the structural barriers that disadvantage particular groups. The movement emerged in the United States during the 1980s as a response to environmental problems that featured distributional and social justice dimensions.[3] In Canada, both research and practice under the umbrella of environmental justice remain in a nascent stage. However, the work of two Canadian theorists – Dianne Draper and Bruce Mitchell – is particularly useful for conceptualizing environmental justice struggles in the Toronto context. Draper and Mitchell (2001) argue that environmental justice's focus on distributive justice takes two forms: demands for substantive rights and demands for procedural rights. "Substantive" refers to "human safety and survival, access to and use of natural resources and the environment." "Procedural" refers to "access to information, fair hearings, and equal participation" (ibid., 93). Substantive and procedural rights, as two distinct environmental justice objectives, require differing types of environmental and social policies, involve differing stakeholders, and rely on differing notions of citizenship. These two concepts provide an important framework for understanding how urban development conflicts – such as brownfields redevelopment – arise between groups that have unequal access to rights and claims in the city.

Resolving distributional injustices in substantive rights requires changes in the regulation of public goods (water, air, and soil) such as restrictions on the practices of industries, businesses, or service providers that affect communities' health, safety, and environmental rights. For example, the Ontario provincial government requires environmental impact assessments and soil cleaning for brownfields. This type of regulatory framework provides guidelines for corporations and developers, and sets out a top-down process (ibid., 96). The state's role in ensuring substantive rights is consistent with a legal or formal sphere of citizenship (Bryan Turner 1992). Here, the state takes a role in ensuring environmental rights through legislating community entitlements to nature and/or community responsibility for nature (Gilbert and Phillips 2003). As Bryan Turner (1992, 44) notes, formal citizenship is passive since rights and privileges are simply handed down to citizens by the state. Formal citizenship and, by extension, substantive environmental justice rights often incorporate aspects of liberalism such as the belief that everyone should be treated equally regardless of group differences. Contradictions between formal political equality and substantive inequality explain why, for instance, a universal right to housing is not a component of formal citizenship rights (ibid.). Liette Gilbert and Catherine Phillips (2003, 319) argue that "citizenship is not only a set of formal rights and practices such as voting or claiming public benefits; rather, it is a continual process of creation and transformation of both society and nature." In this sense, environmental rights and citizenship rights are often the outcomes of struggles.

Environmental justice objectives that involve procedural rights require changes to decision-making processes at all levels to ensure the representation of all stakeholders, regardless of their power or political economic interests. Low-income inner-city residents, for example, demand the right to participate in public municipal meetings about land-use changes in their community. In this sense, in comparison to substantive rights, egalitarian procedural rights are much more difficult for governments to regulate through generalized standards since this involves addressing power imbalances in society. In the United States, these obstacles have led the environmental justice movement to focus on substantive rights.[4]

Procedural rights often involve informal conceptions of citizenship that have the potential to dismantle economic power imbalances. They permit various stakeholders to voice their perceptions of risks and their framings of the same problem. Therefore, environmental justice theory and practice needs to move beyond formal definitions of citizenship to respond to the diversity of interests and the multicultural nature of the polity (Goldberg 1994; Gilbert and Phillips 2003). Informal spheres of citizenship are constituted by the practices of people in the city who are making claims on public space and fostering new social movements and group identities (Isin and Siemiatycki 2000). Citizenship from below, as described by Bryan Turner (1992, 44), is a "consequence of social struggles over resources." This often involves ad hoc associations that arise due to material conflicts between stakeholders.

Iris Marion Young's (1999) work on the politics of difference and citizenship rights provides additional useful insights for the environmental justice movement. For Young (1990, 1999), the public sphere should be accessible to all, and the democratic polity should provide mechanisms (such as support for self-organizing) that enable oppressed groups to enter the decision-making process. Young's approach to multicultural citizenship is also an important step in acknowledging issues of difference in terms of both substantive and procedural rights. Young (1999, 237) argues that citizenship involves people living "together with differentiated others in a common polity."

In a multicultural and global city such as Toronto, the prevention of environmental injustices and the meaningful acknowledgment of diversity depend on political processes that give fair consideration to all claims to spaces. An absence of either substantive or procedural rights, or both, relating to brownfields transformation, can result in environmental injustices. This case study of brownfields redevelopment in Toronto highlights the significance of local struggles for urban survival and environmental justice as struggles for citizenship rights.

Toronto: The Global City

By the early 1990s, Toronto had become a global city in which local urban development was increasingly linked to processes occurring globally (Sassen 1991). Toronto's rising global status was due in part to three important political and economic processes simultaneously at play: these included the city's enhanced role as a financial and service centre for the new global economy and the economic and spatial inequalities emerging from economic restructuring, its population growth and increasing racialization via new immigration, and the stress on its housing market and on the municipal government's ability to provide affordable housing for all its residents. Toronto has become central in coordinating the international division of labour, which involves multinational corporations with various locations of production and distribution, and the global movement of financial capital. The head offices of many top Canadian corporations, as well as foreign-owned corporations, are now located in Toronto, resulting in a commercial real-estate boom in the downtown core from the 1980s onward (Caulfield 1994, 84).

Due to these changes, low-income residents and migrants to the city have been squeezed out of real-estate markets. However, unequal access to housing has also taken on increased racial dimensions. Since the liberalization of Canadian federal immigration policy in the late 1960s, new immigrant groups have been drawn to Toronto from developing nations. These new waves of immigrants are integrated into a racialized division of labour, which ranges from the highly skilled and highly paid professional strata (largely white) to the (largely racialized) low-skilled and low-paid service-sector strata, on the other (Sassen 1991). Today, immigration accounts for a large proportion of the population growth in Toronto. The Ontario government estimates that 4 million people will move to southern Ontario's Golden Horseshoe region within the next thirty years; many of them will be new immigrants to Canada (Ontario. Ministry of Public Infrastructure Renewal 2004).

Income and housing tenure divide recent immigrants into different groups (Murdie and Teixeira 2003). Many new immigrants who cannot afford home ownership in Toronto are restricted to locating affordable housing in the rental market, where vacancy rates have been very low since the 1990s (Murdie 2003; Murdie and Teixeira 2003).

Record low interest rates in recent years have redefined the Toronto real-estate market and have allowed middle-income earners to purchase their first homes, many in trendy inner-city lofts and condominium developments. This booming real-estate market has given the appearance of better times in Toronto's rental market because the recent home purchasing has created more availability at its high end. However, the supply of decent low-rent housing is still very limited (Canada Mortgage and Housing Corporation 2004). Robert Murdie and Carlos Teixeira (2003) report that rental vacancy rates are lowest for the large units that would accommodate immigrant families.

Since the 1980s, globalization and the move toward cleaner technology-driven "new economy" sectors (including information technology, high finance, and new media) have reshaped the demographic geography of Toronto, as increasing numbers of white-collar workers have moved downtown, an area already dominated by white-collar residents. Increasingly, professional classes and groups with capital have been drawn to employment and housing in Toronto's inner city (Sassen 1991; Caulfield 1994; Ley 1996). Meanwhile, the less-skilled strata, which are heavily populated by new racialized immigrants, end up with the remaining and less desirable amenities in the city or, worse yet, in the suburbs, where few government-based social supports exist to offset the inequalities. In addition, since the 1990s, neoliberal agendas have resulted in federal, provincial, and municipal government cutbacks to subsidized housing, health care, education, and the environment (see Chapter 14, this volume).

Low-income marginalized residents, who have few opportunities to "vote with their feet," are disproportionately affected by unregulated growth in real-estate markets and have few resources to practise NIMBYism (not-in-my-backyard syndrome). In competitive Toronto (Kipfer and Keil 2002), various stakeholders struggle for access to scarce resources, and ultimately the benefits and costs of those resources are distributed unevenly. Stakeholders with the greatest access to resources are able to protect their own interests and deprive others of the same access.

Brownfields Redevelopment

Like those of many other deindustrialized cities, Toronto's municipal and provincial policy makers are turning to former industrial sites, or brownfields, as solutions for economic and social needs. An analysis of various urban stakeholders' perspectives regarding the merits of brownfields redevelopment highlights the competing interests involved and suggests the potential impact of such development on social diversity in Toronto.[5]

Communities near Brownfields

Communities situated in the vicinity of abandoned brownfields, which typically consist of low-income and working-class residents, are most vulnerable to the soil and water contamination remaining from industrial use. In the US, problems associated with spatial proximity to brownfields have been framed as environmental justice issues due to the argument that brownfields disproportionately impact marginalized low-income and racialized communities (Byrne and Scattone 2001). Unfortunately, in Canada – and in Toronto in particular – inventories of brownfields locations are lacking (De Sousa 2003), and little information exists about their possible contaminants.

The risks from brownfields are largely unknown to their adjacent communities. The Ontario Ministry of Municipal Affairs and Housing (2004) estimates that as much as 10 to 15 percent of the Greater Toronto Area could be categorized as brownfields. Ontario's Record of Site Condition Regulation (Ontario Regulation 153/04) establishes requirements relating to soil cleanup. However, the registry, which came into existence on 1 October 2005, is voluntary. Property owners are encouraged – but not required – to file a record of site condition before a property's use is changed from industrial, commercial, or community to a more sensitive one, such as residential (Ontario. Ministry of Environment 2005). Nearby communities are potentially subject to environmental injustices because they are unaware of both their substantive rights and their harder-to-access procedural rights.

For these communities, investment due to brownfields redevelopment could lead to such benefits as job creation, improved infrastructure, reduced soil and water contamination, enhanced public safety, and an increased municipal tax base following construction on abandoned lots. However, as John Byrne and Raymond Scattone (2001) note, only when such communities are actively involved as interested long-term partners – from planning through to implementation – is brownfields redevelopment likely to be socially sustainable and inclusive of the affordable housing and infrastructure needs of a broad range of stakeholders. Community participation in brownfields redevelopment is a model that could increase the availability of affordable housing. To date, however, most redeveloped brownfields sites in Toronto – with the exception of the mixed-income and mixed land-use St. Lawrence community – have been geared toward middle-to-high income ownerships (Ontario Coalition Against Poverty 2002a). This is particularly the case in the much-sought-after lakefront properties.

The Ontario Ministry of Municipal Affairs and Housing (2004) defines "strategic" brownfields sites as those with existing transportation and infrastructure facilities (roads, schools, parks, sewers, libraries) that are close to or within downtown cores. Remediations of brownfields that are not in "strategic" locations are most likely to include a mix of tenures and incomes. In short, whether nearby communities realize the benefits of brownfields redevelopment is heavily contingent on their active participation and the location of the brownfields.

Affluent Residents

Downtown Toronto's restructured economy, desirable locations, and trendy amenities have prompted increasing numbers of affluent residents to settle there. Although the downtown core has been subject to previous waves of gentrification, it is important to distinguish the recent movement of affluent residents into its redeveloped brownfields as signifying a new type of gentrification.

As mentioned above, gentrification is normally associated with the movement of elite stakeholders into *existing* working-class and low-income inner-city neighbourhoods. Theorists (Flanagan 2002; Neil Smith 1987) who have taken a structural approach to explaining gentrification have conceptualized it as involving a two-step process. This process can be loosely applied to Toronto and distinguished from the gentrification of former urban brownfield sites.

The first stage of urban gentrification normally involves risk takers, such as artists, who move into rundown neighbourhoods due to a combination of the low property values and the draw of adventurous lifestyles (Ley 1996). According to Jon Caulfield (1994), during the 1970s, these first-wave gentrifiers – who moved into downtown Toronto's working-class and immigrant neighbourhoods rather than settling in suburban areas – were influenced by progressive urban-planning initiatives. These initiatives, reaching a critical mass in the early 1970s, sought to preserve inner-city communities and affordable housing. Toronto urban reformists, motivated by egalitarianism, argued for the need to include local residents in determining the fate of their inner-city neighbourhoods. First-wave gentrifiers did not seek to change these neighbourhoods: rather, they contributed to the inner-city vitality and social diversity that characterized multiculturalism in Toronto (ibid.; Kipfer and Keil 2002). These first-wave newcomers were an important component of desirable urban social mixing that limited economic segregation in Toronto (Rose 2004). In comparison to American cities of the same size, Toronto still has relatively low levels of residential segregation by income (Fong 1996).[6] This is significant because such segregation increases the potential of environmental risks for segregated low-income and marginalized groups.

The second stage of urban gentrification involves a different type of gentrifier (Neil Smith 1987). These newcomers move into existing working-class neighbourhoods in response to the recommendations of real-estate agents, developers, and speculators who predict which residential areas will experience an increase in property values. Gentrification at this stage is not a spontaneous process. In Toronto in the late 1980s, the movements for progressive planning that influenced the first-wave gentrifiers were superseded by a more powerful and pervasive neo-liberal agenda that reoriented city development toward economic growth and a diminished welfare state (Ley 1996; Filion 2001). David Ley (1996) observes that progressive neighbourhood movements, such as cooperative housing initiatives, were abandoned once the federal and Ontario provincial governments were no longer reliable sources of funding. In this climate, second-wave gentrifiers can be characterized as risk-adverse and interested in protecting their own property values, regardless of the housing needs of earlier residents in their neighbourhoods. The second-stage gentrifiers are often relatively affluent and belong to the professional class. They are able to exercise their procedural rights by connecting themselves to developers, real-estate agents, or government actors or simply by having the time and resources to participate in Ontario Municipal Board hearings in order to protect their communities from undesirable land uses and environmental risks.

Spatial conflicts result from the desire of second-wave gentrifiers to upgrade neighbourhood amenities and from the collective moral panic of local businesses and newcomers regarding crime and urban decline (Kipfer and Keil 2002). Eventually, NIMBYism and the upscaling of amenities by newcomers (who dominate decision making in these gentrified communities) lead to increased rental prices and displaced low-income and immigrant residents who can no longer afford to live in their own communities. In Toronto, displaced residents are increasingly being forced to relocate to more affordable suburban areas that lack adequate equivalent social services or settlement services (Teelucksingh 2001). In these cases of second-stage gentrification, changes to the social and economic dynamics of former working-class neighbourhoods occur over time and are in part motivated by real-estate prices and speculation trends. In working-class and low-income areas that are also home to specific ethnic groups, ethnic-owned businesses will eventually be displaced. In downtown neighbourhoods, the eventual displacement of working-class and ethnic populations by middle- and upper-middle-income populations reduces income diversity and ethnic diversity (Sewell 1993). However, while this transformation is taking place – when the old and new residents exist together, although in conflict – the external image of diversity and multiculturalism is preserved. As considered in more detail later, "selling diversity" is still a major component of marketing renovated brownfield sites.

Unlike the gradual two-step process described above, the movement of elite residents into redeveloped brownfields is a form of "instant gentrification" (Rose 2004, 7), one that involves only a single stage. Housing and infrastructure development is aimed, from the outset, to meet the needs of affluent stakeholders, even though – due to the presence of adjacent low-income neighbourhoods and the perception that Toronto's inner-city communities are multicultural – gentrified lakefront brownfield sites may be advertised as spaces of urban social mixing. In reality, many operate as gated communities. New condominium and loft developments currently espouse a commodified version of multiculturalism, with "ethnic culture" and "ethnic restaurants" close by, even while they promote the virtues of physical and symbolic exclusion associated with good security systems and private lakefront walkways (Ley 1996). Brownfield gentrifiers, who enjoy relatively plentiful resources and are skilled at competing in the global economy, are able to reclaim lakefront properties that were formerly reserved for industrial use. They demand the social regulation of perceived low-income behaviours, such as skateboarding and street vending. Such brownfields gentrification effectively reduces multiculturalism and diversity in downtown Toronto

(Filion 2001). At the same time, the interests of these affluent waterfront dwellers are packaged as being good for the city – as environmentally and economically "smart" growth (see below).

Brownfield gentrifiers benefit from brownfields redevelopment due to Toronto's role in the global economy. Gentrification is the logical outcome of the type of investment capital flowing into the city. Increased economic activity linked to globalization has allowed for development in the downtown core to accommodate both new economy businesses and housing for new economy workers. Meanwhile, the increasing numbers of new immigrants, who constitute the other side of the global labour market coin, are not accommodated in terms of affordable housing and well-paying jobs. The city's affluent and poor may be interdependent, but they are spatially separate.

Liberty Village, which is located just west of downtown Toronto and north of the Gardiner Expressway, is a good example of brownfield gentrification aimed at housing new economy workers.[7] This new community, which dates from the late 1990s, is part of the broader resurgence of urban mixed land-use development along both King Street West and the Toronto waterfront. In a relatively short time, Liberty Village has transformed a manufacturing brownfield into a self-contained live-and-work community for affluent residents, who expect it to be lifestyle-oriented and to include amenities such as yoga studios, upscale restaurants, boutiques, and health spas (Hutsul 2004). However, the social and economic needs of these new gentrifiers transform the previous role of the site and set them apart from residents in nearby neighbourhoods. Terrance Belford (2001) aptly sums up the changes in Liberty Village: "Gone are the factories, which provided locals with jobs for half a century. Moving in are those who work in high tech, film and digital entertainment. They want to live near where they work, and their clamouring for new housing is having its effect on the whole area."

According to James Lemon (1985), the area now known as Liberty Village was very important to local mid-twentieth-century manufacturing and industrial companies due to its proximity to railway and harbour facilities in Toronto. Massey Ferguson (formerly Massey-Harris), Inglis Manufacturing, and later, Irwin Toys were major employers (ibid.). Employment opportunities at the time created the need for nearby low-income housing in the Toronto community known as Niagara, just north of the current Liberty Village area, and, following the Second World War, in the Parkdale neighbourhood to the west.[8]

Following the deindustrialization that began in the 1970s, Liberty Village became a brownfield site. However, during the 1980s and early 1990s, artists used its abandoned industrial units as cheap housing and studio spaces. When neighbourhood upscaling commenced in the late 1990s, many of these residents were forced to move. Formal conversion of unused industrial space in Liberty Village began with the renovation of an old carpet factory complex by York Heritage Properties into "a 300,000 square foot, multi-building complex of modern office and commercial space" (Lighthall 2003). At about the same time, plans were established to redevelop the former Massey Ferguson site into a mix of upmarket row houses and a commercial park (Caulfield 1994). Relaxed zoning bylaws in the surrounding areas along King Street West have led to the renovation of many former manufacturing and industrial buildings that are now occupied by predominantly knowledge-based industries, including those devoted to film and video, website design, information technology, and high finance (Lighthall 2003), all part of Toronto's recent global economy orientation.

Liberty Village focuses on jobs, housing, and a lifestyle for the young and the affluent middle class, who are already well positioned, professionally, to compete in the global economy and in the local housing market. Low-income and marginalized inner-city residents, who need both affordable housing and well-paying jobs, are simply ignored and excluded. Many currently compete for affordable rental accommodations in the adjacent Niagara and Parkdale communities. These communities, too, have suffered from deindustrialization but have not experienced the economic boom that characterizes Liberty Village (Teelucksingh 2002).

Public- and Private-Sector Interests

The case of Liberty Village reveals that various levels of government in Toronto and private-sector interests are also key stakeholders in the brownfield gentrification process. The former can play conflicting roles here. On the one hand, they are responsible for safeguarding citizens from the potential health and ecological problems associated with contaminated brownfields – that is, to ensure the substantive rights of all citizens. In terms of procedural rights, the Ontario Ministry of Environment amended Ontario Regulation 681/94 in 2005 to ensure that public consultation would be a component of its risk assessment approach to safeguard against brownfield contaminants. However, it is important to note that brownfield risk assessment does not take social factors and social risks into account, such as displacement from affordable housing or power imbalances that favour elite stakeholders. On the other hand, governments, in the era of neo-liberalism, are increasingly tied to corporate and global economic interests (Filion 2001).

The policies of the federal, provincial, and municipal governments all have an impact on urban development in Toronto. The scope of the City of Toronto's actions and its ability to develop legislation are limited by the authority of, and financial support provided by, the province. [9] That said, brownfields redevelopment is advocated by both the Ontario Ministry of Public Infrastructure and Renewal – in *Places to Grow: Better Choices, Better Future* (2004) – and by the Toronto municipal government in its *City of Toronto Official*

Plan (2002).[10] At both levels, brownfields redevelopment is seen as a way to address housing demand and to foster economic development.

Places to Grow (Ontario. Ministry of Public Infrastructure and Renewal 2004) outlines the Ontario government's proposed strategy to accommodate population growth and promote economic expansion in the Greater Golden Horseshoe. With a strong emphasis on land-use intensification and compact development, it designates pockets along the Golden Horseshoe lakefront as areas for future population and economic growth. In 2000 the provincial Ministry of Municipal Affairs and Housing also implemented a showcase program to assist municipalities in the planning and redevelopment of their brownfields. Similarly, at the municipal level, components of the City of Toronto Official Plan, as articulated in City of Toronto (2001), emphasize the regeneration of Toronto's central waterfront, vacant former industrial sites, and undeveloped land. The overall objectives of this plan are to take advantage of the city's opportunities in the global economy "while preserving its neighbourhoods and natural areas, and providing a good place to live" (ibid., 1). But a good place for whom?

It is important to note that community consultation meetings were a component of the processes for both Places to Grow and the City of Toronto Official Plan. Public consultation for the latter took the form of four town-hall meetings in June 2002. Residents, non-governmental organizations (NGOs), businesses, and ratepayer associations were invited to the meetings where they were asked to comment on the Official Plan, including targeted land-use intensification initiatives (Lura Consulting 2002). The public was also notified of the town-hall meetings through advertisements in Toronto's major daily, community, and ethnic newspapers (ibid.). However, of the sixty-six bodies that were formally invited, only six were ratepayer associations (City of Toronto 1998-2004c), the majority of whom represented comparatively affluent areas of the city. The provincial consultation process for Places to Grow took the form of invitations to the public to mail or e-mail written comments, as well as seven public information sessions held in various southwestern Ontario locations during the last two weeks of July 2004 (Urban Development Institute/Ontario 2004). In short, both cases featured limited strategies to involve marginalized communities as active participants in planning.

Because brownfields redevelopment can be located within the socially acceptable discourses of "smart growth" or sustainable development, it can be a useful tool for governments wishing to repackage the social inequalities associated with urban development that is directed toward the global economy (Bunce 2004). "Smart growth" refers to a critique of low density urban sprawl and to the alternatives of preserving greenfields and agricultural lands, reducing automobile dependency, and making more efficient use of existing inner-city infrastructure. Smart growth is certainly environmentally necessary, but smart-growth initiatives will not be socially sustainable if targeted growth and development further marginalize and exclude low-income populations and deepen racial segregation.

The rising problem of homelessness and the fact that many new immigrants initially need rental accommodation have heightened concerns about affordable housing in Toronto. Yet, since the 1990s, affordable housing has been a low priority for the Ontario and federal governments. In that decade, according to Pierre Filion (2001, 95-96), "developers moved from apartment to condominium building to reap quicker profits and avoid the higher property taxes imposed on apartments." The lack of government involvement in affordable housing has resulted in the privatization of affordable rental housing. Most of the brownfields redevelopment plans for the downtown Toronto waterfront have been based on private-public partnerships that focus only on "economic development" (De Sousa 2003). Developers are offered incentives to develop brownfields, including tax incentives for the remediation of contaminated soil and deregulated zoning to facilitate land-use conversion (Rose 2004). In September 2004, the Ontario government launched its Brownfields Financial Tax Incentive Program, which allows municipalities to provide tax assistance to landowners to encourage environmental rehabilitation.[11] This assistance may take the form of a tax credit applied to the cost of site remediation. The tax incentive program supports the Ontario government's Places to Grow initiative.

Redeveloped brownfields offer new business potential for the real-estate market, private businesses, and lending institutions. The existence of a business sector and the proximity to downtown Toronto and transportation corridors help to draw other new economy businesses to brownfield redevelopments. Overall, the active interest of the private sector is seen as a way to rejuvenate rundown communities and improve the city's overall economic profile. The Portlands area, just east of downtown Toronto, is a good example of the challenges of public-private ventures. For years, it has been mired in the conflicting agendas of private developers and public trustees. At present, no clear or consistent development policy or plan has been adopted or implemented. In 2001 the Toronto Waterfront Revitalization Corporation (TWRC) was established as a private-public partnership to direct the development of unused portions of the waterfront (Morris 2005). The eventual fate of the Portlands area will be a harbinger of the future of Toronto's brownfields development.

In Liberty Village, both public- and private-sector economic and political interests have been concealed through the use of certain discourses. The provincial and municipal governments, as well as private investors, have positioned Liberty Village as a good environmental and social use of reclaimed brownfields, while downplaying the ulterior economic objectives. For instance, transformations in Liberty Village are consistent with the objectives and discourse of the City of Toronto Official Plan (City of Toronto 2001). Sites such as Liberty Village are targeted as urban growth areas with a focus on new employment and residential development. In order to encourage urban development, population densities are not predefined, and mixed-use designations have been assigned to certain growth areas (ibid.). However, pre-existing planning controls are removed without regard for current uses or users, or for the need to facilitate economic opportunities for low-income communities. David Hulchanski notes that the increased development in Liberty Village will lead to displacement for Parkdale's low-income residents: "Rents go up because it's becoming more trendy ... And these days that adds to homelessness" (quoted in Oleniuk 2005). These kinds of regulatory changes, moreover, are made "over the heads" of the people who will be affected by them.

Private developers are drawn to Liberty Village by liberalized zoning and tax incentives. Evidence reveals that they are also marketing and repackaging brownfields redevelopment in new and unanticipated ways. Barry Lyon, president of Lyon Consultants, observes, "In many cities, an area like this [Liberty Village] – a former industrial area – would just lie vacant for many, many years. The area has moved from a functionally obsolescent (industrial) wasteland to a smart-growth showpiece" (quoted in Lighthall 2003). Smart-growth discourse, used by developers such as Lyon, positions urban development in opposition to urban sprawl by advocating high-density inner-city communities that integrate mixed land use. Market-savvy developers portray Liberty Village as taking advantage of renovated living spaces in historical industrial and manufacturing buildings with easy access to transportation and Lake Ontario. According to Linda Mitchell, vice-president of sales and marketing at Monarch, another urban developer in the area, "Liberty Village is going to be a master-planned community featuring a well integrated mix of residential homes, commercial offices, retail amenities, open parkland and landscaped walkways giving the area a community feeling" (quoted in Lighthall 2003). Constructing Liberty Village as a planned smart-growth community has helped to market it to the youthful, creative, and comparatively affluent members of Toronto's workforce.

Clearly, the potential benefits of Toronto brownfields redevelopment are not being equally realized by all stakeholders. Neo-liberalism, which emphasizes economic growth and Toronto's role in the new economy, has played a part here in its failure to focus on either accommodating immigrant population growth or the need for housing market diversity in terms of housing type and tenure. Moreover, growth-oriented development is encouraged at the expense of adequate government attention to the procedural rights of all stakeholders. As argued above, when procedural rights are not taken into account with regard to the determination of urban resources, environmental inequalities can arise. The goal of environmental justice highlights the need to question who gains or suffers from specific environmental advantages or risks.

Reclaiming and Democratizing Brownfields

Underlying the discourses regarding the redevelopment and revitalization of city spaces are concerns about who belongs where in Toronto, the global city, and who should have claims to brownfields (Sassen 2000, 2002). An egalitarian perspective would hold that the basic housing rights of all people living in the city should be recognized, as should similar substantive and procedural rights for all citizens. The claims of global capitalist actors should not be seen as more legitimate than those of less powerful actors. Ideally, as Byrne and Scattone (2001) argue, the brownfields redevelopment process would be open to all stakeholders' perspectives and, from the planning stage onward, would encourage the active participation of local residents. Remediating brownfields would be geared toward environmental justice by, first, taking greater account of their potential ecological and health impacts, and, second, by finding means to improve the economic conditions in low-income communities that do not rely on the introduction of middle- and upper-income residents (Rose 2004, 7).

Another potential solution to the imbalance of stakeholder interests in brownfields redevelopment lies in new urbanism, an approach to urban development that may take steps toward achieving environmental justice. Originating in about 1990, new urbanism advocates the renovation of brownfields with the explicit purpose of creating mixed-income and mixed land-use communities (Flanagan 2002). Its principles, to which some architects, planners, and developers adhere, seek to implement alternatives to modern automobile-oriented planning. In contrast to smart-growth discourses, which share similar critiques of urban sprawl, those of new urbanism attempt to incorporate the needs of all citizens, including those with low incomes, into urban development plans. In Toronto, in keeping with brownfield redevelopment in strategic locations, there has certainly been a move toward increased mixed-*use* zoning but not toward the kind of land use that would attract a mix of housing tenures and services for the poor and the marginalized. New urbanism's focus on *both* mixed land use *and* mixed incomes aims to avoid the exclusionary results of gentrification. One example of the new urbanism approach is the St. Lawrence community: developed in the 1970s by the City of Toronto and situated just southeast of the central business district and the historic St. Lawrence Farmers' Market, it is a successfully revitalized brownfield that created new housing for a mixed-income community, with a variety of property and tenure types including both public- and private-sector cooperatives and non-profit apartments.

New urbanism assigns the state an important role in resolving the inequalities associated with brownfield redevelopment. However, neo-liberal states have limited room to manoeuvre with regard to the conflicting interests of globalized capital and local communities. The state cannot be relied upon to address inequalities in access to capital or to provide social citizenship rights (Marshall 1963; Bryan Turner 1992). Thus, in place of a reliance on state actors, there is a new emphasis on a multiplicity of non-formal political dynamics and acts of resistance (Sassen 2000; Isin and Siemiatycki 2000). These include the practices of people in the city who are making claims on public space and fostering new and often ad hoc social movements and groups.

In Toronto, these acts of resistance function to destabilize dominant power relations and to highlight social inequalities. Ultimately, they seek to change the norms and values of dominant stakeholders and to influence formal processes. This process of destabilization includes revealing how discrimination operates and how dominant power relations threaten Toronto's self-created image of the "multicultural city that works" (Kipfer and Keil 2002). Expressions of resistance and attempts to claim space, such as those associated with Toronto's Tent City, are informal acts of citizenship. A collection of makeshift huts on the former site of an iron foundry in the Portlands, Tent City was erected in 1998 by a group of squatters (Ontario Coalition Against Poverty 2002b). In response to local activism in support of their rights, the City of Toronto implemented an emergency transitional housing project in 2002 to address the needs of those whom Home Depot (the landowner) had evicted from Tent City. In terms of brownfields redevelopment, Tent City is an example of one identifiable group's resistance to the direction of urban redevelopment and of activists' attempts to claim housing for homeless people. According to Gilbert and Phillips (2003, 315), the events surrounding Tent City "are illustrative of how environmental rights and urban rights overlap and of how citizenship is not limited to a status granted by law, but rather is expressed as the political mobilization of citizens around a cause and/or against the governance of a state."

Damaris Rose (2004) raises the important point that a perception of social diversity and harmony in urban communities is important to a city's success in the global arena. Toronto's ability to compete in a knowledge-based economy and to draw capital is tied to its ability to sell its "image of the liveable city" and the desirability of its own culture (Florida 2003, quoted in ibid., 9). Reacting to the negative attention generated by Tent City, former mayor Mel Lastman remarked, "It's [Tent City is] an ugly sight. It's not Toronto. And Toronto should not be portrayed this way" (quoted in Lawlor 2002). Threats to Toronto's image of diversity resulting from spatial inequalities also highlight the city's inability to meet the needs of the significant and growing number of new immigrants who arrive every day.

Redevelopment at Liberty Village has been contested by local residents who question a process that so greatly privileges developers. Residents of the adjacent Parkdale community organized a forum to voice their concerns about the plans of Urbancorp HighRes to erect condominium towers. They focused on "increased density, traffic, pedestrian and public transit congestion, economic and environmental viability, and the aesthetic impact of the buildings" (Glossop 2003, 3). Residents were also concerned about the lack of consultation and the possibility that Urbancorp was planning a "suburban style gated community" in the heart of the Liberty Village area (ibid.). Their point, simply put, was that the lack of community consultation denies democracy in the brownfields redevelopment process.

IBI Group, a multidisciplinary consulting organization that is involved with the development of a proposed live-work building in the Liberty Village area, raises the point that, even when efforts are made to obtain input from residents via community consultation processes, their turnout and response are small (Alicia Turner 2005). Lynn Clay, executive director of the Liberty Village Business Improvement Area, aptly summarizes the consequences of poor community involvement: "We don't have a strong community group here, so we get taken advantage of ... It's the ratepayers who are responsible for keeping these developments honest and we don't really have a residential component" (quoted in ibid., 1). As described above with regard to the City's Official Plan, participation by marginalized groups in the formal decision-making processes has been lacking (City of Toronto 1998-2004c; Lura Consulting 2002). Clearly, time limitations and power imbalances play a role here, but more research needs to be done to determine *why* low-income residents and grassroots-based organizations do not engage in the formal consultation process. Meaningful procedural democracy requires governments to identify the obstacles to citizenship participation and to remove these obstacles.

In terms of informal acts of citizenship, Toronto grassroots groups such as Citizens for Democracy (Boudreau 2000), the Ontario Coalition Against Poverty (OCAP), and the Common Front in Defense of Poor Neighbourhoods (1999) are noteworthy examples of organizations that have influenced formal processes and institutions. In Toronto, local tenants' groups and anti-poverty associations are emerging in public spaces to contribute their perspectives and their protestations regarding urban development plans. The Pope Squat, held in Parkdale during July 2002, consisted of OCAP-led groups of homeless people and activists who occupied the vacant property of 1510 King Street West, which had been abandoned but was still owned by the Ontario government (Ontario Coalition Against Poverty 2002a). The Pope Squat, named in reference to the visit of Pope John Paul II, highlighted the problem of homelessness and the need for affordable housing in the city. The membership of many Toronto grassroots groups reflects the cross-fertilizing of activist networks and community organizations. These intersections suggest the possibility for new identities to emerge – such as the environmental justice movement – that challenge traditional power structures.

Conclusion

Grassroots resistance to neo-liberal visions of urban development is making itself felt in Toronto. Informal citizenship, which strives to protect both the substantive and the procedural rights of all citizens, may involve acts ranging from squatting and protesting to violence in response to proposed brownfields uses. Persistent pressure on the state from grassroots advocacy groups highlights the reality that affordable housing cannot simply be left in the hands of the market. Ultimately, Toronto's urban policy needs to return to a focus on social housing where all levels of government play a significant role in ensuring substantive and procedural rights to housing and other opportunities in the city.

Notes

1   Brownfields are abandoned industrial or commercial properties that remain unused due to known or suspected environmental contamination.
2   My research was conducted between April 2004 and September 2005. Changes to public- and private-sector regulations and policies made since January 2005 may not be reflected in this essay.
3   Landmark American studies that helped to define the field of environmental justice include Robert Bullard (1983); United States General Accounting Office (1983); Bunyan Bryant and Paul Mohai (1992); and United Church of Christ Commission for Racial Justice (1987).
4   Most noteworthy, in terms of American policy, is the Clinton administration's Executive Order 12898, issued in February 1994 to all federal agencies to ensure that federal programs could not intentionally or unintentionally inflict environmental harm on the poor and visible minorities.
5   Damaris Rose (2004, 9) notes that social diversity, or the desired social mix, is measured not only in terms of income, but also according to race, ethnicity, housing type, tenure, and age group.
6   High levels of racial residential segregation are often correlated with high levels of vulnerability to environmental injustices, such as environmental racism.
7   More specifically, Liberty Village lies north of the Gardiner Expressway and south of King Street West, between Dufferin Street to the west and Strachan Avenue to the east.
8   Niagara, which is located south of Queen Street West and north of King Street West, lies east of Strachan Avenue and west of Bathurst Street. Further to the west, Parkdale is near the lakeshore, west of Dufferin Street, east of Roncesvalles Street, and south of Dundas Street.
9   This discussion focuses on the municipal and provincial governments; however, at the federal level, in February 2003, the National Round Table on the Environment and the Economy launched the national brownfield redevelopment strategy, which is outlined in Cleaning Up the Past, Building the Future: A National Brownfield Redevelopment Strategy for Canada.
10   The Places to Grow Act, which provides a legal framework for growth planning in Ontario, received royal assent on 13 June 2005.
11   For information about the Brownfields Financial Tax Incentive Program, consult http://www.rev.gov.on.ca/english/credit/brownfields/ftipe.html.

17

Taking a Stand in Exurbia: Environmental Movements to Preserve Nature and Resist Sprawl

Gerda R. Wekerle, L. Anders Sandberg, and Liette Gilbert

When a thousand enraged citizens showed up for a 12 January 2000 evening town council meeting in Richmond Hill, fifteen miles north of Toronto, they made front page headlines in Toronto's major daily newspapers (Swainson 2000). They demanded a stop to rezoning proposals that would convert twenty-eight hundred hectares of agricultural lands and greenspace on the Oak Ridges Moraine (ORM) into seventeen thousand new housing units. The rallying cry "Save Don't Pave the Oak Ridges Moraine" became a protest sign for what the media described as a "landmark battle" between Richmond Hill Town Council and environmentalists. Certain environmental non-governmental organizations (ENGOs) were vocal in their opposition. These included Earthroots, best known for its direct-action campaigns to stop clear-cuts in Temagami (northern Ontario); Save the Rouge Valley System (SRVS), a group focused on the creation of a national park on the eastern edge of the Greater Toronto Area; Save the Oak Ridges Moraine (STORM), a coalition of a dozen citizen groups; Environmental Defence Canada, a national legal defence organization; and local environmental associations.

The phenomenon of an uprising in which environmentalists made common cause with middle-class exurban homeowners in a town that had overwhelmingly voted Conservative in two previous provincial elections marked a turning point in an environmental and land-use conflict that had been simmering for more than a decade. After protracted organizing on the part of social movements, this resistance culminated in provincial legislation: the Oak Ridges Moraine Conservation Act in 2001, implemented by the Oak Ridges Moraine Conservation Plan in 2002.

Our chapter examines the ways in which urban development on the growing edge of Toronto has been simultaneously framed as both an issue of sprawl and a problem of environmental degradation.[1] We draw on a four-year study of environmentalist challenges to sprawl on the Oak Ridges Moraine.[2] The research is based on participant observation at public consultation meetings organized by the provincial and other levels of government, public presentations made by social movements, press releases, flyers and websites, in-depth interviews with key environmental activists, and analysis of planning documents and newspaper articles.

We focus on the role played by ecological processes in the production of exurban space and how the uses of nature are politically contested.[3] We contend that ecological movements pose a bioregional challenge to a planning system that promotes growth, sprawl, and the destruction and commodification of nature. They use the planning process as a tool to resist these developments and to participate in decision making about the future of the ORM. Thus, the roles played by environmental organizations may include a continuum from protest to consultation, policy making, and program implementation. We examine how movement politics are integrated with institutionalized politics, where "persons and organizations frequently engage in both protest and conventional political actions" (Goldstone 2003, 4).

Environmental Contention in an Exurban Region

The Oak Ridges Moraine is a geological formation of porous rock, sand, and gravel created twelve thousand years ago when the glaciers retreated from southern Ontario. The ORM, which lies directly north of the Greater Toronto Area, extends 160 kilometres from the Niagara Escarpment in the west to the Trent River in the east (see Figure 17.1).

One of the last continuous corridors of greenspace left in south-central Ontario, the ORM includes aquifers on which 250,000 people depend for water; kettle lakes created by the glaciers; the headwaters of sixty-five streams and rivers; the habitats of endangered species; forests; greenspace that includes golf courses, hiking trails, and land trusts; century-old farms; and long-established hamlets and villages. But it is directly in the path of the Greater Toronto Area's urban expansion, offering potential greenfield sites for residential and industrial development and for the infrastructure so vital to capitalist expansion (such as new highways and mega-sewer pipes). The moraine crosses thirty-four municipalities and three regions, and approximately 65 percent of it lies in the Greater Toronto Area. Pressures on the ORM include rapid population growth – the Greater Golden Horseshoe Area is expected to add 3.7 million people by 2031 (Ontario. Ministry of Public Infrastructure Renewal 2005, 4) along with the relatively low density of single-family housing developments, the expansion of employment lands onto greenfield sites, and the greater demand for greenspace and recreational facilities for urban residents. Applications to rezone land on the moraine for all these uses have raised the prospect of environmental destruction on a regional scale.

Exurban residents are characterized as moving to the country to gain what they consider a superior residential environment, "to enjoy unexploited resources" (Davis, Nelson, and Dueker 1994, 46) and "to become part of a

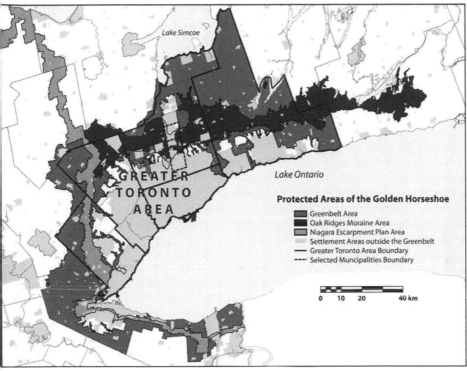

Figure 17.1 Protected areas of the Golden Horseshoe. Map by Gerda R. Wekerle,
L. Anders Sandberg, and Liette Gilbert, based on Schedule 1: Greenbelt Plan Area Map,
produced by the Ontario Ministries of Agriculture and Food, Municipal Affairs and Housing,
and Natural Resources (28 February 2005); additional graphics work by Khyati Nagar.

culture of rurality" (Gerald Walker 2000, 108). When this quiet countryside
is threatened by further development that may bring in more people and
undermine existing pastoral landscapes, urbanites who have moved to exurbia often engage in political action to protect it (ibid.). In Canada, the
United States, and Britain, the new middle class in rural areas often exercises
its political influence through the planning system, which translates middle-
class representations into actual spatial forms (ibid.; Peter Walker 2003;
Murdoch and Day 1998). Some scholars have labelled such activity a defensive politics oriented to the protection of economic, cultural, and property
assets, and designed to exclude any unwanted development through using
the power of the state (Murdoch and Day 1998).

Another position suggests that, when homeowner activism is linked with
campaigns for environmental preservation, as in the case of the Oak Ridges
Moraine, new issues and stakeholders are introduced. The conflict is refocused
on the intersections of nature, society, economics, and politics at a regional

scale, and on stakes that surpass benefits to any particular locality or set of property owners. In this chapter, we explore this phenomenon in more detail. Undoubtedly, like suburban homeowners who attempt to control development in their own backyards, residents of the ORM are motivated by self-interest to prevent the destruction of the nature that enticed them to the countryside in the first place. However – like Susan Lee in her Chapter 13 discussion of the River Road residents – we question the wholesale stereo-typing of *all* exurban activism as protection of property values or quality of life, as this is only part of the story. The battle to preserve the Oak Ridges Moraine also engaged environmental groups, which framed it as a fight to preserve the "bioregion," particularly the aquifers and rivers that run through it (Gilbert, Sandberg, and Wekerle forthcoming). These environmentalists identified land-use planning as a means to preserve and even enhance the ecosystems that formed the bioregion. By expanding the agenda beyond the protection of local environmental features or opposition to unwanted land uses, environmentalists have sought to cultivate among residents an iden-tification with the ORM as a bioregion.

First articulated in the 1970s, the concept of bioregionalism "places the human community within natural communities and recognizes vital links between human terrains of consciousness and geographical terrains" (Berg and Dasmann 1977, 76-77). Central to bioregionalism is the goal of a better relationship (more just and equal, less exploitative and destructive) between human communities and their natural environments. Reinhabiting or re-claiming one's relation to a particular bioregion entails a "process of trans-formative social change at two levels – as a conservation and sustainable strategy [sic] (for nonhuman species) and as a political movement which calls for devolution of power to ecologically and culturally defined bio-regions" (McGinnis 1999, 4).

Land-use planning is about the power to shape place and the meanings and discourses attached to specific places. Contestations over land uses and natural resources such as aquifers and watersheds focus attention on state regulation of land uses and spatial planning, the mediation of conflicting policies of economic growth and environmental conservation, and neo-liberal governance strategies that create a range of new agencies and mech-anisms for decentralizing environmental management and policy implementation. In the environmental planning field, there is a new em-phasis on structured collaborative processes that bring together public, private, and ENGO interests in the creation of multi-stakeholder plans to protect vulnerable natural resources such as watersheds (Foster 2002). How property is to be used engages civil society, quasi-autonomous government agencies, and the private sector through processes of public consultation mandated by planning and environmental legislation. In the planning and management of ecosystems and watersheds, neo-liberal arguments for less state regulation have favoured a greater reliance on multi-stakeholder public-private collaborative efforts, which offer both new opportunities for ENGOs and new complications in their relationship with the state and private sector (Cortner and Moote 1999, 22). State-organized public consultation processes and opportunities for citizen intervention in legal hearings on land use afford new opportunities for environmental movements to advance alternative policy options and to participate in collaborative planning processes with a range of key stakeholders. The downloading of responsibility for policy implementation on non-state agencies may also engage ENGOs in new roles, some of which may meet objectives of local capacity building and democratic participation.

As Hanna Cortner and Margaret Moote (ibid., 42) remark, "Ecologists now speak of the need to take entire bioregions into account to sustain the biological diversity critical to ecological sustainability." The focus is on watersheds and large corridors of connected greenspaces at a regional scale and across mixed ownerships. The ecosystem approach is radical in its challenge to the pro-growth initiatives associated with conventional land-use planning. It involves a "recasting of human/nonhuman relations and the recognition that social and ecological goals are interdependent" (Gibson, Alexander, and Tomalty 1997, 30). Ecosystem management is not just about science. It forges "political connections among humans, nature, science and government" (Cortner and Moote 1999, 1). There is an emphasis on involvement by "communities of place" and on citizen stewardship of the land (ibid., 49).

Environmentalist Strategies and Networks

Throughout the 1990s and to the present, anti-sprawl and pro-environment campaigns emerged in response to accelerated urban sprawl and environmental degradation of exurban areas in fast-growing regions of North America, at times spawning counter-movements focused on property rights (Peter Walker 2003; McCarthy 1998; Wekerle and Abbruzzese forthcoming). Major national environmental organizations such as the Sierra Club in the US and the David Suzuki Foundation in Canada (Gurin 2003) have mounted anti-sprawl campaigns. In the Greater Toronto Area (GTA), environmental organizations operating at a range of scales and with differing histories and agendas adopted a regional approach to anti-sprawl campaigns, linking their core issues of environmental preservation with challenges to neo-liberal urban growth strategies. Applying Manuel Castells' (1997) ideal types of environmental organizations (which cross scales), we find that "conservation of nature" groups include local naturalist bodies, as well as the provincial Federation of Ontario Naturalists (now Ontario Nature), Conservation Authorities, and land trusts. Groups focused on "defence of their own space" include both homeowners' associations, such as the Jefferson Forest Residents Association and the Concerned Citizens of King Township, as well as a large array of environmental organizations focused on protecting specific rivers, watersheds, kettle lakes, and river valleys. "Global save the planet" ENGOs include the World Wildlife Fund, as well as the national representatives of international organizations, such as the Sierra Club. In addition, there are national and international organizations such as Environmental Defence Canada that are public-interest groups specializing in legal challenges or in wider policy issues. Falling under Castells' category of "counter-cultural movements," or Timothy Luke's (1997) "restorational ecologists" – the eco-centric movements rooted in deep ecology and bioregionalism – is Earthroots, founded in 1989 to protest the clear-cutting of Temagami's forests and distinguished by its use of direct action.[4] ENGO initiatives such as the NOAH project and Algonquin to Adirondacks (A2A), based on principles of restorational ecology, seek to implement ecosystem conservation on a continental scale.[5] Although these ideal types are useful for classificatory purposes, environmental movements are flexible, and considerable crossing over from one category to another occurs as circumstances and political contexts change.

Social movement organizations have identified with the moraine struggle, forming coalitions that are regional in scale (Wekerle, Sandberg, and Gilbert forthcoming). One of these, Save the Oak Ridges Moraine (STORM), was founded in 1989 in response to threats of development in the Ganaraska Forest. Other ENGOs included the Federation of Ontario Naturalists (FON, now Ontario Nature), which has a membership of thirty thousand across the province. Save the Rouge Valley System (SRVS), although initially focused on preserving the Rouge River as a national park, became a regional player when it gained "standing" in 2000 to participate and intervene in an Ontario Municipal Board (OMB) hearing related to development plans in Richmond Hill.[6] Toronto-based Earthroots turned its attention to environmental issues closer to home, where most of its membership lives. These key players often coordinated the production of flyers, mass mailings, and events. They appeared as deputants at public consultation meetings and as invited participants in provincially organized stakeholder meetings. Each group tended to take on specialized roles. Some, such as the FON and STORM, were more conservation-focused and willing to work with the Conservative government in power; others were more radical in their critiques of government policies and in their proposals of alternatives (these included SRVS, Environmental Defence Canada, and Earthroots).

Framing the Moraine
Over the course of several years, the movements to preserve the ORM gained visibility, gathered momentum, and changed the way in which the public saw the moraine. They accomplished this by connecting multiple frames

into a broader structure of concepts and policies based on ecosystem planning and bioregionalism. These collective action frames, according to Sidney Tarrow (1998, 89), are "action oriented sets of beliefs and meanings that inspire and legitimate social movement activities and campaigns. The most effective frames direct action, organize discourse, and construct meaning." The alliances and mobilizations that formed around the campaigns to save the ORM from development sought to *reframe the conflict,* transforming it from the simple polarization of necessary growth versus the environment and quality of life to a struggle between reckless and destructive growth, on the one hand, and responsible ecosystem management and land-use policies, on the other. Adequate supplies of safe drinking water and the preservation of rurality and urban greenspace became priority issues. The activists were successful in influencing the mass media to disseminate their message to a vast audience in southern Ontario and beyond.

Initially, citizen groups on the moraine utilized a range of frames, from sprawl to ecosystems, but over time, several of these converged. Protecting endangered species, forests, and headwaters became the focus of campaigns by conservation groups with a large membership base, such as the Federation of Ontario Naturalists, and of smaller naturalist groups, but this was also frequently a theme in public presentations by homeowners' associations. Although residents' associations voiced their concerns for protecting their own amenities and quality of life, and implicitly their property values, over time, their public presentations also raised concerns about the disappearance of aquifers, forest lands, and greenspace. Environmental framing drew on multiple issues and discourses, thereby ensuring continuous media coverage in numerous newspaper articles with changing foci (Martin 2004). Running through the newspaper coverage of environmental devastation on the moraine were themes of destruction of habitats of specific amphibians and animals, preservation of natural heritage for future generations, the security of water quality, and food security.

The initial choice to frame the Oak Ridges Moraine in terms of sprawl and the planning changes needed to combat it (Immen 2001b; Gillespie 2003) built upon ongoing debates in the province regarding the negative impacts of sprawl and the need for smart growth. Combining the environment and sprawl into one campaign provided a wider base of support, thereby avoiding the charge of NIMBYism often levelled at suburban homeowner associations and engaging the resources of Toronto-based environmental groups and the City itself – organizations that were more concerned with environmental issues than with sprawl.

Initially, water was an important frame, linking watersheds, aquifers, and users of water. Both STORM and SRVS hired scientists to conduct studies of the impact of development on the ORM's water resources. Focusing on the moraine as the "rainbarrel" of the region, Earthroots, STORM, and the FON formulated intensive media campaigns about the aquifers that provided drinking water for 250,000 people and the threats posed by moraine development applications to the aquifer recharge and the headwaters of major river systems. An SRVS flyer produced in early 2000 combined an emotional appeal regarding clean water and the needs of children and future generations (through a picture of a young child wading in a stream) with calls for political action at the Ontario Municipal Board hearing on developers' proposals to build housing on the moraine in Richmond Hill:

> Stand up for clean water! Stop the destruction of the Oak Ridges Moraine. Children depend on us for so many things, but one of the most important is water. Right now, the future of this precious resource – and the future of our children – is being decided in a small room filled with lawyers at a place called the Ontario Municipal Board. Most of these lawyers are there, not to represent children, but to speak for developers – developers who want to pave over a crucial section of the Oak Ridges Moraine, located in Richmond Hill. At stake are more than a hundred fragile wetlands, huge underground aquifers, pristine lakes and sensitive headwater streams that feed our rivers. Save the Rouge is at this hearing fighting for future generations, but we can't do it alone. We need your help. For the first time in history, we want thousands of citizens to show up at an OMB hearing and do something very simple: speak for the children.

Although water remained an important theme in the ORM campaigns, the dominant frame gradually became that of ecosystem planning and bioregionalism, with a focus on natural connections, corridors, and watershed protection. This was most clearly articulated by bodies such as STORM, SRVS, the FON, Earthroots, and various place-based stewardship groups. Many of the ENGO campaigns relied on eco-centric narratives that enlarged the boundaries of the community to include water, plants, and animals. In a full-colour flyer produced by SRVS and distributed as an insert with the Saturday edition of the *Toronto Star* to almost half a million households in the region, a picture of a helpless fawn is accompanied by the caption "Homeless. Just like you, this baby fawn needs a place to live. Yet very soon, thousands of animals, birds and amphibians will be driven from the only home they've ever known – the Oak Ridges Moraine. As the bulldozers move to make way for houses, factories and gravel pits, these defenseless creatures will literally run, fly or crawl for their lives."

The Jefferson salamander, an endangered species on the moraine, became an iconic figure, pictured nestled in the hand of SRVS activist Glenn De Baermaeker in a half-page colour spread in the *Toronto Star* (Swainson 2001).

Its fate in human hands, its vulnerability emphasized, the salamander is described as a "tiny, curiously cute animal with its long tail and pointy toes" (Immen 2001a). This fragile amphibian contrasts with the Osgoode Hall law courts in the background, an image of institutionalized power where Environmental Defence Canada was involved in a legal challenge to stop expansion of a roadway into salamander habitat in the Jefferson Forest.

These emotionally charged images in both environmental publications and newspapers communicated a powerful subtext of environmental loss and the devastation of the land wrought by urban development. Via websites and listservs, environmental groups sent out urgent notices of public meetings and protest events, and made their images, flyers, and position papers widely available. Environmentalist control of the media framing managed to stay focused on the issues and avoided the conventional media typecasting of social movement actors as violent and "out of control" anti-government protesters. The choice of multiple and convergent frames pulled together such a diverse group of activists, including those representing state agencies at various levels of government, that the resistance to development on the moraine could not easily be ignored; nor could it be dismissed as the product of self-interest or "special-interest groups."

Pivotal Moments and Political Opportunities

In explaining the *when* of social movement activism (why now and not at other times), social movement scholars utilize the concept of political opportunity structure. According to Sidney Tarrow (1996), the character of a movement's political opportunities is determined by such factors as the openness of institutions to public participation or movement demands, shifting alignments, the availability of influential allies, and cleavages within and among elites. For movements resisting the development of the ORM, political opportunities became more favourable in 1989 when the multiplication of development pressures at diverse sites provoked growing citizen resistance, the Liberal provincial government established a formal process of citizen consultation on the future of the moraine, scientific and planning reports were commissioned, and proposals were made for an ecosystem approach to planning for the moraine as a whole. These developments precipitated the founding of a new regional-scale environmental organization: the STORM coalition.

The election of a neo-liberal Conservative provincial government in 1995, and the relaxation of planning and environmental controls, caused a shift in existing political alignments in the 905 exurban region, which had voted overwhelmingly Conservative.[7] As development applications were made to build tens of thousands of new homes, as well as golf courses and other projects on greenfield sites across the moraine, even Conservative-voting homeowners (in towns such as Caledon, Uxbridge, and Richmond Hill) began to challenge this runaway growth and developers' influence on planning decisions. Naturalist groups active in these exurban towns and in the countryside allied themselves with homeowners and with non-local environmental organizations to focus not only on how sprawl negatively affected the quality of life for humans, but also on the destruction of ecosystems. They gained access to influential allies, including legal expertise from the Sierra Legal Defence Fund and Environmental Defence Canada, the environmental critics of the opposition parties in the provincial legislature, Toronto city councillors, as well as politically well-connected and affluent local exurban residents. They drew upon scientific knowledge from sympathetic scientists and regional quasi-autonomous agencies such as the Conservation Authorities.

Certain events were pivotal, providing a dynamic conjuncture for social actors on the ORM to create their own political opportunities. Environmental groups seized the political moment when the scandal of contaminated drinking water in the town of Walkerton, Ontario, became public on 23 May 2000. This event, in which a bacterial outbreak in contaminated groundwater killed seven people and made another twenty-three hundred ill (Ali 2003), focused the public's attention on water quality, groundwater, and human health issues. Although it was not associated with the moraine, the Walkerton scandal drew attention to the Conservative government's downsizing and privatization of environmental monitoring (see Chapter 14, this volume). For environmentalists active on the ORM, this provided a "wedge issue" with which to attract public and media attention.

Several months of intense organizing preceded a second pivotal moment in the Oak Ridges campaign – the record-breaking crowd of a thousand small-town residents and environmental activists that jammed a Richmond Hill council meeting on 12 January 2000. This emotionally charged meeting was the culmination of a history of contested government approvals for the development of environmentally sensitive local sites (Gilbert, Wekerle, and Sandberg 2005). On 26 August 1998, York Region and the provincial Ministry of Environment and Energy had approved the construction of a road (the Bayview Avenue extension) through the moraine. In 1999 developers disagreed with the York Region's policies limiting development on the moraine and appealed directly to the OMB to build eleven thousand houses on the Yonge Street East and West properties in Richmond Hill (Mallen 2004). When residents met with no success in lobbying Richmond Hill Town Council in fall 1999 to reject applications to develop housing on the moraine, region-wide environmental groups became more prominent in this local battle. On 6 October 1999, SRVS, Richmond Hill Naturalists, STORM, and others organized a joint press conference to call for the protection of the most threatened and biologically rich section of the moraine, a two-kilometre-wide strip called the "pinch point," which was located in Richmond Hill.

Media attention to the ORM struggles intensified. In just one month, November 1999, Toronto newspapers ran fifty-four news items on contestation and resistance to development on the Oak Ridges Moraine. On 15 November, environmental groups organized a "funeral for the moraine" to draw attention to the cutting of the Jefferson Forest to make way for a road extension. Two weeks later, SRVS, Richmond Hill Naturalists, and homeowners set up a blockade to stop the bulldozing of trees for the Bayview extension. Twelve people were arrested.

Developers that had made applications to rezone and develop sites on the moraine became concerned that citizen pressures on Richmond Hill Town Council would delay or stop development. They bypassed council and appealed directly to the Ontario Municipal Board for approvals. On 19 November 1999, the OMB started a pre-hearing to consider proposals to build two thousand houses around Bond and Phillips Lakes, two kettle lakes in Richmond Hill. The Ministry of Natural Resources had identified these lakes as areas with dozens of rare plant and animal species that could be degraded by fertilizers, pesticides, and pollutants from housing developments (Immen 1999). The Jefferson Forest Residents Association and SRVS, represented by Environmental Defence Canada, gained legal standing in the OMB hearing to examine witnesses and bring in experts for testimony. On 8 December 1999, the City of Toronto (finding another way to intervene in the decisions affecting its water supply) allocated $1.6 million to SRVS and the Toronto and Regional Conservation Authority to oppose housing developments on the ORM in the towns of Richmond Hill and Uxbridge. This funding provided the resources for environmental groups to challenge developers with deep pockets, their own local politicians, and the provincial government.

In the new year, the organizing and media campaigns intensified in Richmond Hill, gaining daily coverage in Toronto and regional newspapers. The moraine campaign was taken directly to exurban residents. SRVS started the "Wilderness not Wood Chips Project" – packing garbage bags with wood chips from the remains of moraine trees and dumping these in the offices of Richmond Hill councillors. SRVS volunteers delivered thousands of flyers door-to-door in the town of Richmond Hill to publicize the 12 January council meeting. When, despite the intense opposition, Richmond Hill Town Council voted to move forward with its rezoning plans to approve development on the moraine, the environmental campaigns were redoubled. Save the Rouge Valley System organized a massive drop of twenty thousand small wood-chip bags and sixty-five thousand flyers to Richmond Hill residents. These actions culminated in another council meeting, 23 February 2000, held in a large hotel ballroom to accommodate the sixteen hundred people who attended it. Richmond Hill council finally voted to suspend its plans to urbanize the moraine and asked the provincial government to step in and pass legislation to regulate land use there.

However, development proposals to rezone lands on the moraine also continued to be forwarded to local councils, and environmental groups continued their organizing efforts. On 24 March 2000, taking another legal route, the City of Toronto, the FON, and STORM, funded by the Sierra Legal Defence Fund, attempted to protect the moraine by using the provincial Environmental Bill of Rights to challenge existing provincial land-use planning laws and policies. Another strategy, a public protest in front of the Ontario legislature, was chosen by Earthroots in an effort to shift public attention from local councils to the provincial government, which was responsible for the Planning Act. This drew a relatively small number of supporters and limited media attention.

By April 2000, the Ontario Municipal Board had before it eighteen applications for developments on the Oak Ridges Moraine. On 29 May, a lengthy OMB hearing commenced to consider developer proposals to build housing around two kettle lakes in Richmond Hill. This was the first OMB hearing at which conservation biology was used as a defence for protecting land from development. The months of testimony from developers and experts, as well as environmentalist responses to these positions, were covered on an almost daily basis by the major newspapers in Toronto and the region. Save the Rouge Valley System presented its first expert witness, Dr. Reed Noss, a well-known US conservation biologist and deep ecologist, on 15 January 2001. He "explained his view that the Oak Ridges Moraine represented the best opportunity to maintain and enhance a life-supporting, continentally significant system of natural and agricultural areas within the Lower Great Lakes Region" (Public Spaces 2001). Noss' success in championing large ecosystem preservation plans in Florida prompted environmentalists to advocate linking the moraine to other large natural systems in Ontario and New York, leading to the formation of the NOAH project and the Algonquin to Adirondacks (A2A) project. This was followed on 12 February 2001 by a mass mailing organized by STORM to labour and conservation groups, environmental organizations, politicians, and thirty-five thousand households. Bowing to public pressures, the provincial government passed the Oak Ridges Moraine Protection Act (Bill 55) on 17 May 2001, temporarily freezing all development on the ORM for six months while pursuing a longer-term solution.

As Sidney Tarrow (1996, 54) has argued, "state-centered opportunity structures" and "partially opened access to [political] participation encourages protest." Environmental groups saw the OMB hearing as an opportunity for their positions to gain public exposure through the reports of the media. But the provincial government curtailed this opportunity by summarily

Figure 17.2   A "Live in Harmony with Nature" sign fronts a bulldozed landscape at the site of McLeod's Landing housing development, 2003. Photo Gerda Wekerle.

shutting down the hearing. However, the creation of new structures for public consultation opened up other opportunities for environmental movements to organize and be heard. On 28 June 2001, Ontario's Tory government appointed an Oak Ridges Moraine Advisory Panel, the composition of which would represent developers, politicians, scientists, academics, STORM, and the FON. Notably, STORM and the FON subscribed to "defence of own space" and "nature conservation" discourses; no representatives from the more radical eco-centric or "save the planet" ENGO categories were invited to participate in the advisory panel.

   This body was intended to take the heat off the provincial government by holding public consultations and stakeholder group meetings, and making recommendations for future land uses on the moraine. Over the summer months, it held four public meetings in different locations across the ORM, each of which attracted several hundred people. Those who wished to voice their comments or make presentations were limited to five minutes. Four half-day sessions were set aside for in-camera meetings with invited stakeholders, including developers, aggregate industry representatives, environmental organizations, and resident associations, none of whom were publicly

named. Although represented by the provincial government as *public* consultations, these processes were exclusive, carefully managed, and constructed
so as to achieve the ends defined by the government and its supporters.

The public meetings, extensively covered by print and electronic media,
did, however, provide an opportunity for both ordinary citizens and social
movement organizations to publicly make their claims and outline their
visions for the future of the moraine. In meeting after meeting, representatives of key environmental groups came prepared with their analyses of
planning documents and scientific findings on the ecological significance
of the ORM and with their stories of the moraine as home place for both
residents and non-human nature. Long-time residents related their personal
histories of living on the ORM, its past natural history, and their fears for
its future. Children came forward with their stories about the importance
of the greenspace and nature. The public hearings provided a platform for
the creation of a shared identity and shared myths about the moraine's history and possible future.

On 14 August 2001, the Ontario Ministry of Municipal Affairs and Housing (2001) released the *Share Your Vision* draft recommendations for land-use
planning on the moraine and allocated thirty days for public consultation.
This involved further deputations from citizen groups and subsequent
changes to the recommendations. On 1 November 2001, the province introduced draft versions of its Oak Ridges Moraine Conservation Act (Bill 122)
and Oak Ridges Moraine Conservation Plan (ORMCP), which recommended
protection of natural core areas and natural linkage areas of the moraine,
the establishment of an Oak Ridges Moraine Trust, and a land swap of north
Pickering sites to compensate developers for Richmond Hill land frozen from
development.[8] The province also committed to establishing an Oak Ridges
Moraine Foundation with an endowment of $10 million to fund stewardship
initiatives. The legislation included provision for a ten-year review. Although
environmentalists initially celebrated a victory, they had reservations about
the ten-year review provision and the power of the minister of municipal
affairs and housing to override the provisions of the ORM Conservation
Plan. This proved prescient, as subsequent to the 13 December 2001 passage
of the Oak Ridges Moraine Conservation Act, a Minister's Order of 27 June
2002, which was final and binding, allowed development of the Richmond
Hill lands that were still before the OMB. This permitted an additional sixty-
six hundred houses to be built on the moraine, prompting SRVS to call the
public consultation process "a sham" (Swainson 2002).

Taking Stock

As Tarrow (1996, 44) notes, the "strategies of movements are affected by
their institutional policy environments." In the moraine struggles, political opportunities have often been policy-specific; that is, "the policy and

institutional environment channels collective action around particular issues and with [specific] consequences" (ibid., 42). Social movements tended to react to proposals to develop specific sites on the ORM by forwarding alternative proposals, lobbying local councils, and when that did not work, trying to raise public awareness through the media, protests, and direct action. Perhaps one of the most important shifts was the engagement and expertise developed by local, regional, national, and global environmental organizations in the processes of land development and regional land-use planning. By linking sprawl with environmental degradation and ecosystem preservation, environmental organizations staked their claim as interested parties and mobilized movements that contested growth politics and land-use decision making in a growing urban region. When the provincial Ministry of Municipal Affairs and Housing took on the problem of resolving land-use conflicts on the moraine, environmental and homeowner groups utilized the opportunities for public participation to buy time until public opinion coalesced against development.

Social movement actors played a dual role: organizing and protesting against government land-use policies and agreeing to work collaboratively with the provincial government on legislation and a land-use plan for the whole ORM. Although much of the social movement literature focuses on protest strategies, less attention has been directed to the ways in which social movements operate within or through the state. Jack Goldstone (2003, 7, emphasis in original) suggests that social protest and institutionalized political actions are often complementary: "It may be the ability of groups to combine *both* protest and conventional tactics for influencing government actors that best conduces to movement success." For Goldstone, "social movement activity and conventional political activity are different but parallel approaches to influencing political outcomes, often drawing on the same actors, targeting the same bodies, and seeking the same goals." He argues that social movement actors should not be portrayed so much as insiders or outsiders but as enjoying dynamic and contingent access to and influence on the state. This describes social movement activism on the moraine, where environmental movements sought to influence the state even as state agencies sought to control and channel social movement activities through incorporating them into advisory bodies and public consultation processes.

On the moraine, the lengthy process of protest, resistance, and, ultimately, negotiation and partnership of environmental movements with state agencies raises questions about the changing roles of the state, planning, and social movements. State-organized consultation processes both shaped and constrained citizen participation. Environmental groups such as STORM and the FON, appointed to the advisory panel and actively engaged in negotiating the terms of the legislative framework, moved from an outsider position to that of partner with the state. This gave them new legitimacy and prominence in the public eye but also opened them to charges of "selling out" from other actors in the environmental movement. Through their participation in the advisory panel, STORM and the FON became insiders in secret negotiations between the developers and the provincial government. Thus, they had a stake in the ultimate acceptance of the panel's recommendations, whereas environmental groups on the margins of this process voiced greater skepticism. When the Oak Ridges Moraine Foundation and Trust were established, groups involved in the advisory panel were invited to participate on management boards, thereby continuing their involvement through the implementation of provincial government policy.

The active participation of environmental organizations in stakeholder meetings and their detailed submissions to the public consultations and to the Environmental Bill of Rights Registry meant that these groups proactively engaged in making and amending proposals, and in contributing to the wording of draft acts and final planning legislation. These activities effected a substantial reframing of regional land-use policies relating to the Oak Ridges Moraine. However, by engaging with the state and its system of land-use regulation, ENGOs have also become enmeshed in neo-liberal governance strategies through their roles in task forces and advisory bodies of newly created QUANGOs (quasi-autonomous government agencies) such as land trusts that implement state policies (Logan and Wekerle 2008). This raises questions about how these processes may privilege certain ENGOs, particularly groups that are less strident or overt in their criticism of the state than are their more radical counterparts. Such participation also raises questions about the representation of place-based ENGOs and stewardship groups focused on the preservation of specific habitats, watersheds, or greenspaces should the state choose as partners those groups that it sees as having a broader mandate. The subsequent movement of key environmental actors into government as politicians, staff, and advisors also warrants closer attention to insider-outsider strategies and the institutionalization of social movement agendas within state regulation.[9]

Although media coverage often focused on the social movement opposition to state policies regarding the moraine, as Goldstone (2003, 21) suggests, "states may actually embrace movements and seek their support," particularly when doing so forwards their own agenda. Had the mobilization of citizens not occurred, it is questionable whether legislation to preserve large parts of the ORM would have been passed by a Conservative government. Social movement contention over the Oak Ridges Moraine gave the provincial government the leverage and the rationale to institute region-wide land-use regulation despite the previous strong opposition of the four regional governments and many local municipalities to any region-wide structures of governance (Wekerle et al. 2007). In the absence of a planning authority for the region as a whole, the Oak Ridges Moraine Conservation Act (along with the Greenbelt Plan, passed by the provincial Liberals in February 2005, and a growth management plan, Places to Grow) firmly established the provincial government as *the* regional planning authority for the Greater Golden Horseshoe in southern Ontario. The public attention on the ecological preservation aspects of these policies tended to background the regional coordination and planning outcomes. The institutionalization of environmentalist agendas into legislation and regulatory policies meant that some environmental agendas were now the responsibility of provincial ministries and enforced by provincial legislation rather than being identified with protest politics.

If, as Allan Scott (1990) suggests, social movement success consists of integrating previously excluded issues and groups into the "normal" political process, the "Save the Moraine" movements far exceeded the expectations of environmental and homeowner groups. Provincial legislation and the ORM Conservation Plan clearly set out the ground rules for where development could or could not occur on the moraine. Yet, subsequent development approvals sanctioned by minister's orders saw a whittling away of some of these "saved" lands, and new environmental conflicts surfaced on the eastern edges of the ORM, where provincially owned land was given to developers in a swap and lands protected as an agricultural preserve were threatened by development due to municipal challenges of the provincial legislation. These changes, plus the prospect of a review of the moraine protections after only ten years, made it difficult for environmentalists to claim an unequivocal victory. Nevertheless, the provincial government did pass new planning legislation for the moraine in its entirety, which established a precedent for thinking about the big picture of growth and land use in the Greater Golden Horseshoe. More importantly, perhaps, the discourses of bioregionalism and ecosystem planning approaches were extended into this new planning exercise.

The movement to save the Oak Ridges Moraine no longer exists. In 2005, after reframing its mission, STORM partnered with Citizens Environmental Watch and the Community Mapping Center to develop a "Monitor the Moraine" project, which is funded by the provincial government and the ORM Foundation. Save the Rouge Valley System refocused its efforts on the Rouge Valley after its chief spokesperson, Glenn De Baermaeker, was elected as a Toronto city councillor. As Allan Scott (ibid., 10) notes, "Success is thus quite compatible with, and indeed overlaps, the disappearance of the movement as a movement." However, "success" can also be measured in other ways. Learning from the struggle for the moraine, social movements were able to translate their tactics into more sophisticated contestations over the Greenbelt. They formed a Greenbelt Alliance early in the process and ensured the successful passage of a Greenbelt Act and Plan. The Greenbelt Plan of 2005 acts as a growth boundary that protects 1.8 million acres of environmentally sensitive and agricultural lands from development in an area stretching from the Niagara Peninsula in the southwest to Rice Lake in the east. Despite organized challenges from both developers and farmers – groups that had access to substantial resources to oppose legislation that would freeze developable lands – controls on development were implemented. Perhaps, in hindsight, the greatest gains were made by environmental groups, including local, regional, and national organizations that firmly staked their claims to be heard in environmental policy making, not only in wilderness areas, but also in urban regions dealing with growth management, land use, and sprawl. By successfully linking these planning and economic development issues to ecosystem approaches to planning and to visions of bioregionalism, they reshaped popular understandings and the discourse of planning in the province.

Bioregional discourse introduced new understandings of both nature and development, and generated alternative models for responding to the pressures of urbanization and sprawl. By focusing on the preservation of the Oak Ridges Moraine, a regional natural feature, environmentalists shifted their ground from regulatory changes at the provincial or federal levels to "land-based activism."[10] This meant linking policies to protect aquifers formed in the Ice Age with protecting habitats for amphibians and planning for human habitation. The work of reinhabiting or reaffirming one's belonging to a bioregion encompasses restoring and sustaining local ecosystems but also – as seen on the Oak Ridges Moraine – creating the institutional capacity of communities and co-governance arrangements to ensure environmental conservation. A conservation-driven institutional capacity is inevitably oppositional, rejecting or questioning the current regulation of resources and articulating place-based knowledge that may contradict the expert opinions of managers, scientists, and politicians. Reinhabiting a bioregion both challenges and resists the dominant approaches to socio-economic development and resource use.

Notes

Stephanie Rutherford and Teresa Abbruzzese assisted in developing a chronology of the moraine.

1 Sprawl "is characterized by low overall densities, a rigid specialization of land uses and a near total dependence on the automobile" (Filion 2003, 50).
2 This study was funded by SSHRC grant 410-2002-1483.
3 Exurban areas have been described as "a city's countryside, characterized by varying densities of dispersed and low density urban populations" (Gerald Walker 2000, 108). Within exurbia, there are suburban subdivisions, estate developments, large farms, hobby farms, small towns and hamlets, forests, streams, lakes, and natural heritage areas that include unique ecosystems and habitats.
4 Eco-centric philosophy posits that all entities that self-produce have intrinsic value and deserve moral consideration (Eckersley 2004a; Naess 1983).

5 NOAH identifies significant land forms and natural habitats across southern Ontario and northern New York State. Its acronym stands for Niagara Escarpment, Oak Ridges Moraine, Algonquin to Adirondack Heritage System.

6 The Ontario Municipal Board is a provincially appointed planning appeal board. "Standing" means that groups were able to actively participate in the OMB hearing, their lawyers cross-examining witnesses for developers and the government, as well as presenting the testimony of their own expert witnesses. SRVS was prevented from calling its own witnesses due to the termination of the hearings, with the exception of Reed Noss, who was called out of turn due to scheduling conflicts. The City of Toronto was refused standing on the grounds that Toronto lay beyond the boundaries of the moraine. The City argued that Toronto would suffer the downstream effects if the headwaters of the city's rivers were degraded by development (Bragg 1999).

7 In public debate, the region is commonly identified as 905, a form of shorthand referring to its telephone area code. Planning controls on development were relaxed by the neo-liberal Tory government. Developers put pressures on local councils to fast-track approvals for new development. Although applications for rezoning greenfields are controlled by municipal Official Plans and subject to council approvals, new provincial legislation stipulated that developers could appeal to the Ontario Municipal Board (OMB) if a council had not made a decision within ninety days. The OMB had been stacked with Tory government supporters and was labelled as pro-development when its decisions supported development approvals against the rulings of local councils.

8 As required by law, Bill 122 and the draft land-use plan were posted on the Environmental Bill of Rights Registry for thirty days. Over six hundred responses were received from the public.

9 For example, Erin Shapero, active in SRVS, was elected to Markham council; Glenn De Baermaeker, SRVS president, was elected as a Toronto city councillor; Josh Matlow, moraine organizer for Earthroots, ran as a Liberal candidate in a provincial by-election where he failed to defeat provincial Tory leader Ernie Eves but was subsequently elected as a Toronto school trustee; and John Mackenzie, environmental planner for SRVS, became a planner with the Ministry of Municipal Affairs and Housing, and was then hired as the assistant to Minister of Municipal Affairs and Housing John Gerretsen when the Liberals were elected in 2003.

10 Debbe Crandall, executive director of STORM, used this term, arguing that it better described their work than "environmental activism" (interviewed by Gerda Wekerle and L. Anders Sandberg April 2004).

18
Democracy from the Trenches: Environmental Conflicts and Ecological Citizenship
Laurie E. Adkin

Environmental conflicts are not only about the health of particular communities, the future of non-human species, the aesthetic quality of life, or even long-term ecosystem devastation, although they are, of course, about all of these. When citizens become involved in "environmental" struggles, they very quickly find themselves enmeshed in a much broader web of relationships and issues. Often, they come to question the democratic character of their local governments: Whose interests do they represent? Whose voices do they disregard? How do the consultation processes express biases regarding legitimate concerns or "stakeholders"?

In these ways, environmental demands such as the regulation of GMOs, the implementation of sustainable fisheries management, the banning of forest clear-cutting, the protection of provincial parks, the adoption of sustainable alternatives to the dumping of waste, the preservation of a valley from dissection by an auto-route, the ending of industrial pollution, and countless others lead to democratic demands. These include calls for citizen participation in decision-making processes (planning, approval, monitoring, priority setting), transparency of these processes, resources for citizens' groups to "level the playing field" in conflicts involving well-funded corporations, independent regulatory agencies, more funding for "independent" scientific research, compensation for environmental harms, government and corporate accountability and responsibility, stricter regulation of corporate activities, community autonomy, and self-determination.

In other words, the criticisms and demands emerging from the environmental struggles of citizens, communities, and Aboriginal peoples constitute demands for the democratization of the state and the economy, and for the transformation of citizenship. The citizenship implications of these struggles may not be evident to all of the participants; the democratizing import of their demands may not be explicitly recognized. Yet, elements of democratic discourse link together quite disparate struggles and express with remarkable consistency a shared understanding of the political and social roots of the environmental crisis.

The Political Economy of Environmental Conflicts

The case studies in this book demonstrate the ways in which environmental conflicts have been generated and intensified by contemporary developments in global capitalism and its underpinning neo-liberal ideology. A number of the authors make use of David Harvey's (2006, 153) account of neo-liberalism in the advanced capitalist countries as "an ongoing process of accumulation by dispossession." Harvey identifies four principal practices that "restored class power to capitalist elites" (ibid.). These include privatization, financialization, the management and manipulation of international debt crises, and state redistributions of income from the lower to the upper classes. Of these, it is privatization that receives the most attention from our authors as a cause of worsening degradation of the environment as well as growing social insecurity and inequality. Under the rubric of "privatization," Harvey includes (ibid.) "the corporatization, commodification and privatization of hitherto public assets ... [whose] primary aim has been to open up new fields for capital accumulation in domains hitherto regarded off-limits to the calculus of profitability."

Various writers have documented such processes in regard to public utilities, social welfare (or entitlements, such as to public health care or government pensions), or public institutions (such as universities). Political ecologists draw attention to the enclosure of the commons (wild fish stocks, forests, parks) by corporate interests, aided by governments. In Chapter 10, Jim Overton joins David Marquand (2004), David Harvey (2005, 2006), Colin Leys (2001), and others in viewing such processes not only as a form of hollowing out of the state's regulatory functions, but also as a hollowing out of citizenship. Citizens are losing rights (political, social, and environmental) that were hard-fought for by previous generations. In this context, the authors in this book call not only for the restoration and protection of social citizenship rights, but also for the extension of citizenship rights to environmental goods (clean, safe water, clean air, healthy ecosystems) and for the meaningful recognition of Aboriginal rights.

The studies in the first part of the book demonstrate the environmental and social costs of the hyper-industrialization of food production, which is linked to intense competition for global markets. In Chapter 2, Peter Andrée and Lucy Sharratt describe a Canadian regulatory framework geared to promoting the commercialization of biotechnology products. The effects of corporatization and intensive resource exploitation on communities that depend on the wild fisheries for livelihoods are illustrated in Chapters 4 and 5, by Donna Harrison and Martha Stiegman respectively. In British Columbia, employment in the commercial fisheries dropped massively, by more than two-thirds in five years (1996-2001). We see similar effects in Quebec, with the industrialization of meat production: production for global markets increases scale, drives mergers and the centralization of production, and exerts downward pressure on wages, as well as imposing environmental costs on local communities. Plants have been closed, and unionized workers have been laid off. Ethical questions regarding the welfare of pigs don't seem to have entered the public debate. As for salmon, concerns have been voiced regarding the extinction of species. In both cases, the industries are, in Harrison's words, "fiercely competitive," "characterized by global overproduction," and dominated by large international corporations (56). Global overproduction and competition from low-wage countries with minimal environmental standards have pushed down the prices of salmon and pork, and this has in turn driven further "industrial rationalization."

The chapters in the third part of the book draw our attention to the consequences of deregulation, privatization, and corporatization in relation to urban infrastructure and development. In Chapter 14, Michael Mascarenhas recounts First Nations' experiences with the provincial deregulation of water services, combined with new auditing, or accountability practices, which download responsibilities while withholding needed resources. In Chapter 16, Cheryl Teelucksingh provides the important insight that the drive to make Toronto a global city – a financial and service centre for the new global economy – has meshed with the smart-growth discourse about urban planning. The city's changing class structure and developers' interests are fuelling a real-estate boom and forms of gentrification that are squeezing low-income groups out of affordable housing. At the same time, Toronto's development affects the exurban communities who, as Gerda Wekerle, L. Anders Sandberg, and Liette Gilbert describe in Chapter 17, are fighting urban sprawl. The City of Hamilton, meanwhile, seeks to attract investment to generate employment and to avoid becoming a bedroom community for commuters who work in Toronto. As Jane Mulkewich and Richard Oddie report in Chapter 15, Hamilton's economic growth coalition argues for the necessity of an expressway through the Red Hill Valley, one of the region's most significant greenspaces.

These competitive pressures become the "economic realities" to which governments, corporations, or producers' associations refer when justifying anti-ecological policies or policies that undermine communities' livelihoods, security, and connection to place. Due to such "economic realities," deregulation (to attract investment capital or to download costs), commercialization, increased concentration of ownership (as in the case of fishing quotas), the "development" of green areas (such as the Red Hill Valley), the production of new environmental risks (GMOs), and the relentless exploitation of forests and fisheries are all said to be in the "public interest." The challenge for environmentalists and local communities is to identify and to defend alternatives that can provide sustainable livelihoods.

Connecting Struggles to Create Alternatives

As long as global commodity chains and global market forces determine the economic options available to local communities, social actors will be squeezed to conform to or accept the profit-maximizing strategies of corporations.[1] One social group is pitted against another – very often, environmentalists against workers in the resource or manufacturing sectors. Or particular groups of workers (such as the meat-processing-plant workers in Quebec) are forced to bear the burden of job losses due to the rationalization of production. The ideological structuring of conflict around oppositional identities such as workers versus environmentalists, or fishing communities versus environmentalists, is perpetuated by media constructions, corporate discourses, the "stakeholder" organization of policy-making processes, and the strategic decisions made by citizens' groups, unions, ENGOs, and other actors.

In Chapter 3, Nathalie Berny, Raymond Hudon, and Maxime Ouellet reveal how conflict among agricultural producers' associations and environmentalists has contributed to the *absence* of an alternative (ecologically and socially sustainable) development model for the sector. Instead, the government and industry owners are using "globalization imperatives" as the pretext for imposing intensified wage competition and a larger scale of production. The unions are weakened, and the environmental costs ignored. Half of Quebec's pork is exported, requiring the long-distance transportation of meat products, with its environmental costs. The industrial production of pork entails large concentrations of animals, with the associated costs in animal suffering, disease (and disease control), and management of wastes. A more environmentally sustainable system would adjust production to regional market demand and might require compensatory tariffs or fair trade certification to be imposed on imported food products. Which social forces in Quebec might stand behind a slow food/ecological farming movement?

Chapters 5 and 17 both provide glimpses of more successful coalition building among subordinate actors. The former chapter reveals how, using community-based management, Nova Scotia fishers established their own council and created a working relationship with a First Nation to assert local autonomy against regulatory control by Fisheries and Oceans Canada (DFO). The latter chapter, focusing on Ontario's Oak Ridges Moraine (ORM), discusses how "the alliances and mobilizations that formed around the campaigns to save the ORM from development sought to *reframe the conflict*, transforming it from the simple polarization of necessary growth versus the environment and quality of life to a struggle between reckless and destructive growth, on the one hand, and responsible ecosystem management and land-use policies, on the other" (285).

Alternatives arise organically, from local struggles, as well as from networking with, and learning from, other groups in other places. Ecological economics and the programs of Green Parties around the world offer many ideas for economic restructuring, and these can help to turn conflicts among subordinate actors into ecological alternatives. Yet, these connections must be actively developed by "organic intellectuals" capable of bridging struggles and creating new counter-hegemonic identities.[2]

The State of Environmental Conflicts

As the above discussion suggests, the most recent phase of capitalist globalization has been accompanied by a number of transformations in the roles of states vis-à-vis economic and other forms of regulation. States have ceded authority – along with policy levers and planning capacity – to international financial institutions, trade organizations and their tribunals, transnational corporations, and unregulated financial markets that may destabilize national economies overnight. The hollowing out of the state in these ways has had enormously negative consequences not only for areas of social welfare, but also for environmental protection.

In Canada, the inadequacies of environmental regulation are compounded by the federal structure, which assigns most environmental responsibilities, along with natural-resource management, to the provinces and territories (Kathryn Harrison 1996). The result is a patchwork of provincial policies and standards. At the same time, in areas where the federal government has constitutional jurisdiction, such as the fisheries, its policies in recent decades have promoted the concentration of ownership and enclosure of the commons (Chapters 4 and 5). This experience has led some to argue that "devolution," in the form of community-based management of resources, will result in more ecologically sustainable practices. When it comes to Aboriginal communities, such as the Ahousaht, the Nuxalk, the Chippewa, the Nishnawbe Aski, the Lubicon, the Haudenosaunee, or the Mi'kmaq, the recognition of Aboriginal and treaty rights by both levels of government is critical to their cultural survival and may result in a more sustainable use of land and resources.

It is important to note, however, that decentralized governance may serve hegemonic interests when it takes the form of consultation processes largely constructed and controlled by neo-liberal governments and corporate interests. In the Alberta cases documented by John Parkins in Chapter 11, network governance served to isolate "tiny publics" from larger public spheres, making them easier for corporate and government interests to manage, or dominate, and preventing local conflicts from connecting to the resources and discourses of broader social movements. As Parkins observes, rural communities are often dependent for employment on the companies proposing the "development." This characterization of network governance in the Alberta case is confirmed in Chapter 8 by Colette Fluet and Naomi Krogman in their analysis of the provincial government's Northern East Slopes Sustainable Resource and Environmental Management Strategy (NES). The restricted terms of reference of the NES Strategy process also functioned to privilege pro-development interests and to exclude the perspectives and alternatives that may have been advanced by actors outside of small resource-dependent communities. Network governance may exclude urban and other interests from having a say in consultation processes by constructing such interests as "outsiders," or non-stakeholders. In Chapter 6, Patricia Ballamingie reports this kind of exclusion of urban populations in the north-south opposition constructed in Ontario's Lands for Life process.

One of the most consistent findings of these case studies is the existence of a pro-development bias in government ministries, sometimes characterized as "captured" by corporate interests. In Chapters 4 and 5, both Harrison and Stiegman see a clear commitment on the part of DFO to the concentration of quota ownership in the West and East Coast fisheries, and Harrison illustrates the provincial and federal governments' promotion of aquaculture. The Alberta Ministries of Environment, Sustainable Resource Development, and Economic Development may all be viewed as captured by corporate interests, given the government's neo-liberal orientation, which leads it to equate corporate exploitation of natural resources with the public interest. In Chapter 2, Peter Andrée and Lucy Sharratt note the close relationship between Agriculture and Agri-Food Canada and the biotechnology industry; in Chapter 3, Berny, Hudon, and Ouellet describe the corporatist relationship between Agriculture Quebec and the Union des producteurs agricoles; in Chapter 6, Ballamingie characterizes the government-industry relationship in Ontario as "collusion"; and, in Chapter 7, Darren Bardati depicts the BC Ministry of Forests as biased toward the lumber industry. In these processes, corporations are typically insiders, involved in setting the criteria, goals, and policy options of a given consultation process. As an Alberta business representative puts it, "If you're involved up front, you can really help form the process or regulation that's going to come out. That's a big thing because then you can kind of build it how it will work for you" (134).

It is therefore not surprising that citizens engaged in environmental conflicts perceive their governments as being in the pocket of private interests and as failing to protect public goods. Ahousaht chief Angus Campbell concluded, in the context of the BC government's decision to lift the four-year moratorium on fish farm expansion without consulting First Nations, "This is a slap in the face to our people ... Their own studies show how this poisons the ocean and everything around it, but they have bowed to the big-money lobby to put our health last" (quoted in Danard 1999, A3). Susan Lee concludes in Chapter 13, regarding the Fundy region waste management decision, "Citizens who tried to participate in this process found it to be an insurmountable challenge; many said that, in retrospect, it seemed more like a 'done deal' than a democratic process" (210).

Such sentiments are echoed by citizens involved in many other policy processes or regulatory conflicts. A participant (Gordon Taylor 2005) in the 2005 public review of the Great Lakes Water Quality Agreement protested,

> Being a past 5 year member of the BPAC [Bi-national Public Advisory Committee] for the Detroit River I can state that this plan has been a failure. Those of us advocating on behalf of our environment were constantly being outvoted on remediation initiatives by governments and their corporate partners ... The people responsible for the impairment [of the environment] should not be part of the decision making process of how to remediate the impairment ... Governments and their advisors are responsible to protect the rights of all ecosystem inhabitants not just the ones who finance the political campaigns of elected representatives.

Another participant (Gilbertson 2005) in the same review, identifying himself as a biologist who had worked for the Canadian federal government as well as for the International Joint Commission for thirty-four years, stated,

> For the past twenty-five years, it was a precondition for promotion within the federal governments that managers in environmental departments were there to 'manage the issue,' not to address and solve the cause of the injury to health and property from water pollution. Part of that process of managing the issue was managing the outraged public. Fortunately for them, within this democracy, there were enough developers and other corporate interests represented on the RAP [Remedial Action Plan] committee that the environmental activists and scientists never won the votes, and the managers never had the ignominious experience of having to return to their offices to seek funding for remedial action.[3]

A Calgary wildlife scientist who has struggled for a long time to intervene in the dealings between the oil and gas companies and the Alberta Energy and Utilities Board (AEUB) protests that the never-ending stream of industry requests for exemptions from environmental regulations cannot be monitored and responded to by the public concerned with wilderness protection: "The public simply doesn't have the resources and the time to remain vigilant enough to catch the many assaults on both the process and on our public lands. It's beyond any individual or even any group ... The only systematic way of ... protecting the public's rights is to have an honest, honourable and effective government system in place. Neither the EUB nor SRD [Sustainable Resources Development] are providing that ... We need some sort of a structure here that protects the public resources, assets and processes" (Brian Horejsi, quoted in Boschman 2007, A4).

In these circumstances, citizens have essentially two choices: to retreat, exhausted and defeated, to private life (as some of the women in Lee's Chapter 13 case study did), or to reframe their particular struggles as struggles to democratize political decision making – to empower themselves and others as citizens. Thus, many citizens' groups, ENGOs, and other organizations have put forward proposals to open up decision-making processes to meaningful citizen participation.

The Science of Environmental Conflicts

Science plays contradictory roles in policy conflicts, as suggested by the terms "citizen science," "independent science," "public science," "post-normal science," "traditional ecological knowledge," "expert knowledge," and others. "Scientific data" are produced and used by international institutions such as the UN intergovernmental scientific panel on climate change, governments, corporations, academics, NGOs, and private consultants, whereas citizens' groups, farmers, fishers, industrial workers, First Nations, and others generate "knowledge" of various other kinds. The New Denver residents in the Slocan Valley (Chapter 7) had knowledge of their watershed that called into question the expert knowledge of the Ministry of Forests and Slocan Forest Products. In other cases, citizens' groups have informed themselves about hydrogeology (Chapter 13) or the effects of toxic chemicals (Adkin 1998b). Women are often the first to recognize the incidence of environmentally caused illness among their neighbours (MacGregor 2006). Aboriginal elders observe changes in their environment over time, which inform their causal explanations of deformations of wildlife or illnesses in their communities. For example, in Dianne Meili (2007), Fred and Margaret MacDonald, members of the Fort McKay First Nation in Alberta, describe changes in the Athabasca River in the wake of oil sands exploitation.

Science plays an important – and sometimes pivotal – role in many decision-making processes. Some conservation biologists and resource management experts call for environmental policy to be "science-based," thinking that this condition will free policy from the influence of various economic interests – that policy will, in other words, be grounded in "objective" criteria rather than biases (pro-development or otherwise). Their expectation is that the findings of scientists will be respected and used as the foundation for decision making. Such beliefs are, of course, somewhat naive, especially insofar as they accept the possibility of scientific knowledge that is free of any social determination.

First, scientific method embodies a particular set of assumptions about what constitutes evidence or legitimate knowledge. Other ways of knowing are viewed as unscientific and hence less legitimate. In Chapter 12, Jason Found and Michael M'Gonigle describe the way in which universities fragment knowledge among specialized disciplines and avoid examination of the reality shared by their constituents, as well as their connections to the outside world. Observing that local knowledges "are not only ignored, but often disqualified by the university," these authors propose an alternative planning strategy for universities that they call "comprehensive local innovation" (205).

In other cases, we see that local, "lay," or "traditional" forms of knowledge are often treated as "merely" anecdotal or alarmist, or as unsubstantiated by "real data." Indeed, governments and corporations often use scientific criteria to dismiss or downplay risks, as Andrée and Sharratt's Chapter 2 study suggests. Citizens, on the other hand, assert the precautionary principle, testifying to their mistrust of scientific claims and assurances.[4] William Leiss and Michael Tyshenko (2002, 335) note that the Canadian government takes a "highly restrictive perspective on risk ... Risk is restricted only to those hazards that may be characterized with precision by scientific practice ... However, it has also been known for some time now that among the public ... there is a quite different perspective on risk, one that is much broader in scope than that of the regulators and that is permeated by values different from those that govern scientific practice." Given the closeness of government and the private sector, Leiss and Tyshenko (ibid., 337) conclude that, in the case of new biotechnology, what is needed is "a unique, specialized regulatory body at a national level, working in parallel (but entirely independent of) the existing departmental regulatory agencies charged with assuring product safety."

Second, it is increasingly less the case that research carried out in universities can be viewed as independent of particular interests, particularly in the science and engineering fields, where a considerable portion of research is funded or co-funded by private corporations. At issue is not so much the compromising of scientific method, but the very framing and selection of the questions that will – or will not – be asked or investigated. "Public" science (that which is performed in government-funded departments, research centres, or universities) may nevertheless have a certain degree of autonomy from private interests. Indeed, a number of our studies highlight the important role of independent science or public science in assisting environmentalists and citizen groups to make their cases.

However, as other instances show, little may distinguish the ideological orientations of captured governments from the goals of their corporate clients. Government-employed scientists may be subject to various political pressures (such as the ruling party's desire not to close a major fishery just prior to an election), and their findings and recommendations may be repressed. For example, in the case of Alberta's NES Strategy (Chapter 8), we see that government ministries and companies determined which environmental impact data would be considered and which would not – *whose* science to use. Ministry data were suppressed; industry data were substituted.

The characterization of policy as science-based conveys the ideas of impartiality and superior rationality, and may serve to circumscribe debate. Ted Schrecker (2001, 52) describes the hegemonic uses of science:

> Firms or industries that view environmental conservation initiatives as an economic threat are likely to contend that all policy and regulatory decisions should be 'purely science-based,' or to use some similar phrase. They invoke the cognitive authority of science (who, after all, can be against science?) for strategic purposes, in order to avoid discussing concepts like standards of proof in terms that are appropriate. Governments may pursue the same course of action in order to defend decisions that have been reached primarily on grounds unrelated to either science or the environment, such as trade-policy implications or the avoidance of major economic disruptions. Thus, the idea of science-based environmental policy is superficially attractive, yet ultimately pernicious.

On the basis of their analysis of the making of the regulatory framework for GMOs, Andrée and Sharratt conclude in Chapter 2 that

> The notion that environmental and health regulation must be science-based has become a central tenet of neo-liberal governance. However, it is problematic on several levels – and not simply because examples have shown that science-based decisions are influenced by specific economic interests (see Sharratt 2001). Science-based decision making is often structured to recognize certain issues as worthy of consideration while ignoring others, whether intentionally or through the tacit judgment of certain groups of "experts" (Wynne 1994) ... A science-based process also validates only some expert knowledge as legitimate. Consequently, the risk evaluations of farmers, consumers, and other lay people are often discredited, even though these groups have valuable experience and insights to offer the process, and they will be affected by decisions made. (30)

Andrée and Sharratt note, in particular, the exclusion of socio-economic or ethical issues from "science-based" policy considerations. They argue that the question of who benefits from a new technology or a development project should be part of the criteria for governmental approval. That is, socio-economic and ethical criteria should be incorporated in a statutory definition of a review process. Science produces knowledge, but only partial knowledge. It is important, also, to consider the distinction offered by a Cree elder, quoted in *Rethinking History*, a mixed-media collage by artist Jane Ash Poitras: "Knowledge is about the past. Wisdom is about the future."[5]

In Chapter 7, Bardati argues for the acceptance of greater epistemological pluralism as well as dialogue between "experts" and citizens in decision-making processes, suggesting that such "civic science" can help to resolve the following questions: "Which data sources should be used? What level of information quality is appropriate? And who is competent enough to interpret these data sources and information types?" (121). Civic science, he argues, democratizes the hierarchical model of knowledge production and "transmission" from experts to citizens; it respects "knowledge based in practice" and actively "engages local communities and workers in the production and use of knowledge" (120).

Science – like all forms of knowledge – is important to democratic deliberation; hence, making knowledge transparent is a condition of a democratic process. In the case of GMO regulation, the lack of transparency of information has been a major problem. Andrée and Sharratt's Chapter 2 study shows that proprietary rights or interests may be put before citizens' rights to information. In Canada, commercial interests predominate over citizens' rights to know which food products contain genetically modified ingredients. In many public consultation processes, information is withheld from non-corporate participants on the grounds of proprietary interests. As Bardati reveals in Chapter 7, Slocan Forest Products asserted this prerogative with regard to the Slocan Valley community resource board.

In Chapter 11, Parkins shows how expert knowledge is used to pacify citizens in one forest certification public advisory committee, and how citizen input is nullified, or vetoed, by company representatives in another. The problem here is that the procedures permit corporate interests to dominate – to veto conditions they dislike (such as foregoing the exploitation of part of a forest). Parkins argues that post-normal science can encompass citizen knowledge – diverse knowledges – better than "normal" or positivist science can; it is not citizens who must be transformed into scientists before their knowledge can be accepted as legitimate, but rather science that needs to be democratized and viewed critically as socially embedded. The incorporation of a variety of knowledges into the discursive engagement will, of course, generate or clarify conflicts, but this, in Parkins' view, is a good thing – a democratic approach. The current trend is to exclude, suppress, or silence knowledge that does not conform to Western, positivist, science. Such "scientific" criteria function to obscure what Parkins describes as the "vested interests that support the scientific enterprise" (190) and to rule out-of-order counter-hegemonic discourses.

First Nations and Environmentalism

The environmental justice movements in the United States and globally have made it very clear that environmental risks are not distributed equally. Many of the chapters in this book document the ways in which First Nations are being severely undermined by the destruction and enclosure of nature. For Aboriginal communities everywhere, cultural survival is at stake.

The responses of Aboriginal communities to corporate encroachment on their lands and fisheries (often approved or promoted by governments) are necessarily shaped by the limited economic and political opportunities available to them. Environmentalists have sometimes lamented the agreements made between particular bands and companies in the forestry, oil, fishery, or other industries for joint exploitation of these resources. For example, in BC, some bands have signed agreements with salmon farm companies. In Alberta, some bands have made contracts with oil companies to supply production sites. The Athabasca Chipewayan First Nation (ACFN) supported an application by Suncor to the Alberta Energy and Utilities Board (AEUB) to double its production, in exchange for "an agreement for employment and contracts for its members and affiliated companies" (Buehler 2007). The Fort McKay First Nation owns a group of seven companies that service corporate clients in the oil sands, pipeline, and forestry sectors. As Dianne Meili (2007) explains, "Each band member is awarded substantial payouts from profits brought in by the Fort McKay Group of Companies owned and managed by the band." The chief of the Fort McKay First Nation recognizes that "the environmental cost [of such agreements] has been great" but concludes that "there is no other economic option; hunting, trapping, fishing is gone" (quoted in Struck 2006). Harrison's sympathetic treatment of the Ahousaht dilemma (Chapter 4) keeps ever in the foreground the reality that other means of livelihood have been destroyed or enclosed for big capital to exploit. In Bella Coola, where logging threatens the salmon and other wildlife upon which the Nuxalk have depended for subsistence, William Hipwell (Chapter 9) notes their high level of unemployment (nearly 80 percent), the loss of their traditional land base, and their poverty and social problems. Racism against First Nations also characterizes many conflicts, further isolating and marginalizing them in the competition with non-Natives for scarce natural resources.

As a result of such intense pressures and limited options, First Nations are often divided internally regarding development projects and government consultations. The Ahousaht were deeply divided over the idea of making a business agreement with Pacific National Aquaculture. The "holdout" bands in Nova Scotia suffered internal tension because of the economic pressures to come to an agreement with DFO (the MacKenzie agreements). The Nuxalk were split between the radical-sovereigntist House of Smayusta and the elected Nuxalk Nation Band Council. In the Red Hill Valley case, differences existed within the "traditional" Haudenosaunee Confederacy and the elected Six Nations Band Council.

The Mikisew Cree First Nation in Alberta opposed the Suncor application referred to above and has rejected the Government of Alberta's Water Management Framework for the Athabasca River (Buehler 2007; Pembina Institute 2007). A committee of elders from the Athabasca Chipewayan First Nation (ACFN) and the Mikisew Cree First Nation joined with the Pembina Institute in March 2007 to protest that "the oil sands industry and Government are placing oil ahead of Albertans' fresh water" (Pembina Institute 2007). As a First Nations representative put it, "We're talking about the survival of the Athabasca River, but more than that this is about the survival of our people" (Pat Marcel, chair of the ACFN Denesuline Traditional Environmental Knowledge Elders Committee, quoted in ibid.). In October 2006, a gathering of representatives from First Nations and Metis communities produced a "Keepers of the Water Declaration" aimed at protecting the waters of the Mackenzie Basin (Norwegian and Willson 2006).

The positions adopted by First Nations are not reducible to "environmental virtue" or "environmental sell-out." In Chapter 4, Harrison points out that signing an agreement does not necessarily mean that, with regard to their environmental concerns, Aboriginal groups cease to monitor, criticize, or lobby. The Ahousaht, she says, "remain active and honest critics of the farms in their area, and they attempt to hold regulatory bodies and industry accountable" (64). Likewise, the Fort McKay First Nation is "advocating for environmental improvements," says its chief, Jim Boucher (quoted in Meili 2007). Since signing an agreement with Suncor, the chief of the Athabasca Chipewayan First Nation has repeatedly expressed concern about rates of cancer in the area, which local people believe are linked to the pollution of the Athabasca River by the oil sands operations (Meili 2007; Buehler 2007; Weber 2006).[6] In Chapter 15, Jane Mulkewich and Richard Oddie offer an eloquent and thoughtful reflection on the limitations of mainstream environmentalism's failures to recognize the "unique needs, concerns, and life experiences of Aboriginal peoples" (250), and, more generally, "the interconnections between ecological degradation, socio-economic disparity, systemic injustice, and the uneven relationships of power that operate within modern social, political, and economic institutions" (257).

Both non-Native environmentalists and First Nations might consider alliances between them an important potential resource in influencing government policy. However, attempts to form such alliances have been both uncommon and problematic in Canada, for reasons that several of the chapters in this book unravel. In the Ahousaht case study (Chapter 4), support from the Friends of Clayoquot Sound and the David Suzuki Foundation helped the Ahousaht to defend the wild salmon stock from the environmental harms of the commercial fish farms. As noted above, the Pembina

Institute has allied with First Nations in Alberta to protect the Athabasca River. On the other hand, Hipwell's study of the Nuxalk relationship with the Forest Action Network (Chapter 9) and Mulkewich and Oddie's analysis of the struggle to preserve the Red Hill Valley (Chapter 15) highlight the complexities and tensions of such alliances, and Ballamingie's account documents a complete breakdown of solidarity (Chapter 6).

Chapters 6, 10, and 15 provide further insight into a significant difference between non-Native and indigenous understandings of human-nature relationships. The Nishnawbe Aski understanding of the appropriate relationship of humans to nature stands as a radical challenge to the instrumental exploitation of nature that has predominated in European societies since the sixteenth century (Merchant 1980). Whereas the ENGO coalition Partnership for Public Lands aimed to secure the setting aside of 10-15 percent of Ontario Crown lands for conservation, a member of the Anishnabek First Nation was mystified by the distinction between land that deserves ecological consideration and land that does not. In the wake of the Forest Accord, the Nishnawbe Aski Nation has continued to oppose the creation of protected areas within its traditional territory and is advocating its own "community-based, land use planning process for determining the allocation of land for traditional uses (including wildlife and habitat protection) and industrial development" (NRTEE 2004).[7] In Chapter 10, Overton found that the Protected Areas Association of Newfoundland and Labrador was interested only in protecting areas of "ecological importance." These cases raise an important question for non-Native environmentalists: what view of the appropriate relationship of humans to nature underpins conservation battles? Mulkewich and Oddie (Chapter 15) observe that, as a result of connections made between environmental and aboriginal activists in the Red Hill Valley conflict, foundations have been laid for replacing wilderness/urban or nature/humans oppositions with an environmental justice framing of such conflicts. Indeed, these authors argue that "bringing together an analysis of ecological degradation with attention to the systems of oppression" (258) is a condition for the formation of broad-based political ecology coalitions.

Institutional Resources for Ecological Democratization
As every environmental conflict is embedded in multiple spatial scales, actors have recognized that reforms are needed from local to international levels: from municipal governments (the Toronto and Hamilton case studies in this book give some indication of the problems with their "growth coalition" politics and planning processes) to the international agreements and institutions (such as NAFTA and the WTO) that have eroded the regulatory powers of elected governments.[8] Environmental lawyer David Boyd (2003, 259) argues that, as a result of having lost NAFTA cases brought under Chapter 11 of the agreement, the Canadian government is afraid to enact

new environmental laws. WTO free trade rules have squashed national attempts at environmental protection on a number of occasions. The following discussion proceeds with the acknowledgment that some of the reforms advocated by Canadian ENGOs, environmental lawyers, and others will have to confront such obstacles.

In terms of constitutional rights and responsibilities, and in terms of opportunities to meaningfully participate in decision making, ecological citizenship is weak in Canada. Our federal constitution and our environmental laws do not recognize the rights of citizens to public goods such as a clean (healthful) environment, clean water, or clean air; nor do they effectively protect biodiversity or entrench a precautionary principle with regard to environmental and health risks.[9] Environmental rights, embedded constitutionally, would be new resources for citizen action – would help to assert collective rights to health and ecosystem integrity over private property rights. Thus, the constitution could be amended to include substantive (collective and individual) environmental rights that have the status of human rights. According to environmental lawyers Elaine Hughes and David Iyalomhe (1998-99, 242), when a government fails to act to protect these rights, citizens should have access to a judicial review of its conduct. In their view, "complainants should be able to bring a claim for any act or omission which degrades the present level of environmental quality, impairs essential biological processes or ecosystem function, undermines collective or individual human health or survival, impairs the supply of basic resources, denies fair access to such resources, or undermines the ability of the environment to support human culture" (ibid., 243).

Environmental bills of rights have been adopted at subnational levels: in the Northwest Territories in 1990, the Yukon in 1991, and Ontario in 1993. None, however, has met the criteria of "strong" environmental rights as set out by such legal scholars as Hughes and Iyalomhe (ibid.) – including that citizens should be able to hold their governments accountable for failure to protect environmental rights. Importantly, citizens are not empowered to bring cases.

The existence of rights recognized in law does not, of course, guarantee their effective protection or implementation, as the historical experience of Aboriginal peoples in Canada has shown. In Chapter 8, Fluet and Krogman suggest that the Alberta government makes a greater effort to consult Aboriginal groups than it does to consult environmental organizations because it has a constitutional obligation to do so and wishes to avoid legal conflicts. In Chapter 9, Hipwell notes that, for the Nuxalk, Aboriginal title could be used to "counteract state power" (156). In these cases, constitutional rights that have been affirmed through litigation are viewed as important resources for First Nations' struggles. On the other hand, an estimated nine hundred land claims remain unsettled, Aboriginal rights are not respected (Foot 2007), and most Aboriginal communities lack the financial and personnel resources to engage in endless litigation and multiple consultation processes.

Canadian environmental law is notorious (among environmental lawyers and ENGOs) for its weaknesses, including its largely discretionary nature, reliance on voluntary compliance agreements, absence of rights for citizens to initiate prosecutions, and patchwork nature (Boyd 2003).[10] The chapters in this book provide examples of the largely discretionary nature of environmental regulation in Canada. In the Oak Ridges Moraine case (Chapter 17), Ontario's municipal affairs and housing minister introduced the Oak Ridges Moraine Conservation Act and Conservation Plan, in which he retained the power to override the latter. After passage of the act, he issued a ministerial order allowing development of lands that were still being reviewed by the Ontario Municipal Board. His decision was final and binding; there was no appeal. Moreover, the capacity of environmental ministries to monitor or enforce existing regulations is generally poor. Monitoring of the state of the environment by governments and transparent, regular reporting to citizens of the findings are typically inconsistent or non-existent. Finally, as noted earlier, Canadian federalism presents some particular problems for the setting of national goals and standards, and for their implementation and enforcement.[11]

Institutional opportunities for citizen resistance to capitalism's accumulation by dispossession, or to the generation of new environmental risks, vary from case to case. For example, as Chapter 3 shows, in Quebec the Bureau d'audiences publiques sur l'environnement (BAPE) provides an important forum for public deliberation, albeit one subject to many pressures. Unique in Canada, the BAPE is an institutional model of deliberative democracy with a "sustainable development" mandate that could usefully be studied by social movement actors.[12] Apart from the BAPE, public consultation processes inside and outside of Quebec are typically managed more directly by governments. Provincial and federal legislation requiring environmental impact assessments provides some opportunity for citizen participation. As Jim Overton's study of parks privatization in Newfoundland and Labrador (Chapter 10) shows, however, governments do not always respect either the spirit or the letter of such environmental impact assessment laws. Other types of consultation (over land or water use, for example, or a regulatory initiative) are often ad hoc, having no legal "trigger."

Within the consultation processes studied in this book, various practices or strategies were employed by government authorities and corporate representatives to predetermine the outcomes while maintaining the appearance of an open-ended and inclusive decision-making process. Wekerle, Sandberg, and Gilbert conclude in Chapter 17 that, "although represented by the provincial government as *public* consultations, these processes were exclusive, carefully managed, and constructed so as to achieve the ends defined by the government and its supporters" (291). In most instances, moreover, the conclusions of such consultative processes are not binding on governments. Environmentalists viewed this as a major flaw in a number of cases, including the Alberta NES Strategy (Chapter 8). In Alberta, ENGOs have participated in a series of consultation processes, some producing hard-negotiated compromises, only to have the recommendations of the deliberative bodies rejected or ignored by government.

In a recent report for the Canadian Policy Research Networks (CPRN), Amanda Sheedy, Mary Pat Mackinnon, Sonia Pitre, and Judy Watling (2008, 12) observe that in very few jurisdictions around the world are there institutionalized public consultation processes. They describe (ibid., 12-13) four criteria for institutionalized "citizen engagement" that are taken from the work of Lori Turnbull and Peter Aucoin (2006). These are

- that public involvement should be a "core element" of the policy process
- that public input is given "substantive consideration throughout the planning and execution stages" (rather than having only token value)
- that the commitment to institutionalized public involvement be government-wide rather than assigned only to certain departments
- that both the public service and elected representatives play a role in the institutionalization of public involvement.

Additional necessary procedural conditions for meaningful citizen participation have been identified by citizen activists, ENGOs, and other actors. Their demands include that processes be open (not limited to economic stakeholders or "tiny publics"), that their terms of reference not be predetermined, that they be neutrally administered, that adequate notice be given of questions to be decided, that all participants have equal access to information, that there be adequate resources for input from the public and Aboriginal peoples, that a civic-science-type model of knowledge production and sharing be used, and that indigenous (or traditional ecological) knowledge be included in the deliberation. These criteria imply a very different conception of citizenship than that which prevails in liberal democracies today – one that facilitates and "normalizes" meaningful citizen participation in policy making.

A number of the authors in this book refer to the existence of insiders and outsiders in policy making. Insiders participate in the public processes but also have privileged access to government decision makers in private contexts, and their interests tend to be privileged in the ideological framework of values and goals that determines policy outcomes. Where exceptions occur, or these prerogatives are threatened, firms and business associations are quick to defend their positions. In the Quebec case (Chapter 3), we see that business associations are strategizing about how to reign in the BAPE. If the recommendations of Yves Rabeau (2006), Louis Bernard (2007), André Caillé (cited in Boisvert 2006), and others are followed, the main deliberative forum for citizen participation will see the curtailment of its autonomy and the restriction of the scope of issues or interests that it is entitled to consider. It is clear that opportunities for citizen participation in decision-making processes, along with deliberative democratic criteria, must be actively and permanently defended against powerful private interests.

Public consultation processes are failing to include affected Aboriginal communities in meaningful and effective ways, as illustrated in Chapters 6 and 8. Nor does the "duty to consult" translate into effective power for First Nations to *refuse* development in their territories. In the Ahousaht case (Chapter 4), ministries with pro-development biases assess the "validity" of First Nations claims that their rights are affected by salmon farm licensing. Moreover, the Ahousaht are not allowed to veto, or to reject, any form of industrial development in territories that they claim. This lack of veto power was also evident in the Ontario Lands for Life process (Chapter 6) – that is, *no development is not an option.* Thus, the deliberative democracy criteria of (government) impartiality and inclusiveness (of all interested parties, as well as of all perspectives and alternatives) are violated. In Chapter 8, Fluet and Krogman identify some of the obstacles to First Nations representation in land-use consultations, such as linguistic and cultural differences as well as the multiple demands on very small, poorly resourced communities to provide representatives to such bodies. Also significant are the divisions within bands and competing claims to representation. Governments can certainly assist by providing funding for expenses, adapting timelines to the needs of communities that practise collective decision making, translating documents, and taking other measures. And yet, some would argue that the real solution for First Nations is not more participation in Canadian (federal or provincial) government-managed and -initiated consultations, but self-government with territorial sovereignty.

Like Aboriginal groups, low-income populations in general are often marginalized in and by public consultation processes. Lacking the middle strata's social capital and financial means to participate, the low-income working class, the unemployed, and the homeless – groups containing disproportionate numbers of persons with disabilities, urban Aboriginals, racialized recent immigrants, and, in some contexts, other racialized minorities – are often the most poorly represented and the most affected by industrial or urban development. In Chapter 16, Teelucksingh argues, "Meaningful procedural democracy requires governments to identify the obstacles to citizenship participation and to remove these obstacles" (277). Val Plumwood (2005, 625) calls for "communicative and participatory ideals and institutions that not only permit but actively solicit the voice from below." Yet, it is clear that procedures alone cannot level a playing field characterized by enormous inequalities of wealth and power. Thus, Plumwood (ibid.) argues that "deep democracy" requires "a justice dimension as redistributive equality."

Another aspect of Canada's current citizenship deficit is the underfunding of independent media. Some of the chapters in this book report the important role of publicly funded investigative journalism and broadcasting in informing the public of various risks or injustices. Likewise, the corporatization of universities, as recounted in Chapter 12, is eroding crucial resources for citizenship. As many of our case studies have illustrated, ecological citizenship requires publicly funded university research and insistence on its availability to the public.

Lastly – though there is so much more to explore – we should consider other conditions that prevent people from becoming informed about policy debates, from attending or organizing meetings, or from participating in planning and consultation processes (in other words, all the activities that constitute citizenship in action). This question immediately draws our attention to such factors as work norms and lack of free time (Adkin 1998a, 2003; Lipietz 1989), the absence of publicly funded child care, poor or expensive public transportation, and the lack of free public spaces in which to meet (Horton 2006). These examples suggest the ways in which environmental citizenship is inseparable from the defence and the extension of social citizenship.

Consensus, Conflict, and Ecological Democracy

Democratic reforms are needed not because they will bring about a non-exclusive consensus among societal interests, but because environmental politics are inherently *conflictual*. In the context of existing power relationships, the best deliberative democratic processes cannot produce "consensus" outcomes that reflect the highest aspirations toward ecological integrity and social justice. As argued in Chapter 1 of this book, the consensus achievable when opposed interests are in play will reflect the resources (institutional, organizational, monetary, knowledge-related, moral, and political) that the protagonists are able to mobilize. Conflicting interests (such as those of capitalist accumulation versus those of ecology and equality) cannot be dissolved into a common, generalized interest, despite the pretensions of sustainable development discourse. Citizens groups, NGOs, and Aboriginal peoples engaging in policy-making processes very quickly identify the democratic shortcomings of existing institutions as well as the nature of the interests at stake. In this sense, the criteria of deliberative democracy constitute an effective and societally rooted critique of liberal democratic institutions.

However, the problems that these actors identify extend well beyond the rules governing the deliberative forums of public consultation processes. They touch upon the hugely unequal relationships of wealth and power that characterize our society and in which governance frameworks are embedded. To begin to level *this* playing field – to overcome the obstacles to mobilizing green publics and realizing societal alternatives – calls for a much broader range of democratizing reforms. Here we move into the territory of the conditions needed to make ecological alternatives socially equitable and attractive to a substantial majority of the population. These include transitional strategies for workers in industrial and resource sectors, community involvement in regional (or bioregional) economic planning, income redistribution, and changing work-life norms.

Demands for substantive and procedural rights interact in complex ways that alter the conditions for the achievement of both. Procedural reforms allow previously excluded interests to gain representation and put alternatives on the public agenda. At the same time, the mobilization of informed (and "sustainable") green publics is indispensable to ensuring that procedural reforms occur and that their outcomes are implemented. The only solution to this chicken-and-egg problem is the one that social movements already practise: to be active at *multiple* sites and scales, crossing the boundaries of state and civil society, law and culture, public and private. The democratizing reforms identified in this book constitute a substantial challenge to hegemonic interests and institutions. To the extent that they are achieved, they will alter the landscape of the "trench warfare" – the war of position – waged on a daily basis by counter-hegemonic actors. That is, they will enhance the resources available to such actors and allow their alternatives to be more seriously considered. Such demands are both goals of ecological citizenship and evidence of ecological citizenship in action.

Notes

1 Jean-Paul Rodrigue, Claude Comtois, and Brian Slack (2006, ch. 3 concept 3) define a commodity chain as "a functionally integrated network of production, trade and service activities that covers all the stages in a supply chain, from the transformation of raw materials, through intermediate manufacturing stages, to the market. The chain is conceptualized as a series of nodes, linked by various types of transactions, such as sales and intra-firm transfers. Each successive node within a commodity chain involves the acquisition or organization of inputs for the purpose of added value."

2 This term was used by Antonio Gramsci to refer to those individuals who are "the thinking and organizing element of a particular fundamental social class" (Hoare and Smith 1971, 3). I use the term here to refer to the individuals who are engaged in the formation of affinities among diverse, subordinate, subject positions.

3 The IJC was created by the 1909 *Boundary Waters Treaty*, signed by the United States and Great Britain, with a mandate to prevent and resolve disputes involving Canada-US boundary waters. It oversees the implementation of the Great Lakes Water Quality Agreement, first signed by Canada and the United States in 1972. For more information about the IJC, see its website http://www.ijc.org/. This characterization of bureaucratic practice is confirmed

by Ann Dale (2001), in her analysis of the obstacles within Canadian government institutions to the implementation of sustainable development policies.

4 Ted Schrecker (2001, 55) observes that "scientific standards" of proof often trump the precautionary principle, as was the case in the WTO ruling on the legitimacy of the EU ban on the import of beef produced with growth hormones.

5 Ash Poitras' work was displayed at the Edmonton Art Gallery in March-April 1992, as part of the group exhibition titled "Rethinking History."

6 Like the observations of local residents, environmental toxicology has supported the likelihood of causal links between toxins in the river and cancers in humans, and between toxins and the deformities seen in fish and wildlife. See Kevin Timoney (2007).

7 In November 2006, the chiefs of the Nishnawbe Aski Nation (NAN) signed the "Northern Table," a bilateral partnership with the Government of Ontario to jointly develop a process to resolve questions of consultation and accommodation, resource development, mining, parks, and licensing permits within NAN traditional territory. Earlier that year, the NAN had protested drilling within its traditional territory by Platinex Mining, whereupon Platinex sued it for $10 billion in "damages." In March 2008, six NAN leaders, from the Kitchenumahykoosib Inninuwug (KI) First Nation, were jailed for preventing Platinex Mining employees access to their territory. Coverage of this conflict has been provided by a coalition supporting the "KI6" (http://www.freeki6.ca/index.php) and by the Wawatay Native Communications Society (http://www.wawataynews.ca/search/node/KI6).

8 I am aware that this discussion of governance leaves out a lot. For example, indigenous actors have demanded the transformation of the system of governance imposed on First Nations by the Indian Act. In Chapter 12, Found and M'Gonigle look at the ways in which institutions such as universities could be democratized, and we could also talk about reforms to corporate governance or trade union organizations.

9 In their June 2006 joint submission to the parliamentary review of the Canadian Environmental Protection Act (CEPA), Environmental Defence Canada and the Canadian Environmental Law Association observed that, "Although CEPA requires the federal government to apply the precautionary principle, more weight is given in practice to social, economic, and legal considerations than to protecting health of the environment. The Act does not operationalize the principle by setting out how it shall be explicitly used at every stage of the decision-making process." See also PollutionWatch (2006, 27-29).

10 Private prosecutions brought by individuals may be "taken over" by provincial attorneys general, then dropped before they get to trial. Citizen suits are either not yet permitted in Canadian law or have not been effective (Boyd 2003, 247-48). One problem is that such rights are not accompanied by intervenor funding. According to Boyd (ibid., 248), the citizen suit provisions that exist in the Canadian Environmental Protection Act "have never been used successfully."

11 Like other environmental lawyers, David Boyd (2003) argues that, given the signals that the Supreme Court has been sending, the Government of Canada could go much further in asserting its jurisdictional authority in this field but has chosen not to test its constitutional powers. Instead, the federal Liberal government opted for "harmonization" agreements that further decentralized regulatory responsibilities to the provinces.

12 I do not mean to suggest here that the BAPE is an ideal model. For example, for the most part, it does not initiate public consultations but does so at the request of the Quebec minister for sustainable development, environment, and parks, and its recommendations to the minister are merely advisory. The BAPE's executive is appointed by the government, raising questions about the impartiality of those leading the consultation processes. Public consultation periods are only forty-five days. The definition of sustainable development and the objectives of the BAPE itself also invite critical examination. Detailed information regarding the BAPE's mandate and actions may be found at http://www.bape.gouv.qc.ca/sections/documentation/Rap_annuel_2007-2008.pdf.

Photo Yzobela Hyett

References

AAFC (Agriculture and Agri-Food Canada). 2005. Telephone interview with an AAFC representative by Peter Andrée, 27 July.

Abells, Susan, and Michael Henry. 2001. *The Northern East Slopes (NES) Sustainable Resource and Environmental Management Strategy: Stakeholder Consultations October to December 2000, Final Report.* Edmonton: Alberta Environment.

–. 2002. *The Northern East Slopes (NES) Sustainable Resource and Environmental Management Strategy: Stakeholder Consultations February 2002, Final Report.* Edmonton: Alberta Environment.

Aberley, Doug. 1999. Interpreting Bioregionalism: A Story from Many Voices. In *Bioregionalism*, ed. M.V. McGinnis, 13-42. London: Routledge.

Abers, Rebecca. 2000. *Inventing Local Democracy: Grassroots Politics in Brazil.* London: Lynne Rienner.

Adkin, Laurie E. 1992. Counter-Hegemony and Environmental Politics in Canada. In *Organizing Dissent: Contemporary Social Movements in Theory and Practice*, ed. William K. Carroll, 135-56. Toronto: Garamond Press.

–. 1998a. Ecological Politics in Canada: Elements of a Strategy of Collective Action. In *Political Ecology: Global and Local*, ed. R. Keil, D. Bell, L. Fawcett, and P. Penz, 292-322. New York: Routledge.

–. 1998b. *Politics of Sustainable Development: Citizens, Unions, and the Corporations.* Montreal: Black Rose Books.

–. 2003. Ecology, Political Economy, and Social Transformation. In *Changing Canada: Political Economy as Transformation*, ed. W. Clement and L.F. Vosko, 393-421. Montreal and Kingston: McGill-Queen's University Press.

Agamben, Giorgio. 2005. *State of Exception.* Chicago: University of Chicago Press.

Agrawal, A., and C. Gibson. 1999. Enchantment and Disenchantment: The Role of Community in Natural Resource Conservation. *World Development* 27, 4: 629-49.

Agyeman, Julian, Robert Bullard, and Bob Evans. 2003. *Just Sustainabilities: Development in an Unequal World.* Cambridge, MA: MIT Press.

Agyeman, Julian, and Bob Evans. 2006. Justice, Governance, and Sustainability: Perspectives on Environmental Citizenship from North America and Europe. In *Environmental Citizenship*, ed. Andrew Dobson and Derek Bell, 185-206. Cambridge, MA: MIT Press.

Alberta. Aboriginal Affairs and Northern Development. 2000a. *Strengthening Relationships: The Government of Alberta's Aboriginal Policy Framework.* Edmonton: Government of Alberta.

–. 2000b. Indian Land Claims – Backgrounder. 31 March. http://www.assembly.ab.ca/lao/library/egovdocs/alaa/2000/149532.pdf.

Alberta. Aboriginal Relations. 2003. *Best Practices Handbook for Traditional Use Studies.* Edmonton: Government of Alberta. http://www.aboriginal.alberta.ca/documents/tsu_BP_Internet_Handbook.pdf.

–. 2007. *Consultative Guidelines – An Update.* http://www.aboriginal.alberta.ca/571.cfm.

Alberta. Department of Energy. 2005. The Government of Alberta's First Nations Consultation Policy on Land Management and Resource Development. 16 May. http://www.energy.alberta.ca/AbRel/pdfs/Approved_Consultation_Policy_-_May_16.pdf.

Alberta. Energy Resources Conservation Board. 2007. Public Involvement. (Under the "Public Safety" area of "Public Zone.") http://www.ercb.ca.

Alberta. Ministry of Environment. 2006. A Sustainability Strategy for Southern Alberta. http://www3.gov.ab.ca/env/regions/southern/strategy.html.

Alberta. Sustainable Resources Development. 2008. Land-Use Framework. 3 December. http://www.landuse.alberta.ca/.

Alberta Wilderness Association. 2003. Letter to Abells Henry Public Affairs, 26 March.

Alfred, T. 1999. *Peace, Power, Righteousness: An Indigenous Manifesto.* Toronto: Oxford University Press.

Ali, S. Harris. 2003. A Socio-Ecological Autopsy of the *E. coli* 0157:H7 Outbreak in Walkerton, Ontario, Canada. *Social Science and Medicine* 58, 12: 2601-12.

Allen, Patricia, Margaret FitzSimmons, Michael Goodman, and Keith Warner. 2003. Shifting Plates in the Agrifood Landscape: The Tectonics of Alternative Agrifood Initiatives in California. *Journal of Rural Studies* 19: 61-75.

Allen Wood Consulting. 2001. *Report on Native Involvement in Commercial Fisheries.* Vancouver: Native Fishing Association.

Anderson, Erin. 1996. Crane Mountain Gets Landfill: Cabinet. *Saint John Times Globe,* 11 September, A1.

Anderson, J. 2000. Four Considerations for Decentralized Forest Management: Subsidiary, Empowerment, Pluralism and Social Capital. In *Decentralization and Devolution of Forest Management in Asia and the Pacific,* ed. T. Enters, B. Patrick, P. Durst, and M. Victor, 11-22. Bangkok: Food and Agricultural Organization of the United Nations and RECOFTC.

Anderson, Terry L., and Alexander James, eds. 2001. *The Politics and Economics of Park Management.* Lanham, MD: Rowman and Littlefield.

Andersson, K., C. Gibson, and F. Lehoucq. 2004. The Politics of Decentralized Natural Resource Governance. *Political Science and Politics* (American Political Science Association) 37, 3: 421-26.

Andrée, Peter. 2002. The Biopolitics of Genetically Modified Organisms in Canada. *Journal of Canadian Studies* 37, 3: 162-91.

–. 2005. The Genetic Engineering Revolution in Agriculture and Food: Strategies of the 'Biotech Bloc.' In *The Business of International Environmental Governance,* ed. D. Levy and P. Newell, 135-66. Boston: MIT Press.

Angus, M. 1992. Comprehensive Claims: One Step Forward, Two Steps Back. In *Nation to Nation: Aboriginal Sovereignty and the Future of Canada,* ed. D. Engelstad and J. Bird, 67-77. Concord: Anansi Press.

Anonymous. 2004a. The Conflict. Percy Schmeiser website. http://www.percyschmeiser.com/conflict.htm.

–. 2004b. Industry and Government Continue to Work toward GM Acceptance Process. *Canola Link* 35, 22 April. http://archives.foodsafety.ksu.edu/agnet/2004/4-2003/agnet_april_23.htm#story1.

–. 2004c. Ottawa to Examine Economics of GM Wheat. Canadian Broadcasting Corporation, 20 April. http://winnipeg.cbc.ca/regional/servlet/View?filename=mb_wheatboard20040420. Accessed 26 November 2005; page now discontinued.

APCFNC (Atlantic Policy Congress of First Nations Chiefs). 2001a. The Management of Fisheries on Canada's Atlantic Coast: A Discussion Document on Policy Direction and Principles. http://www.rism.org/isg/dlp/bc/perspectives/mmp1.htm.

–. 2001b. Mi'kmaq Maliseet Passamaquoddy Integrated Natural Resource Management Policy. http://www.rism.org/isg/dlp/bc/perspectives/mmp2.htm.

Apostle, R., B. McCay, and K. Mikalsen. 2002. *Enclosing the Commons: Individual Transferable Quotas in the Nova Scotia Fishery.* St. John's: Institute of Social and Economic Research (ISER), Memorial University of Newfoundland.

Arnold, Ron. 1996. Overcoming Ideology. In *A Wolf in the Garden: The Land Rights Movement and the New Environmental Debate*, ed. P.D. Brick and R. McGreggor Cawley, 15-26. Lanham, MD: Rowman and Littlefield. Also available at http://www.cdfe.org/center-projects/wise-use/.

Arnstein, S. 1969. A Ladder of Citizen Participation. *Journal of the American Institute of Planners* 35, 4 (July): 216-24.

Arquette, M., M. Cole, B. LaFrance, M. Peters, J. Ransom, E. Sargent, V. Smoke, and A. Stairs. 2002. Holistic Risk-Based Environmental Decision Making: A Native Perspective. *Environmental Health Perspectives* 110: 259-64.

Assembly of First Nations. 2005. Federal Budget Fails to Address First Nations Health Care Crisis. http://www.afn.ca/article.asp?id=181.

–. 2008. The $9 Billion Myth Exposed: Why First Nations Poverty Endures. http://www.crr.ca/divers-files/en/publications/reports/diversReports/mythExposed.pdf.

Atleo, Richard E. (Umeek). 2004. *Tsawalk: A Nuu-chah-nulth Worldview*. Vancouver: UBC Press.

Auditor General of Quebec. 2007. *Report to the National Assembly for 2007-2008.* Vol. 2, *Report of the Sustainable Development Commissioner, Highlights.* Quebec City: Gouvernement du Québec. http://www.vgq.gouv.qc.ca/publications/rapp20072008/Rapport2007-2008-T2.pdf.

August, Denise. 2005a. Senate Standing Committee Visits Ahousaht, NCN Send Clear Message to DFO: Your Policies Are Destroying the Resource! *Ha-Shilth-Sa*, 3 November, 16.

–. 2005b. Uu-a-thluk Council of Ha'wiih Prepare for Policy Discussions with DFO. *Ha-Shilth-Sa*, 15 December, 4.

Auld, G., and G.Q. Bull. 2003. The Institutional Design of Forest Certification Standards Initiatives and Its Influence on the Role of Science: The Case of Forest Genetic Resources. *Journal of Environmental Management* 69: 47-62.

Babbie, E. 1995. *The Practice of Social Research*. Toronto: Wadsworth.

Bäckstrand, Karin. 2003. Civic Science for Sustainability: Reframing the Role of Experts, Policy-Makers and Citizens in Environmental Governance. *Global Environmental Politics* 3, 4: 24-41.

BAPE (Bureau d'audiences publiques sur l'environnement). 2003. *L'inscription de la production porcine dans le développement durable, rapport d'enquête et d'audience publique.* Quebec City: Gouvernement du Québec.

Barcelo, Yan. 2007. Agriculture: un combat extrême. *Forces* 149 (March): 40-45.

Bardati, Darren. 2003. Land Use Planning in British Columbia, Canada: Evaluating a Participatory Experiment. *The Land: Journal of the International Land Use Society* 7, 1: 57-81.

Baril, Jean. 2007. *Le BAPE devant les citoyens. Pour une évaluation environnementale au service du développement durable.* Quebec City: Les Éditions PUL-IQRC.

Barker, D. 1986. Core, Periphery and Focus for Geography? *Area* 18: 157-60.

Barnes, T.J., and R. Hayter, eds. 1997. *Troubles in the Rainforest: British Columbia's Forest Economy in Transition.* Canadian Western Geographical Series 33. Department of Geography, University of Victoria. Victoria: Western Geographical Press.

Barron, Jennifer. 2000. In the Name of Solidarity: The Politics of Representation and Articulation in Support of the Labrador Innu. *Capitalism, Nature, Socialism* 11, 3 (September): 87-112.

Barron, Tracy. 2001. Golf Course Endangers Park: Activist. *St. John's Evening Telegram*, 23 June, 1-2.

–. 2002. Windmill Bight Released from Full Assessment. *St. John's Evening Telegram*, 30 April, A4.

Barry, John. 1996. Sustainability, Political Judgement and Citizenship: Connecting Green Politics and Democracy. In *Democracy and Green Political Thought*, ed. Brian Doherty and Marius de Geus, 115-31. London: Routledge.

–. 1999. *Rethinking Green Politics: Nature, Virtue, and Progress.* London: Sage.

–. 2006. Resistance Is Fertile: From Environmental to Sustainability Citizenship. In *Environmental Citizenship*, ed. Andrew Dobson and Derek Bell, 21-48. London: MIT Press.

Barton, Gregory. 2002. *Empire Forestry and the Origins of Environmentalism.* Cambridge: Cambridge University Press.

Beder, Sharon. 2002. *Global Spin: The Corporate Assault on Environmentalism.* Foxhole, Devon, UK: Green Books.

–. 2006. The Changing Face of Conservation: Commodification, Privatization and the Free Market. In *Gaining Ground: In Pursuit of Ecological Sustainability,* ed. David M. Lavigne, 83-97. Guelph, ON: International Fund for Animal Welfare.

Bedford, David, and Danielle Irving. 2001. *The Tragedy of Progress: Marxism, Modernity and the Aboriginal Question.* Halifax: Fernwood.

Belford, Terrance. 2001. New Digs. *National Post* (Toronto), 10 November.

Benjamin, R., and S. Carroll. 1998. The Implications of the Changed Environment for Governance in Higher Education. In *The Responsive University: Restructuring for High Performance,* ed. W.G. Tierney, 92-119. Baltimore: Johns Hopkins University Press.

Berg, Peter, and Richard Dasmann. 1977. Reinhabiting California. *The Ecologist* 7, 10: 399-401.

Berger, T. 1977. *Northern Frontier, Northern Homeland: The Report of the Mackenzie Valley Pipeline Inquiry.* Ottawa: Supply and Services.

Bergman, Hans. 2002. Interview by Peter Andrée. Den Hague, Netherlands, 22 April.

Berkes, F. 2002. Cross-Scale Institutional Linkages: Perspectives from the Bottom Up. In *The Drama of the Commons,* ed. Elinor Ostrom, T. Dietz, N. Dolsak, P.C. Stern, S. Stonich, and E.U. Weber, 293-321. Washington, DC: National Academy Press. http://www.nap.edu/catalog/10287.html.

–. 2003. Alternatives to Conventional Management: Lessons from Small-Scale Fisheries. *Environments* 31, 1: 5-20.

Berkes, F., D. Feeny, B.J. McCay, and J.M. Acheson. 1989. The Benefits of the Commons. *Nature* 340: 91-93.

Berkes, F., R. Mahon, P. McConney, R. Pollnac, and R. Pomeroy. 2001. *Managing Small-Scale Fisheries: Alternative Directions and Methods.* Ottawa: International Development Research Centre.

Bernard, Louis. 2007. L'acceptabilité sociale des projets de développement. Paper presented at the 32nd Congress of the Association des économistes québécois (ASDEQ), Quebec City, 18 May. http://www.asdeq.org/congres/pdf/32ieme_congres/louis-bernard-asdeq-congres2007.pdf.

Berny, Nathalie. 1999. L'Union des producteurs agricoles et les interventions gouvernementales sur les problèmes d'environnement d'origine agricole: participation et contribution d'un acteur à un processus de décision publique. Master's thesis, Political Science, Université Laval.

Birnbaum, R., ed. 1991. *Faculty in Governance: The Role of Senates and Joint Committees in Academic Decision Making.* San Francisco: Jossey-Bass.

Blaikie, Piers, and Harold Brookfield. 1987. *Land Degradation and Society.* London: Methuen.

Blomley, Nicholas. 1994. Activism and the Academy. *Environment and Planning D: Society and Space* 12: 383-85.

Bocking, S. 2006. *Nature's Experts: Science, Politics and the Environment.* New Brunswick, NJ: Rutgers University Press.

Boisvert, Michèle. 2006. André Caillé prône l'abolition du BAPE. *La Presse* (Montreal), 4 November, A1.

Bok, D. 2003. *Universities in the Marketplace: The Commercialization of Higher Education.* Princeton: Princeton University Press.

Bombay, Harry. 1996. *Discussion Paper: Aboriginal Forest-Based Ecological Knowledge in Canada.* Ottawa: National Aboriginal Forestry Association.

Borchers, J., and J. Kusel. 2003. Toward a Civic Science for Community Forestry. In *Community Forestry in the United States: Past Practice, Crafting the Future,* ed. M. Baker and J. Kusel, 147-63. Washington, DC: Island Press.

Boschman, Caroline. 2007. Province Not Considering Public Input into Oil and Gas Industry, Says Scientist. *Lethbridge Herald,* 20 February, A4.

Boudreau, Julie-Ann. 2000. *Megacity Saga: Democracy and Citizenship in This Global Age.* Montreal: Black Rose Books.

Bourdieu, P. 1988. *Homo Academicus.* Stanford: Stanford University Press.

Boyd, David R. 2003. *Unnatural Law: Rethinking Canadian Environmental Law and Policy.* Vancouver: UBC Press.

Bradshaw, B. 2003. Questioning the Credibility and Capacity of Community-Based Resource Management. *Canadian Geographer / Le Géographe canadien* 47, 2: 137-45.

Braga, Francesco. 1996. Over-the-Counter Derivatives and Price Risk Management in a Post-tripartite Environment: The Case of Cattle and Hogs. *Canadian Journal of Agricultural Economics* 44: 369-74.

Bragg, Rebecca. 1999. City Joins Movement to Save Moraine – Council Votes to Push Province for 6-Month Development Freeze. *Toronto Star,* 26 November, 1.

Braun, Bruce. 2002. *The Intemperate Rainforest: Nature, Culture and Power on Canada's West Coast.* Minneapolis: University of Minnesota Press.

Brechin, S., P. Wilshusen, C. Fortwangler, and P. West. 2002. Beyond the Square Wheel: Toward a More Comprehensive Understanding of Biodiversity Conservation as Social and Political Process. *Society and Natural Resources* 15, 1: 41-64.

Brennenstuhl, G.M. 1996. Letter to Nancy Colpitts, Committee Chairperson, 28 March. In *Bella Coola Local Resource Use Plan,* ed. LRUP Committee, sheet 17-18 in unnumbered front matter. Hagensborg, BC: BC Ministry of Forests.

British Columbia. BCStats. 2001. College Region 11 – North Island. Statistical Profile of Aboriginal Peoples 2001: With Emphasis on Labour Market and Post-secondary Education Issues. http://www.bcstats.gov.bc.ca/data/cen01/abor/cr11.pdf.

–. 2007a. Aboriginal Peoples Labour Force Survey 2007 – Revised February 1, 2008. http://www.bcstats.gov.bc.ca/data/lss/abor/abo_lfs_2007.pdf.

–. 2007b. Labour Market Characteristics of the Off-Reserve Aboriginal Population in BC. *Earnings and Employment Trends* (April). http://www.bcstats.gov.bc.ca/pubs/eet/eet0704.pdf.

British Columbia. BCStats. Ministry of Management Services. 2002. *British Columbia's Fisheries and Aquaculture Sector, 2002 Edition.* http://www.bcstats.gov.bc.ca/data/bus_stat/busind/fish.asp.

British Columbia. Environmental Assessment Office. 1997. *The Salmon Aquaculture Review: Final Report.* Victoria: Government of British Columbia.

British Columbia. Legislative Assembly. 2006. Special Committee on Sustainable Aquaculture, Minutes (6 June). http://www.leg.bc.ca/cmt/38thparl/session-2/aquaculture/hansard/W60606p.htm.

British Columbia. Ministry of Environment. 2005. *The 2004 British Columbia Seafood Industry Year in Review.* http://www.env.gov.bc.ca/omfd/reports/index.html#STATPUB.

–. 2007. *Salmon Aquaculture in British Columbia.* http://www.env.gov.bc.ca/omfd/fishstats/aqua/salmon.html.

British Columbia. Ministry of Forests. N.d. *The Ecology of the Interior Cedar-Hemlock Zone.* Victoria: BC Ministry of Forests.

Brodie, Janine. 2002. Citizenship and Solidarity: Reflections on the Canadian Way. *Citizenship Studies* 6, 4: 377-94.

Brookymans, Hanneke. 2004. From Enemies to Collaborators: Alliance Is a Breath of Fresh Air. *Edmonton Journal,* 7 March, A2.

Brown, R.G. 1995. Public Influence [sic] on Reforestation in British Columbia. Paper presented at the Third Global Conference on Paper and the Environment, London, UK, March. http://www.for.gov.bc.ca/hfp/publications/00134/index.htm.

Brownson, J.M.J. 1995. *In Cold Margins.* Missoula, MT: Northern Rim Press.

Brulé-Babel, A.L., R.C. Van Acker, and L.F. Friesen. 2003. Issues Related to Release of GM Wheat: Gene Flow and Selection. In *Proceedings of the 2002 Manitoba Agronomists Conference,* 144-51. Winnipeg: Faculty of Agricultural and Food Sciences, University of Manitoba. http://www.umanitoba.ca/afs/agronomists_conf/2002/pdf/brulebabel.pdf.

Bryant, Bunyan, and Paul Mohai. 1992. *Race and Incidence of Environmental Hazards: A Time for Discourse.* Boulder, CO: Westview Press.

Bryant, R., and G. Wilson. 1998. Rethinking Environmental Management. *Progress in Human Geography* 22, 3: 321-43.

Bryant, Raymond, and Sinéad Bailey. 1997. Environmental Non-governmental Organizations. In *Third World Political Ecology*, ed. R. Bryant and S. Bailey, 124-50. New York: Routledge.

Bueckert, Dennis. 2004. Agriculture Canada Puts Brakes on Roundup Ready Wheat Project. *Toronto Globe and Mail*, 10 January, A7.

Buehler, Clint. 2007. Fort Chip Natives Oppose More Oil Business. *First Nations Drum*, Winter 2006-07. http://www.firstnationsdrum.com/Fall%202006/HealthChip.htm.

Bullard, Robert. 1983. Solid Waste Sites and the Houston Black Community. *Sociological Inquiry* 53: 273-88.

Bullis, Connie. 1996. Retalking Environmental Discourses from a Feminist Perspective: The Radical Potential of Ecofeminism. In *The Symbolic Earth: Discourse and Our Creation of the Environment*, ed. J.G. Cantrill and C.L. Oravec, 123-48. Lexington: University Press of Kentucky.

Bunce, Susannah. 2004. The Emergence of 'Smart Growth' Intensification in Toronto: Environment and Economy in the New Official Plan. *Local Environment* 9, 2: 177-91.

Burke, Brenda. 2006. *Don't Drink the Water: The Walkerton Tragedy*. Victoria: Trafford.

Burke, Derek. 1998. Why All the Fuss over Genetically Modified Food? *British Medical Journal* 316, 7148 (20 June): 1845-46.

Burroughs, Rick. 2005. Wheat – The Tradition Is Going Strong. Statistics Canada. http://www.statcan.ca/english/kits/agric/wheat.htm.

Byrne, John, and Raymond P. Scattone. 2001. Community Participation Is Key to Environmental Justice in Brownfields. *Race, Poverty and the Environment* 3, 1 (Winter): 6-7. http://ceep.udel.edu/publications/ej/publications/2001_ej_community_is_key.pdf.

Calhoun, C., ed. 1992. *Habermas and the Public Sphere*. Cambridge, MA: MIT Press.

Canada. Advisory Committee on Reconstruction. 1944. *Final Report of the Subcommittee*. Vol. 2, *Conservation and Development of Natural Resources*. Ottawa: King's Printer.

Canada. Department of Agriculture. 1993. Federal Government Agrees on New Regulatory Framework for Biotechnology. Press release, 11 January. Ottawa (on file with Peter Andrée).

Canada. Department of Agriculture and Agri-Foods. 2004. Profile of the Canadian Wheat Industry. 16 July. http://www.seedquest.com/News/releases/2004/july/9313.htm.

Canada. Department of Fisheries and Oceans. 1982. *Task Force on Atlantic Fisheries. Navigating Troubled Waters: A New Policy for the Atlantic Fisheries: Highlights and Recommendations*. Ottawa: DFO.

–. 2004. A Policy Framework for the Management of Fisheries on Canada's Atlantic Coast. http://www.dfo-mpo.gc.ca/afpr-rppa/link_policy_framework_e.htm.

–. 2005. Fisheries Management, Pacific Region. *Commercial License Reports by Fishery*. http://www-ops2.pac.dfo-mpo.gc.ca/.

Canada. Indian and Northern Affairs Canada. 2003. National Assessment of Water and Wastewater Systems in First Nations Communities, Summary Report. Ottawa. http://www.ainc-inac.gc.ca/enr/wtr/pubs/watw/watw-eng.pdf.

–. 2007. Backgrounder: Made-in-Nova Scotia Process Framework Agreement. http://www.ainc-inac.gc.ca/nr/prs/j-a2007/2-2847-bk_e.html.

Canada. Office of the Auditor General of Canada. 2000. Chapter 30: Fisheries and Oceans – The Effects of Salmon Farming in British Columbia on the Management of Wild Stocks. In *2000 December Report of the Auditor General of Canada*. Ottawa: Government of Canada. http://www.oag-bvg.gc.ca/internet/English/parl_oag_200012_30_e_11217.html.

Canada. Senate Standing Committee on Fisheries. 1998. *Privatization and Quota Licensing in Canada's Fisheries: Report of the Standing Senate Committee on Fisheries*. December. http://www.parl.gc.ca/36/1/parlbus/commbus/senate/com-e/fish-e/rep-e/rep03dec98-e.htm.

Canada Mortgage and Housing Corporation. 2004. *Canadian Housing Observer*. Ottawa: Canada Mortgage and Housing Corporation. http://www.cmhc-schl.gc.ca/en/cahoob/index.cfm. Website now discontinued.

Canada Parks Agency. 2000. *"Unimpaired for Future Generations"? Conserving Ecological Integrity with Canada's National Parks*. Ottawa: Ministry of Public Works and Services.

Canada Revenue Agency. 2003. Income Statistics 2003 (2001 Tax Year), Final Basic Table 3: Sample Data for British Columbia. http://www.cra-arc.gc.ca/gncy/stts/gb01/pst/fnl/pdf/menu-eng.html.

Canadian Broadcasting Corporation. 2005. Boil Water Advisory Tops 9 Years for B.C. Native Community. CBCnews.ca. http://www.cbc.ca/story/canada/national/2005/10/31/kwicksutaineuk051031.html.

–. 2006. Water Still a Problem on 76 Reserves. CBCnews.ca. http://www.cbc.ca/story/canada/national/2006/02/20/aboriginal-water060220.html.

–. 2008. Pipeline Construction Can't Start: Lubicon Cree. CBCnews.ca. http://www.cbc.ca/canada/edmonton/story/2008/10/16/lubicon-cree-pipeline.html.

Canadian Sustainable Forestry Certification Coalition. 2008. Certification Status and Intentions. http://www.certificationcanada.org.

CAPP (Canadian Association of Petroleum Producers). 2002. Coalition Formed to Advance "Made in Canada" Strategy on Climate Change. Press release, 26 September. http://www.cpeq.qc.ca/anglais/apropos/ReleaseCCRES.pdf.

Cartwright, John. 2003. Environmental Groups, Ontario's Lands for Life Process, and the Forest Accord. *Environmental Politics* 12, 2 (Summer): 115-32.

Cashore, B. 2002. Legitimacy and the Privatization of Environmental Governance: How Non-state Market-Driven (NSMD) Governance Systems Gain Rule-Making Authority. *Governance* 15: 503-29.

Cashore, B., G. Auld, and D. Newsom. 2004. *Governing through Markets: Forest Certification and the Emergence of Non-state Authority*. New Haven: Yale University Press.

Cashore, B., G. Hoberg, M. Howlett, J. Rayner, and J. Wilson. 2001. *In Search of Sustainability: British Columbia Forest Policy in the 1990s*. Vancouver: UBC Press.

Cashore, B., and I. Vertinsky. 2000. Policy Networks and Firm Behaviour: Governance Systems and Firm Responses to External Demands for Sustainable Forest Management. *Policy Sciences* 33: 1-30.

Castells, Manuel. 1997. *The Power of Identity*. Oxford: Blackwell.

Castree, Noel. 2003. Environmental Issues: Relational Ontologies and Hybrid Politics. *Progress in Human Geography* 27, 2: 203-11.

–. 2004. Differential Geographies: Place, Indigenous Rights and 'Local' Resources. *Political Geography* 23: 133-67.

Castree, Noel, and Bruce Braun. 1998. The Construction of Nature and the Nature of Construction. In *Remaking Reality: Nature at the Millennium*, ed. Bruce Braun and Noel Castree, 3-24. London: Routledge.

–, eds. 2001. *Social Nature: Theory, Practice, and Politics*. Malden, MA: Blackwell.

Castree, Noel, and M. Sparke. 2000. Introduction: Professional Geography and the Corporatization of the University: Experiences, Evaluations, and Engagements. *Antipode* 32, 3: 222-29.

Castro, P., and E. Nielsen. 2001. Indigenous People and Co-management: Implications for Conflict Management. *Environmental Science and Policy* 4: 229-39.

Caulfield, Jon. 1994. *City Form and Everyday Life: Toronto's Gentrification and Critical Social Practice*. Toronto: University of Toronto Press.

CCCE (Canadian Council of Chief Executives). 2002. *The Kyoto Protocol Revisited: A Responsible and Dynamic Alternative for Canada. A Strategy Prepared by the Canadian Council of Chief Executives*. Ottawa: CCCE.

CCEDC (Central Coast Economic Development Commission). 1995. *Background Fact Sheet – Mid Coast Forest District – 1980 to 1994*. Bella Coola: CCEDC.

CCPFH (Canadian Council of Professional Fish Harvesters). 2001. *Preliminary Response of the Canadian Council of Professional Fish Harvesters to the Atlantic Fisheries Policy Review*. March. http://www.ccpfh-ccpp.org/e_dbViewer.asp?cs=policy&id=25.

CELA (Canadian Environmental Law Association). 1999. Mining "Side Deal" Undermines Lands for Life Proposal, Says Environmental Group. CELA. Press release, Toronto, 21 May. http://www.cela.ca/newsevents/detail.shtml?x=1209.

CFIA (Canadian Food Inspection Agency). 2003. Technical Workshop on Management of Herbicide Tolerant Crops Report. http://www.inspection.gc.ca/english/plaveg/bio/consult/herbtolrepe.shtml. Accessed 22 February 2005; page now discontinued.

Chambers, S. 2003. Deliberative Democratic Theory. *Annual Review of Political Science* 6: 307-26.

Chapin, Mac. 2004. A Challenge to Conservationists. *World Watch Magazine* (WorldWatch Institute), November-December, 17-31.

Chess, Caron, and Kristen Purcell. 1999. Public Participation and the Environment: Do We Know What Works? *Environmental Science and Technology* 33, 16: 2685-92.

Chiefs of Ontario. 1998. Chiefs Reject Ontario Lands for Life Policy. Press release, 31 July. http://www.chiefs-of-ontario.org/bulletins/bt_july31-98.html.

Christiansen Ruffman, Linda. 2002. Atlantic Canadian Coastal Communities and the Fisheries Trade: A Feminist Critique, Revaluation and Revisioning. *Canadian Woman Studies / les cahiers de la femme* 21-22, 1-4 (Spring-Summer): 56-63.

Christoff, Peter. 1996. Ecological Citizens and Ecologically Guided Democracy. In *Democracy and Green Political Thought*, ed. Brian Doherty and Marius de Geus, 151-69. London: Routledge.

City of Hamilton. 2003. Court Order Brings End to Greenhill Avenue Occupation. Press release, 15 August. http://www.myhamilton.ca/NR/rdonlyres/D8F60DA3-0C10-483D-98DF-56D908DCE2D3/0/RHVPMediaReleaseAug152003.pdf.

–. 2004. Joint Stewardship Board between the Six Nations Community and the City of Hamilton, Public Works Report PW05093, 20 June. http://www.myhamilton.ca/NR/rdonlyres/AA666A48-6EC2-462B-B521-DBC5E890D619/0/PW05093.pdf.

City of Toronto. 1998-2004a. Key Facts on Rental Housing in Toronto. http://www.city.toronto.on.ca/torontoplan/op2_fact_sheet_rental.htm Accessed 23 May 2004; website now discontinued.

–. 1998-2004b. Preserving Rental Housing. http://www.city.toronto.on.ca/torontoplan/op2_housing.htm. Accessed 23 May 2004; page now discontinued.

–. 1998-2004c. Toronto Plan – Official Consultation. http://www.city.toronto.on.ca/torontoplan/consult.htm; website now discontinued.

–. 2001. A Land Use Strategy for Toronto. *Toronto Plan Newsletter* 5 (June). http://www.city.toronto.on.ca/torontoplan/toronto_future.htm. Accessed 22 May 2004; website now discontinued.

–. 2002. City of Toronto Official Plan (November, 2002). Toronto: City of Toronto. http://www.city.toronto.on.ca/torontoplan/htm. Accessed May 2004; website now discontinued.

Clairmont, Susan. 2003. New Deputy Police Chief Kept Peace during Red Hill 'Journey.' *Hamilton Spectator*, 9 December, A12.

Clark, F., and D. Illman. 2001. Dimensions of Civic Science. *Science Communication* 23, 1: 5-27.

Coastal Community Network. 2002. The State of BC's Coastal Communities: A Decade of Change, 1993 – 2002. Paper presented at the "Tenth Anniversary Conference of Coastal Communities," Port Alberni, BC, 2-4 May.

Coates, John. 2003. *Ecology and Social Work: Toward a New Paradigm*. Halifax: Fernwood.

Common Front in Defense of Poor Neighbourhoods. 1999. *Parkdale Tenants and Rezoning*. Toronto: Parkdale Legal Services.

Confédération des syndicats nationaux (CSN). 2007. Olymel: La CSN réagit à la stratégie gouvernementale. Press release, 2 February.

Conklin, B.A., and L.R. Graham. 1995. The Shifting Middle Ground: Amazonian Indians and Eco-politics. *American Anthropologist* 97, 4: 695-710.

Connelly, James. 2006. The Virtues of Environmental Citizenship. In *Environmental Citizenship*, ed. Andrew Dobson and Derek Bell, 49-74. Cambridge, MA: MIT Press.

Connor, D.M. 1996. Public Participation in Canada: Development, Current Status and Trends. *Interact: The Journal of Public Participation* 2, 1: 1-5.

Copes, P. 1998. Adverse Impacts of Individual Quota Systems in Small-Scale Fisheries: What Are Positive Alternatives? In *Managing Our Fisheries, Managing Ourselves*, ed. L. Loucks, T. Charles, and M. Butler, 23-30. Halifax: Gorsebrook Research Institute for Atlantic Canada Studies.

CORE (Commission on Resources and Environment). 1995. *The Provincial Land Use Strategy*. Vol. 3, *Public Participation*. Victoria: Government of British Columbia.

Cortner, Hanna J., and Margaret A. Moote. 1999. *The Politics of Ecosystem Management.* Washington, DC: Island Press.

Cosgrove, William J. 2005. Démocratie de participation: où en sommes-nous? Opening comments to the joint congress of the Association québécoise pour l'évaluation d'impacts and the Ordre des urbanistes du Québec on the theme "La participation citoyenne au développement durable," Boucherville, QC, 16 November. http://www.bape.gouv.qc.ca/sections/rapports/conferences/05-11-16_AQEI-Cos.pdf.

Côté, Gilles. 2004. La participation des acteurs sociaux à l'évaluation et au suivi des impacts environnementaux et sociaux: le cas du complexe industriel d'Alcan à Alma. PhD diss., Université du Québec à Chicoutimi and Université du Québec à Rimouski.

Court of Queen's Bench of New Brunswick. 1997. *River Road Action Team v. New Brunswick (Minister of Environment)* CanLII 9476 (NB Q.B.).

Cox, Sarah K. 2004. *Diminishing Returns: An Investigation into the Five Multinational Corporations That Control British Columbia's Salmon Farming Industry.* Victoria: Raincoast Conservation Society.

Cronon, William. 1995. The Trouble with Wilderness or Getting Back to the Wrong Nature. In *Uncommon Ground: Rethinking the Human Place in Nature,* ed. William Cronon, 69-90. New York: W.W. Norton.

Cruikshank, Ken, and Nancy Bouchier. 2004. Blighted Areas and Obnoxious Industries: Constructing Environmental Inequality on an Industrial Waterfront, Hamilton, Ontario, 1890-1960. *Environmental History* 9, 3: 436-63.

CSA (Canadian Standards Association). 2002. CAN/CSA-Z809-02, Sustainable Forest Management: Requirements and Guidance. Mississauga. http://www.csa-intl.org/onlinestore/GetCatalogItemDetails.asp?mat=000000000002415908.

Curran, Deborah, and M. Leung. 2000. *Smart Growth: A Primer.* Victoria: Smart Growth BC and the Eco-Research Chair on Environmental Law and Policy, University of Victoria.

Curtis, M. 1997. Ministry Shuts Down Fish Farm. *Victoria Times Colonist,* 8 August, 1.

Dahl, R. 1989. *Democracy and Its Critics.* New Haven: Yale University Press.

Dalby, S. 2002. *Environmental Security: The Geopolitics of Colonizing Nature.* Minneapolis: University of Minnesota Press.

Dalby, S., and F. Mackenzie. 1997. Reconceptualizing Local Community: Environment, Identity and Threat. *Area* 29, 2: 99-108.

Dale, Ann. 2001. *At the Edge: Sustainable Development in the 21st Century.* Vancouver: UBC Press.

Danard, Susan. 1999. Fish Farm Expansion Upsets Some Native Leaders. *Victoria Times Colonist,* 20 October, A3.

Davidson, D.J., and N.A. MacKendrick. 2004. All Dressed Up with Nowhere to Go: The Discourse of Ecological Modernization in Alberta. *Canadian Review of Sociology and Anthropology* 41, 1: 47-65.

Davis, A. 1991. Insidious Rationalities: The Institutionalisation of Small Boat Fishing and the Rise of the Rapacious Fisher. http://www.stfx.ca/research/gbayesp/insidious_report.htm.

Davis, A., and S. Jentoft. 2001. The Challenge and the Promise of Indigenous Peoples' Fishing Rights – From Dependency to Agency. *Marine Policy* 25, 3: 223-37.

Davis, Judy S., Arthur C. Nelson, and Kenneth J. Dueker. 1994. The New Burbs: The Exurbs and Their Implications for Planning Policy. *American Planning Association Journal* 60,1: 45-59.

Day, Bob. 2002. Parks Not Solution. *Toronto Globe and Mail,* 8 January, A12.

De Angelis, Massimo. 2004. Separating the Doing and the Deed: Capital and the Continuous Character of Enclosures. *Historical Materialism* 12, 2: 57-87.

de la Harpe, Derek, Peter Fernhead, George Hughes, Richard Davies, Anna Spencely, Jonathan Barnes, Jenny Cooper, and Brian Child. 2004. Does 'Commercialization' of Protected Areas Threaten Their Conservation Goals? In *Parks in Transition: Biodiversity, Rural Development and the Bottom Line,* ed. Brian Child, 189-210. London: Earthscan.

De Sousa, Christopher A. 2003. Turning Brownfields into Green Space in the City of Toronto. *Landscape and Urban Planning* 62: 181-98.

Dear, M.J., J.J. Drake, and L.G. Reeds, eds. 1987. *Steel City: Hamilton and Region.* Toronto: University of Toronto Press.

Dearden, Philip, and Jessica Dempsey. 2004. Protected Areas in Canada: Decade of Change. *Canadian Geographer / Le Géographe canadien* 48, 2: 225-39.

Deleuze, G., and F. Guattari. 1987. *A Thousand Plateaus: Capitalism and Schizophrenia.* Trans. B. Massumi. Minneapolis: University of Minnesota Press.

Deloria, Vine. 1983. Circling the Same Old Rock. In *Marxism and Native Americans,* ed. W. Churchill, 113-36. Boston: South End.

Deslauriers, Pierre. 1995. Zonage agricole et utilisation du sol en milieu rural. Quelques exemples québécois. In *The Sustainability of Rural Systems: Le développement durable de systèmes ruraux,* ed. R.C. Bryant and C. Marois, 168-78. Montreal: Département de géographie, Université de Montréal.

Devall, William, and George Sessions. 1985. *Deep Ecology: Living as If Nature Mattered.* Salt Lake City: Peregrine Smith.

Di Chiro, Giovanna. 1995. Nature as Community: The Convergence of Environment and Social Justice. In *Uncommon Ground: Rethinking the Human Place in Nature,* ed. W. Cronon, 298-320. New York: W.W. Norton.

Dion, Stéphane. 2005a. Canada's Project Green. Speaking Notes for the Honourable Stéphane Dion, P.C., M.P., Minister of the Environment, Vancouver Board of Trade, Vancouver, BC, 19 September. http://www.ec.gc.ca/media_archive/minister/speeches/2005/050919_s_e. htm.

–. 2005b. Cutting Megatonnes of GHGs: Making Megaprofits. Speaking Notes for the Honourable Stéphane Dion, P.C., M.P., Minister of the Environment. 15 September. http://www. ec.gc.ca/minister/speeches/2005/050915_s_e.htm.

Dobson, Andrew. 1996. Democratising Green Theory: Preconditions and Principles. In *Democracy and Green Political Thought,* ed. Brian Doherty and Marius de Geus, 132-50. London: Routledge.

–. 2003. *Citizenship and the Environment.* Oxford: Oxford University Press.

–. 2006. Ecological Citizenship: A Defence. *Environmental Politics* 15, 3 (June): 447-51.

Dobson, Andrew, and Derek Bell, eds. 2006. *Environmental Citizenship.* Cambridge, MA: MIT Press.

Donnelly, Peter. 1993. The Right to Wander: Issues in the Leisure Use of Countryside and Wilderness Areas. *International Review for the Sociology of Sport* 28, 2-3: 187-292.

Dorcey, A., and T. McDaniels. 2001. Great Expectations, Mixed Results: Trends in Citizen Involvement in Canadian Environmental Governance. In *Governing the Environment: Persistent Challenges, Uncertain Innovations,* ed. E.A. Parson, 247-302. Toronto: University of Toronto Press.

Dorf, Michael C., and Charles F. Sabel. 1998. A Constitution of Democratic Experimentalism. *Columbia Law Review* 98, 2: 267-473.

Downes, David. 1996. Neo-corporatism and Environmental Policy. *Australian Journal of Political Science* 31, 2 (July). http://www.tandf.co.uk/journals/carfax/10361146.html.

Downey, R.K., and H. Beckie. 2002. Isolation Effectiveness in Canola Pedigree Seed Production. Internal Research Report, Agriculture and Agri-Food Canada, Saskatoon Research Centre, Saskatoon.

Draper, D. 1998. *Our Environment: A Canadian Perspective.* Toronto: Nelson Thomson.

Draper, Dianne, and Bruce Mitchell. 2001. Environmental Justice Considerations in Canada. *Canadian Geographer / Le Géographe canadien* 45, 1: 93-98.

Drews, Keven. 2008. No Fish Farm Renewal; Wild Stocks at "Alarming Levels." Westcoaster. ca, 21 January. http://www.westcoaster.ca/modules/AMS/article.php?storyid=3476.

Dreyfus, H.L., and P. Rabinow. 1983. *Michel Foucault: Beyond Structuralism and Hermeneutics.* Chicago: University of Chicago Press.

Drushka, K., B. Nixon, and R. Travers, eds. 1993. *Touch Wood: BC Forests at the Crossroads.* Madeira Park, BC: Harbour.

Dryzek, John S. 1995a. Political and Ecological Communication. *Environmental Politics* 4, 4: 13-30.

–. 1995b. Strategies of Ecological Democratization. In *Democracy and the Environment: Problems and Prospects*, ed. William M. Lafferty and James Meadowcroft, 108-23. Cheltenham, UK: Edward Elgar.

–. 1999. Global Ecological Democracy. In *Global Ethics and Environment*, ed. N. Law, 264-82. London: Routledge.

–. 2000. *Deliberative Democracy and Beyond: Liberals, Critics, Contestations*. Oxford: Oxford University Press.

–. 2005. Deliberative Democracy in Divided Societies: Alternatives to Agonism and Analgesia. *Political Theory* 33, 2 (April): 218-42.

Ducharme, Jeff. 2006. Trash Talk: Lifespan of Landfill in Dispute. *Saint John Telegraph Journal*, 17 April, B2.

Duderstadt, J.J. 2000. *A University for the 21st Century*. Chicago: University of Michigan Press.

Dufresne, Denis. 2002. Québec a cédé devant la 'psychose porcine.' La Fédération des producteurs de porcs du Québec dénonce les mesures adoptées par le MEF. *La Tribune* (Sherbrooke), 15 June, A12.

Duhamel, Alain. 2007. Un marché commun pour les producteurs de porcs? *Les Affaires* (Montreal), 8 February.

Dumais, Mario. 2006. L'agriculture intense est nécessaire. *Le Devoir* (Montreal), 7 March.

Dunlop, Claire. 2000. GMOs and Regulatory Styles. *Environmental Politics* 9, 2: 149-55.

Eckersley, Robyn. 1992. *Environmentalism and Democratic Theory: Toward an Ecocentric Approach*. Albany: State University of New York Press.

–. 1996. Greening Liberal Democracy: The Rights Discourse Revisited. In *Democracy and Green Political Thought*, ed. Brian Doherty and Marius de Geus, 212-36. London: Routledge.

–. 2004a. Ecocentric Discourses: Problems and Prospects for Nature Advocacy. *Tamkang Review* 34, 3-4 (Spring-Summer): 155-86.

–. 2004b. *The Green State (Rethinking Democracy and Sovereignty)*. Cambridge, MA: MIT Press.

Ecotrust Canada and Ecotrust. 2004. Catch-22: Conservation, Communities and the Privatization of B.C. Fisheries – An Ecological, Social, and Economic Impact Study. Vancouver, November. http://www.ecotrust.ca/files/Catch-22-November2004_0.pdf.

Edge, Marc. 2007. Convergence and the 'Black News Hole': Canadian Newspaper Coverage of the 2003 Lincoln Report. *Canadian Journal of Media Studies* 2, 1 (April): 76-108.

Edwards, Paul. 1990. *The Al-Pac Review Hearings: A Case Study*. Edmonton: Environmental Law Centre.

EEC (Council of the European Economic Community). 1990. Council Directive 90/220/EC on the Deliberate Release into the Environment of Genetically Modified Organisms. *Official Journal of the European Communities* (8 May) L 117/15.

Eisner, M.A. 2004. Corporate Environmentalism, Regulatory Reform, and Industry Self-Regulation: Toward Genuine Regulatory Reinvention in the United States. *Governance* 17: 145-67.

Environment Canada. 2008. Introduction. Freshwater Website. http://www.ec.gc.ca/water/en/info/pubs/mountain/e_intro.htm.

Epstein, Barbara. 1999. Grassroots Environmental Activism: The Toxics Movement and Directions for Social Change. In *Earth, Air, Fire, Water*, ed. J.K. Conway, K. Keniston, and L. Marx, 170-83. Boston: University of Massachusetts Press.

EQ (Environnement Québec). 2003. *Synthèse des informations environnementales disponibles en matière agricole au Québec*. Quebec City: Ministère de l'Environnement, Gouvernement du Québec.

–. 2005. Rapport annuel de gestion 2004-2005. Ministère de Développement durable, de l'Environnement et des Parcs, Gouvernement du Québec. http://www.mddep.gouv.qc.ca/ministere/rapports_annuels/rapport_2004-2005.pdf.

Escobar, Arturo. 1995. Imagining a Post-development Era. In *Power of Development*, ed. J. Crush, 211-27. London: Routledge.

Esping-Andersen, Gøsta. 1999. *Social Foundations of Post-industrial Economies*. Oxford: Oxford University Press.

European Council and Commission. 1999. Minutes of the Council Meeting 24-25 June. Annex: Declaration regarding the Proposal to Amend Directive 90/220/EEC on Genetically Modified Organisms. Brussels, 24-25 June.

Fairley, B., C. Leys, J. Sacouman, and R. Williams. 1990. Restructuring and Resistance in Atlantic Canada: An Introduction. In *Restructuring and Resistance from Atlantic Canada*, ed. B. Fairley, C. Leys, and J. Sacouman. Toronto: Garamond Press.

Fallis, George. 2004. The Mission of the University. Paper submitted to the Postsecondary Review: Higher Expectations for Higher Education, Government of Ontario.

–. 2007. *Multiversities, Ideas and Democracy*. Toronto: University of Toronto Press.

Federation of Canadian Municipalities. 2005. *Building Capacity through Communication: Municipal-Aboriginal Partnerships in Land Management*. Book 1, *Project Overview*. Ottawa: Federation of Canadian Municipalities.

Feinberg, J. 1996. Unusual Protest Had One Message: FAN Go Home. *Hagensborg Coast Mountain News*, 9 May, 1, 3.

Fernandez, M.R., F. Selles, R.M. DePauw, and R.P. Zetner. 2005. Crop Production Factors Associated with Fusarium Head Blight in Spring Wheat in Eastern Saskatchewan. *Crop Science* 45: 1908-16.

Filion, Pierre. 2001. The Urban Policy-Making and Development Dimension of Fordism and Post-Fordism: A Toronto Case Study. *Space and Polity* 5, 2: 85-111.

–. 2003. Towards Smart Growth? The Difficult Implementation of Alternatives to Urban Dispersion. *Plan Canada* 12, 1: 48-70.

Fine, G.A., and B. Harrington. 2004. Tiny Publics: Small Groups and Civil Society. *Sociological Theory* 22, 3: 341-56.

Fine, Robert. 1999. Benign Nationalism? The Limits of the Civic Ideal. In *People, Nation and State: The Meaning of Ethnicity and Nationalism*, ed. Edward Mortimer and Robert Fine, 149-61. London: I.B. Tauris.

Fischer, F. 2000. *Citizens, Experts and the Environment: The Politics of Local Knowledge*. Durham: Duke University Press.

Fiske, J. 1990. Native Women in Reserve Politics: Strategies and Struggles. In *Community Organization and the State in Canada*, ed. R. Ng, G. Walker, and J. Muller, 131-46. Toronto: Garamond Press.

Flamborough Chamber of Commerce. 2005. Advocacy Group Formed to Promote New Corridor. *Bottom Line* (Flamborough Chamber of Commerce newsletter), September, 5-6.

Flanagan, William G. 2002. *Urban Sociology: Images and Structure*. Boston: Allyn and Bacon.

Fluet, Colette. 2003. The Involvement of Aboriginal Groups and Environmental Organizations in a Regional Planning Strategy for the Northern East Slopes of Alberta. Master's thesis, University of Alberta.

Fong, Eric. 1996. A Comparative Perspective on Racial Residential Segregation: American and Canadian Experiences. *Sociological Quarterly* 37, 2: 199-226.

Fonseca, José. 2002. *Complexity and Innovation in Organizations*. New York: Routledge.

Foot, Richard. 2007. Land Claims Frustrations Lead to National Day of Protest: Aboriginals Want Government to Address Nearly 900 Outstanding Grievances. *Edmonton Journal*, 24 June, A2.

Ford, Jennifer, and Ransom Myers. 2008. A Global Assessment of Salmon Aquaculture Impacts on Wild Salmonids. *PLoS Biology* 6, 2 (12 February). http://biology.plosjournals. org/perlserv/?request=get-document&doi=10.1371/journal.pbio.0060033&ct=1.

Ford, Marcus. 2002. *Beyond the Modern University: Towards a Constructive Post-modern University*. Westport: Praeger.

Forest Action Network. 1995. *Forest Action Network Newsletter and Action Log* 1 (May).

–. 1996. *Yellow-Ribbon Protest Video*.

Forester, J. 1980. Critical Theory and Planning Practice. *Journal of the American Planning Association* 46: 275-84.

Foster, Sheila. 2002. Environmental Justice in an Era of Devolved Collaboration. In *Justice and Natural Resources*, ed. K.M. Mutz, G.C. Bryner, and D.S. Kenney, 139-60. Washington, DC: Island Press.

Fothergill, Ed. 2003. Cases for and Against: Vital Link – Expressway Is Crucial to Building Hamilton's Future Prosperity. *Hamilton Spectator,* 25 October, FO9.

Foucault, Michel. 1977. Nietzsche, Genealogy, History. In *Language, Counter-Memory, Practice: Selected Essays and Interviews by Michel Foucault,* ed. D. Bouchard, 139-64. Ithaca: Cornell University Press.

–. 1980. *Power/Knowledge: Selected Interviews and Other Writings 1972-1977/ Michel Foucault,* ed. Colin Gordon. New York: Pantheon Books.

–. 1983. Afterword: The Subject and Power. In *Michel Foucault: Beyond Structuralism and Hermeneutics,* ed. H.L. Dreyfus and P. Rabinow, 208-26. Chicago: University of Chicago Press.

–. 2003. *Society Must Be Defended: Lectures at the Collège de France, 1975-76.* New York: Picador.

Found, Jason, and R. Michael M'Gonigle. 2005. *UnCommon Ground: Creating Complete Community at the University of Victoria.* Victoria: POLIS Project on Ecological Governance.

Four Winds and Associates. 2000. Northern East Slopes Integrated Resource Management Strategy: Report on the Aboriginal Consultation Process. Edmonton.

–. 2002. Appendix I: Community Consultations. Edmonton.

Fox, Warwick. 1990. *Toward a Transpersonal Ecology: Developing New Foundations for Environmentalism.* Boston: Shambhala.

FPPQ (Fédération des producteurs de porcs du Québec). 2001. Rapport sommaire de recherche: Étude de perceptions pour la valorisation de la profession de producteur de porcs. Léger Marketing, Montreal, May.

–. 2005. Pour une production porcine viable et durable. A brief submitted in the context of the public consultations on the Quebec Sustainable Development Plan, Saint-Georges-de-Beauce, QC, 25 April.

–. 2007. Retombées économiques. http://www.leporcduquebec.qc.ca/fppq/prod-4.html.

Francis, Wendy. 2001. *An Analysis of Alberta's Northern East Slopes Sustainable Resource and Environmental Management Strategy.* Calgary: Canadian Parks and Wilderness Society.

Francoeur, Louis-Gilles. 1995. Les agriculteurs sont-ils à l'abri de la loi sur l'environnement? *Le Devoir* (Montreal), 23 August, A1.

–. 1997. Pollution agricole: Une plainte déposée contre Québec en vertu de ALENA. *Le Devoir* (Montreal), 10 April, A4.

–. 2005. Les tas de fumier resteront dans les champs. *Le Devoir* (Montreal), 28 September, A1.

–. 2006a. Le ministère de l'Environnement en désaccord avec Nathalie Normandeau. *Le Devoir* (Montreal), 23 June, A2.

–. 2006b. La Loi sur la protection du territoire agricole nuirait aux régions. *Le Devoir* (Montreal), 23 June, A5.

Fraser, Keith. 1997. Natives Ready to Dismantle Fish Farm. *Vancouver Province,* 28 February, A12.

Fraser, Nancy. 1994. After the Family Wage: Gender Equity and the Welfare State. *Political Theory* 22, 4 (November): 591-618.

–. 1997. *Justice Interruptus: Critical Reflections on the "Postsocialist" Condition.* New York: Routledge.

Freeman, Bill. 2001. *Hamilton: A People's History.* Toronto: James Lorimer.

Freeze, R. Allan, and John A. Cherry. 1989. What Has Gone Wrong? *Groundwater* 27, 4: 458-64.

Friedmann, John. 1993. *Empowerment: The Politics of Alternative Development.* Malden, MA: Blackwell.

Friends of Clayoquot Sound. 2003. Pacific National Aquaculture Charged with Irresponsible Fish Farming. Press release, 3 January.

–. 2008. Cermaq (Mainstream Canada) in Clayoquot Sound. http://www.focs.ca/fishfarming/cermaq.asp.

Friends of the Lubicon. 2007. Chief Rebuts Prentice's Latest Upside-Down Backwards Interpretation of Canadian History, Law and Justice, 6 June. http://www.tao.ca/~FOL/pa/negp/po06/po070606.htm.

Friesen, L.F., A.G. Nelson, and R.C. Van Acker. 2003. Evidence of Contamination of Pedigreed Canola *(Brassica napus)* Seedlots in Western Canada with Genetically Engineered Herbicide Resistant Traits. *Agronomy Journal* 95: 1342-47.

FRSWC (Fundy Region Solid Waste Commission). 2003. Our Fundy Region's History. http://www.fundyrecycles.com/ourhistory.html. Accessed 24 October 2003; page now discontinued.

–. 2005. Commission: History. http://www.fundyrecycles.com/index.php?id=9#fundyregion.

FSC (Forest Stewardship Council of Canada). 2008. Forest Management Standards. http://www.fsccanada.org.

Fulton, James. 1996. National Parks Abandoned: Ottawa's Budget Plans Means No New Parks, Less Protection for Existing Ones. *Alternatives* 22, 4: 9.

Fung, Archon, and Erik Olin Wright, eds. 2003. *Deepening Democracy: Institutional Innovations in Empowered Participatory Governance.* London: Verso.

Funtowicz, S., and J.R. Ravetz. 1985. Three Types of Risk Assessment. In *Risk Analysis in the Private Sector,* ed. C. Whipple and V.T. Covello, 217-31. New York: Plenum.

–. 1993. Science for the Post-normal Age. *Futures* 25, 7: 739-57.

–. 1994. Emergent Complex Systems. *Futures* 26, 6: 568-82.

Furtan, W.H., R.S. Gray, and J.J. Holzman. 2002. The Optimal Time to License a Biotech 'Lemon.' Department of Agricultural Economics, University of Saskatchewan, Saskatoon.

G.E. Bridges and Associates. 1994. *Socio-economic Assessment of Timber Supply Scenarios: Mid Coast TSA.* Victoria: BC Ministry of Forests, Economics and Trade Branch.

Gagné, Jean-Charles. 1996a. Le gouvernement évite la grande manifestation. *La Terre de chez nous* (Longueuil), 16 October.

–. 1996b. Les pressions de l'UPA secouent le gouvernement Bouchard. *La Terre de chez nous* (Longueuil), 3-9 October.

Gale, John. 1997. *Review of Environment Impact Statement: Regional Landfill at Crane Mountain.* Torbay, NL: Fracflow Consultants.

Gallaugher, Brian, and G. Fred Lee. 1997. *Review of Potential Public Health, Groundwater Resource, Financial, and Other Impacts of the Crane Mountain Landfill.* El Macero, CA: G. Fred Lee and Associates.

Galloway, Gloria. 2007. Ottawa Tells Federal Bodies to Reprioritize Spending. *Toronto Globe and Mail,* 31 August, A4.

George, Susan. 1998. Preface. In *Privatizing Nature: Political Struggles for the Global Commons,* ed. Michael Goldman, ix-xiv. New Brunswick, NJ: Rutgers University Press.

Georgia-Pacific Canada and the Unama'ki Mi'kmaq Communities. 1998. *Melford Gypsum Mine Agreement.* Sydney, NS: Union of Nova Scotia Indians.

Gertler, Michael E. 1999. Sustainable Communities and Sustainable Agriculture on the Prairies. In *Communities, Development and Sustainability across Canada,* ed. J.T. Pierce and A. Dale, 121-39. Vancouver: UBC Press.

Gibson, Chris. 1999. Cartographies of the Colonial/Capitalist State: A Geopolitics of Indigenous Self-Determination in Australia. *Antipode* 31, 1: 45-79.

Gibson, Robert B., Donald H.M. Alexander, and Ray Tomalty. 1997. Putting Cities in Their Place: Ecosystem-Based Planning for Canadian Urban Regions. In *Eco-City Dimensions: Healthy Communities, Healthy Planet,* ed. M. Roseland, 25-39. Gabriola Island, BC: New Society.

Gilbert, Liette, and Catherine Phillips. 2003. Practices of Urban Environmental Citizenships: Rights to the City and Rights to Nature in Toronto. *Citizenship Studies* 7, 3: 313-30.

Gilbert, Liette, L. Anders Sandberg, and Gerda R. Wekerle. Forthcoming. Building Bioregional Citizenship: The Case of the Oak Ridges Moraine, Ontario, Canada. *Local Environment.*

Gilbert, Liette, Gerda R. Wekerle, and L. Anders Sandberg. 2005. Local Responses to Development Pressures: Conflictual Politics of Sprawl and Environmental Conservation. *Cahiers de Geographie du Quebec* 49, 138: 377-92.

Gilbertson, Michael. 2005. Public Engagement in Implementation. International Joint Commission Web Dialogue. Intervention posted 2 December 2005. http://www.webdialogues.net/cs/ijc-greatlakes-participants/view/di/77?x-t=participants&ln=g.

Gill, A., and M. Reed. 1999. Incorporating Postproductivist Values into Sustainable Community Processes. In *Communities, Development and Sustainability across Canada*, ed. J.T. Pierce and A. Dale, 166-89. Vancouver: UBC Press.

Gillespie, Kerry. 2003. Sprawl Costs Billions, Sierra Says. *Toronto Star*, 29 July.

GIWGGMW (Grain Industry Working Group on Genetically Modified Wheat). Conditions for the Introduction of Genetically Modified Wheat. Discussion Document (Draft 5) (on file with Peter Andrée).

Glaser, B.G., and A.L. Strauss. 1967. *The Discovery of Grounded Theory.* New York: Aldine.

Glazer, Penina, and Myron Glazer. 1998. *The Environmental Crusaders: Confronting Disaster and Mobilizing Community.* University Park: Pennsylvania State University Press.

Glossop, Erica. 2003. Public Grills HighRes Condo Developers about Planned Towers. *Parkdale Liberty,* June, 3.

GLSL (Great Lakes–St. Lawrence Round Table). 1997a. Round Table Minutes. Record of Public Consultation. 30 September. Ministry of Natural Resources (digital record of all GLSL Round Table minutes on file with Patricia Ballamingie).

–. 1997b. Record of Public Consultations. 11 November.

–. 1998a. Record of Public Consultations. 13 January.

–. 1998b. Record of Public Consultations. 9 June.

Goldberg, David. 1994. *Multiculturalism: A Critical Reader.* Cambridge, MA: Blackwell.

Goldstone, Jack. A. 2003. Bridging Institutionalized and Non-institutionalized Politics. In *States, Parties and Social Movements,* ed. J.A. Goldstone, 1-26. Cambridge: Cambridge University Press.

Goodman, David, Bernardo Sorj, and John Wilkinson. 1987. *From Farming to Biotechnology: A Theory of Agro-industrial Development.* Oxford: Blackwell.

Gordon, K. 2004. *The Slocan: Portrait of a Valley.* Winlaw, BC: Sono Nis Press.

Gordon S. Gislason and Associates. 1998. Fishing for Money: Challenges and Opportunities in the BC Salmon Fishery. Final Report. Prepared for the BC Job Protection Commission, June.

Gosine, Andil. 2003. Myths of Diversity. *Alternatives Journal* 29, 1: 12-15.

Gower, J. 1990. The Impact of Alternative Ideology on Landscape: The Back-to-the-Land Movement in the Slocan Valley. Master's thesis, Department of Geography, University of British Columbia, Vancouver.

Graf, William. 1995. The State in the Third World. In *Why Not Capitalism? Socialist Register 1995,* ed. L. Panitch, 140-62. London: Merlin.

Gramsci, Antonio. 1971. *Selections from the Prison Notebooks.* Ed. Quintin Hoare and Geoffrey Nowell Smith. New York: International.

Green, J. 1995. Towards a Detente with History: Confronting Canada's Colonial Legacy. *International Journal of Canadian Studies* 12: 85-105.

Green, Kenneth, and Sylvia LeRoy. 2005. Let the Free Market Save Canada's Parks. *National Post* (Toronto), 12 May, A26.

Greenpeace. 2003. GM Wheat. http://www.greenpeace.ca/e/campaign/gmo/depth/wheat/archive.php. Accessed 18 July 2005; page now discontinued.

Greenpeace International. 2008a. Great Barrier Reef Saved from Shale Oil Exploitation. http://www.greenpeace.org/international/news/barrier-reef-shale-oil-victory250808.

–. 2008b. US Congress Announces Ban on Toxic Chemicals. http://www.greenpeace.org/international/news/congress-announces-ban-on-toxi.

Grenier, M. 2002. A Discussion Paper on Agronomic Assessment of Roundup Ready Wheat. Canadian Wheat Board.

Grima, A. 1985. Participatory Rites: Integrating Public Involvement in EIA. In *Environmental Impact Assessment: The Canadian Experience,* ed. J. Whitney and V. MacLaren, 33-52. Toronto: University of Toronto Institute of Environmental Studies.

Grindstaff, Laura. 2002. *The Money Shot: Trash, Class, and the Making of TV Talk Shows.* Chicago: University of Chicago Press.

Gunton, T. 1998. Forestry Land Use Policy in British Columbia: The Dynamics of Change. *Environments* 25, 2-3: 8-13.

Gurin, David. 2003. *Understanding Sprawl: A Citizen's Guide.* Vancouver: David Suzuki Foundation.

Gutmann, Amy, and Dennis Thompson. 2004. *Why Deliberative Democracy?* Princeton: Princeton University Press.

Habermas, Jürgen. 1984. *The Theory of Communicative Action.* Vol. 1, *Reason and the Rationalization of Society.* Boston: Beacon Press.

–. 1987. *The Theory of Communicative Action.* Vol. 2, *System and Lifeworld.* Boston: Beacon Press.

–. 1996. *Between Facts and Norms: Contributions to a Discourse Theory of Law and Democracy.* Cambridge: Polity Press.

Hagensborg Coast Mountain News. 1995a. Illegal Blockade on King Island. 28 September, 5.

–. 1995b. Large Force of RCMP Prepare to Move into Fog Creek Area. 28 September, 5.

–. 1995c. A View from the Woods. Forestry and Fog Creek. 28 September, 5.

Halseth, G., and A. Booth. 2003. "What Works Well; What Needs Improvement": Lessons in Public Consultation from British Columbia's Resource Planning Processes. *Local Environment* 8, 4: 437-55.

Hammersley, Martyn, and Paul Atkinson. 1995. *Ethnography: Principles in Practice.* New York: Routledge.

Hammond, H. 1991. *Seeing the Forest among the Trees: The Case for Wholistic Forest Use.* Vancouver: Polestar Book Publishers.

Hanen, Marsha, Tom Austin, and Eric Higgs. 2003. Planning Possibilities: Pathways for Constructive Change. Report of the Planning Review Team, University of Victoria, November. http://web.uvic.ca/vpfin/campusplan/cdcreview/cdcreview-nov03.html.

Hanna, K. 2000. The Paradox of Participation and the Hidden Role of Information: A Case Study. *Journal of the American Planning Association* 66, 4: 398-410.

Hannesson, R. 2004. *The Privatization of the Oceans.* Cambridge, MA: MIT Press.

Hanson, Jeff. 2000. *Privatization Opportunities for Washington State Parks.* Seattle, WA: Washington Policy Center. http://www.washingtonpolicy.org/ConOutPrivatization/PNParksPrivatization2000-01.html.

Haraway, Donna. 1990. A Manifesto for Cyborgs: Science, Technology, and Socialist Feminism in the 1980s. In *Feminism/Postmodernism,* ed. L. Nicholson, 190-233. New York: Routledge.

Harding, Harry. 2001. Can We Survive Our Tourism Department? *St. John's Evening Telegram,* 4 November, A13.

Hardt, M., and A. Negri. 2000. *Empire.* Cambridge, MA: Harvard University Press.

Harrigan, Daniel. 2006. Open Letter to Crane Mountain Host Community. *River Valley News* (Crane Mountain area), 13 April, 2.

Harrison, Kathryn. 1996. *Passing the Buck: Federalism and Canadian Environmental Policy.* Vancouver: UBC Press.

Harrison, Trevor. 1995. Making the Trains Run on Time: Corporatism in Alberta. In *The Trojan Horse: Alberta and the Future of Canada,* ed. Trevor Harrison and Gordon Laxer, 118-31. Montreal: Black Rose Books.

Harvey, David. 1996. *Justice, Nature and the Geography of Difference.* Oxford: Blackwell.

–. 1999. The Environment of Justice. In *Living with Nature: Environmental Politics as Cultural Discourse,* ed. F. Fischer and M.A. Hajer, 153-85. New York: Oxford University Press.

–. 2003. *The New Imperialism.* Oxford: Oxford University Press.

–. 2005. *A Brief History of Neoliberalism.* Oxford: Oxford University Press.

–. 2006. Neo-liberalism as Creative Destruction. *Geographic Annals* 88 B, 2: 145-58.

Ha-Shilth-Sa. 2005. Major Study to Examine Treaty. 8 September, 3.

Hayter, R. 2000. *Flexible Crossroads: The Restructuring of British Columbia's Forest Economy.* Vancouver: UBC Press.

Hayward, Tim. 2006a. Ecological Citizenship: Justice, Rights and the Virtue of Resourcefulness. *Environmental Politics* 15, 3: 435-46.

–. 2006b. Ecological Citizenship: A Rejoinder. *Environmental Politics* 15, 3: 452-53.

Head, L. 1990. Conservation and Aboriginal Land Rights: When Green Is Not Black. *Australian Natural History* 23, 6: 448-54.

Health Canada. 2008. Drinking Water Advisories. Ottawa. http://www.hc-sc.gc.ca/fnih-spni/promotion/water-eau/advis-avis_concern_e.html.

Hemson Consulting. 2003. *Providing Employment Land in Hamilton – Financial Options.* Toronto: Hemson Consulting.

Hessing, Melody, and Michael Howlett. 1997. *Canadian Natural Resource and Environmental Policy: Political Economy and Public Policy.* Vancouver: UBC Press.

Hipwell, W.T. 1997. 'They've Got No Stake in Where They're At': Radical Ecology, the Fourth World and Local Identity in the Bella Coola Region. Master's thesis, Department of Geography, Carleton University.

–. 2004a. A Deleuzian Critique of Resource-Use Management Politics in Industria. *Canadian Geographer / Le Géographe Canadien* 48, 3: 356-77.

–. 2004b. Political Ecology and Bioregionalism: New Directions for Geography and Resource-Use Management. *Journal of the Korean Geographical Society* 39, 5 (Series no. 104): 735-54.

–. 2007. The Industria Hypothesis. *Peace Review* 19, 3: 305-13.

Hoare, Quintin, and Geoffrey Nowell Smith. 1971. General Introduction. In *Selections from the Prison Notebooks,* ed. Quintin Hoare and Geoffrey Nowell Smith, xviii-xcvi. New York: International.

Hornborg, Alf. 1998. The Mi'kmaq of Nova Scotia: Environmentalism, Ethnicity and Sacred Places. In *Voices of the Land: Identity and Ecology in the Margins,* ed. A. Hornborg and M. Kurkiala, 135-71. Lund, Sweden: Lund University Press.

Horrigan, Leo, Robert S. Lawrence, and Polly Walker. 2002. How Sustainable Agriculture Can Address the Environmental and Human Health Harms of Industrial Agriculture. *Environmental Health Perspectives* 110, 5: 445-56.

Horton, Dave. 2006. Demonstrating Environmental Citizenship? A Study of Everyday Life among Green Activists. In *Environmental Citizenship,* ed. Andrew Dobson and Derek Bell, 127-50. Cambridge, MA: MIT Press.

Hospitality Newfoundland and Labrador. 1997. Hospitality Newfoundland and Labrador Presentation to the Honourable Paul Dicks, Minister of Finance. St. John's, 17 February.

House, J.D. 1999. *Against the Tide: Battling for Economic Renewal in Newfoundland and Labrador.* Toronto: University of Toronto Press.

Howlett, Michael, and Jeremy Rayner. 1995. Do Ideas Matter? Policy Network Configurations and Resistance to Policy Change in the Canadian Forest Sector. *Canadian Public Administration* 38, 3: 382-410.

Hudon, Raymond, Christian Poirier, and Stéphanie Yates. 2008. When Ecology Collides with Economy in Infrastructure Projects. *Network Industries Quarterly* 10, 1: 4-6.

Hughes, Elaine L., and David Iyalomhe. 1998-99. Substantive Environmental Rights in Canada. *Ottawa Law Review* 30: 229-58.

Hume, Stephen. 2004. Fishing for Answers. In *A Stain upon the Sea: West Coast Salmon Farming,* ed. S. Hume, A. Morton, B.C. Keller, R.M. Leslie, O. Langer, and D. Staniford, 17-77. Madeira Park, BC: Harbour.

–. 2008. The Hoof and Mouth Disease of the Salmon Farming Industry. *Vancouver Sun,* 7 April. http://www.canada.com/vancouversun/news/editorial/story.html?id=15f5e452-6d8f-4fed-ae54-a0b8917584f5.

Hutsul, Christopher. 2004. Calm: Yoga and Spas Where I Wanted to Be. *Toronto Star,* 18 April, B2.

Immen, Wallace. 1999. Showdown Brewing with Developers over Oak Ridges. *Toronto Globe and Mail,* 8 November, A11.

–. 2001a. Small Salamander Stands in Way of Road Construction. *Toronto Globe and Mail,* 24 October, A9.

–. 2001b. Environmentalists Call for Tougher Checks on Sprawl. *Toronto Globe and Mail,* 19 November, A15.

Immen, Wallace, and James Rusk (with a report from Murray Campbell). 2002. Toronto's Tent City Sealed Off, Squatters Ejected. *Toronto Globe and Mail,* 25 September, A1.

India. Department of Biotechnology. 1998. *Revised Guidelines for Research in Transgenic Plants and Guidelines for Toxicity and Allergenicity Evaluations of Transgenic Seeds, Plants and Plant Parts.* New Delhi: Ministry of Science and Technology, Government of India.

Irwin, A. 1995. *Citizen Science: A Study of People, Expertise and Sustainable Development.* London: Routledge.

Isin, Engin, and Myer Siemiatycki. 2000. Making Space for Mosques: Struggles for Urban Citizenship in Diasporic Toronto. In *Race, Space and the Law,* ed. S. Razack, 186-209. Toronto: Between the Lines.

Jacobs, Jane M. 1996. *Edge of Empire: Postcolonialism and the City.* London: Routledge.

Jacoby, Karl. 2001. *Crimes against Nature: Squatters, Poachers, Thieves, and the Hidden History of American Conservation.* Berkeley: University of California Press.

James, Michelle. 2003. Native Participation in British Columbia Commercial Fisheries – 2003. Prepared for Ministry of Agriculture, Food and Fisheries, Victoria, November.

Jeffs, A. 2003. Wheat Sales Won't Be Risked: Impact on Exports Must Be Studied – Agriculture Minister. *Edmonton Journal,* 13 March, H1.

Jentoft, S. 2000. The Community: A Missing Link of Fisheries Management. *Marine Policy* 24, 1: 53-59.

Jentoft, S., and B. McCay. 1995. User Participation in Fisheries Management – Lessons Drawn from International Experiences. *Marine Policy* 19, 3: 227-46.

Jerrett, Michael, Richard Burnett, Pavlos Kanaroglou, John Eyles, Norm Finkelstein, Chris Giovis, and Jeffery Brook. 2001. A GIS – Environmental Justice Analysis of Particulate Air Pollution in Hamilton, Canada. *Environment and Planning A* 33: 955-73.

Jessop, B. 1997. Capitalism and Its Future: Remarks on Regulation, Government and Governance. *Review of International Political Economy* 4, 3: 561-81.

–. 1999. Narrating the Future of the National Economy and the National State: Remarks on Remapping Regulation and Reinventing Governance. In *State/Culture: State-Formation after the Cultural Turn,* ed. G. Steinmetz, 378-405. Ithaca, NY: Cornell.

–. 2002. Liberalism, Neoliberalism and Urban Governance: A State-Theoretical Perspective. In *Spaces of Neoliberalism: Urban Restructuring in North America and Western Europe,* ed. N. Brenner and N. Theodore, 105-25. Malden, MA: Blackwell.

Kaplan, Gabriel E. 2004. Do Governance Structures Matter? *New Directions for Higher Education* 127: 23-39.

Katz, Cindi. 2001. Vagabond Capitalism and the Necessity of Social Reproduction. *Antipode* 33, 4: 709-28.

Kearney, John. 1998. Alternative Approaches to Management of the Fixed Gear Sector. In *Managing Our Fisheries, Managing Ourselves,* ed. L. Loucks, T. Charles, and M. Butler, 50-53. Halifax: Gorsebrook Research Institute for Atlantic Canada Studies.

–. 2005. Community-Based Fisheries Management in the Bay of Fundy: Sustaining Communities through Resistance and Hope. In *Natural Resources as Assets: Lessons from Two Continents,* ed. M. West Lyman and B. Child. Washington, DC: Aspen Institute and Sand County Foundation.

Keefer, Tom. 2007. The Politics of Solidarity: Six Nations, Leadership and the Settler Left. *Upping the Anti: A Journal of Theory and Action* 4. http://uppingtheanti.org/node/2728.

Kelly, Sandra. 1997. Making Choices. *St. John's Evening Telegram,* 4 March, 6.

Kennett, Steven A. 1999. *Towards a New Paradigm for Cumulative Effects Management.* Calgary: Canadian Institute of Resources Law.

Kennett, Steven, and Monique M. Ross. 1998. *In Search of Public Land Law in Alberta.* Calgary: Canadian Institute of Resources Law.

Kerans, Patrick, and John Kearney. 2006. *Turning the World Right-Side Up: Science, Community, and Democracy.* Halifax: Fernwood.

Kezar, Adrianna. 2004. What Is More Important to Effective Governance: Relationships, Trust, and Leadership, or Structures and Formal Processes? *New Directions for Higher Education* 127: 35-46.

Kipfer, Stefan, and Roger Keil. 2002. Toronto Inc? Planning the Competitive City in the New Toronto. *Antipode* 34, 2 (March): 227-45.

Klein, Naomi. 2000. *No Logo: Taking Aim at the Brand Bullies.* Toronto: Knopf Canada.

Kloppenberg, J. 1988. *First the Seed: The Political Economy of Plant Biotechnology.* New Haven: Yale University Press.

Krajnc, Anita. 2000. Wither Ontario's Environment? Neo-conservatism and the Decline of the Environment Ministry. *Canadian Public Policy* 26, 1: 111-27.

–. 2002. Conservation Biologists, Civic Science and the Preservation of BC Forests. *Journal of Canadian Studies* 37, 3: 219-38.

Krauss, Celene. 1994. Women of Color on the Front Line. In *Unequal Protection: Environmental Justice and Communities of Color,* ed. Robert D. Bullard, 256-71. San Francisco: Sierra Club Books.

Krkosek, Martin, Jennifer S. Ford, Alexandra Morton, Subhash Lele, Ransom A. Myers, and Mark A. Lewis. 2007. Declining Wild Salmon Populations in Relation to Parasites from Farm Salmon. *Science* 318 (14 December): 1772-75.

Krupp, Staci Jeanne. 2000. Environmental Hazards: Assessing the Risk to Women. *Fordham Environmental Law Journal* 12, 1: 111-39.

Laprade, Yvon. 2007. Bouchard parle de "victoire collective," Carbonneau d'un "gros dégât." Le Canal Argent (Groupe TVA), 15 February. http://lca.canoe.com/infos/quebec/archives/2007/02/20070215-064948.html.

Larose, Jean. 2007. Agriculture durable: la nécessaire cohérence. Paper presented at the 34th congress of the Association des économistes québécois (ASDEQ), Quebec City, 18 May. http://www.asdeq.org/congres/pdf/32ieme_congres/Jean%20Larose/jlarose%20mai07.pdf.

Latta, Alex. 2007. Locating Democratic Politics in Ecological Citizenship. *Environmental Politics* 16, 3 (June): 377-93.

Latta, Alex, and Nick Garside. 2005. Perspectives on Ecological Citizenship: An Introduction. *Environments* 33, 3: 1-8.

Lavoie, Judith. 1997. Ahousaht Plans Flotilla to Protest Salmon Farm. *Victoria Times Colonist,* 27 February, 1.

–. 2008. Clayoquot Bands Clash over Mining. *Victoria Times Colonist,* 16 April, A1.

Law Commission of Canada. 1999. *Urban Aboriginal Governance in Canada: Re-fashioning the Dialogue.* Ottawa: Law Commission of Canada and the National Association of Friendship Centres.

Lawlor, Allison. 2002. Squatters Ousted from Toronto's Tent City. *Toronto Globe and Mail (Update),* 24 September. https://lists.resist.ca/pipermail/news/2002-September/000047.html.

Lee, K. 1993. *Compass and Gyroscope: Integrating Science and Politics for the Environment.* Washington, DC: Island Press.

Leggatt, Stewart. 2001. Clear Choices, Clean Waters: The Leggatt Inquiry into Salmon Farming in British Columbia. http://www.davidsuzuki.org/files/Leggatt_reportfinal.pdf.

Leiss, William, and Michael Tyshenko. 2002. Some Aspects of the 'New Biotechnology' and Its Regulation in Canada. In *Canadian Environmental Policy: Context and Cases,* 2nd ed., ed. D.L. VanNijnatten and R. Boardman, 321-44. Don Mills, ON: Oxford University Press.

Lemon, James T. 1985. *Toronto since 1918: An Illustrated History.* Toronto: James Lorimer.

Le Quotidien. 2002. Nouvelles normes du ministre Boisclair. 19 June, 11.

LeRoy, Sylvia. 2004. Can the Market Save Our Parks? *Fraser Forum* (April): 10-11.

Levidow, Les. 1999. Blocking Biotechnology as Pollution: Political Cultures in the UK Risk Controversy. Paper presented at "Alternate Futures and Popular Protest" conference, Milton Keynes, UK, 29-31 March.

Lewis, Paul. 2006. The Swingeing Cuts Casting a £1m Shadow over Wordsworth Country. *Guardian,* 27 January, 9.

Ley, David. 1996. *The New Middle Class and the Remaking of the Central City.* New York: Oxford University Press.

Leys, Colin. 2001. *Market-Driven Politics: Neoliberal Democracy and the Public Interest.* London: Verso.

–. 2002. Public Services: Needed, Some Principles. *Red Pepper* (London, UK), April. http://www.queensu.ca/msp/pages/Conferences/Services/Leys.htm.

Li, T.M. 1996. Images of Community: Discourse and Strategy in Property Relations. *Development and Change* 27: 501-27.

Lien, Jon. 2005. Comments on *Radio Noon*. CBC Radio, St. John's, 4 July.

Lighthall, William D. 2003. Housing for Thousands Coming Online at 'Staggering' Speed from Spadina Ave to Dufferin Str. *Toronto Star*, 24 May.

Lipietz, Alain. 1989. *Choisir L'Audace*. Paris: Editions la Découverte.

Lister, Ruth. 2007. Inclusive Citizenship: Realizing the Potential. *Citizenship Studies* 11, 1 (February): 49-61.

Logan, John R., and Harvey L. Molotch. 1987. *Urban Fortunes: The Political Economy of Place*. Berkeley: University of California Press.

Logan, Shannon, and Gerda R. Wekerle. 2008. Neoliberalizing Environmental Governance? Land Trusts, Private Conservation and Nature on the Oak Ridges Moraine. *Geoforum* 39, 6: 2097-108.

Lolive, Jacques. 1997. La montée en généralité pour sortir de Nimby. La mobilisation associative contre le TGV Méditerranée. *Politix* 39: 109-30.

Long, D. 1992. Culture, Ideology and Militancy: The Movement of Native Indians in Canada, 1969-91. In *Organizing Dissent: Contemporary Social Movements in Theory and Practice*, ed. W. Carroll, 118-34. Toronto: Garamond Press.

Lowery, David, Virginia Gray, and Matthew Fellowes. 2005. Sisyphus Meets the Borg. Economic Scale and Inequalities in Interest Representation. *Journal of Theoretical Politics* 17, 1: 41-74.

LRUP Committee. 1996. *Bella Coola Local Resource Use Plan*. Hagensborg, BC: BC Ministry of Forests.

Ludwig, D., R. Hilborn, and C. Walters. 1993. Uncertainty, Resource Exploitation, and Conservation: Lessons from History. *Science* 260: 17-18.

Luke, Timothy W. 1997. *Ecocritique: Contesting the Politics of Nature, Economy and Culture*. Minneapolis: University of Minnesota Press.

–. 2002. The People, Politics and the Planet: Who Knows, Protects and Serves Nature Best? In *Democracy and the Claims of Nature*, ed. B.A. Minteer and B. Pepperman Taylor, 301-20. Lanham, MD: Rowan and Littlefield.

Lura Consulting. 2002. *Town Hall Meetings on Toronto's Official Plan: Summary of Citizen Feedback Final Report*. Toronto: Lura Consulting.

MacDonald, M. 2005. Lessons and Linkages: Building a Framework for Analyzing the Relationship between Gender, Globalization and the Fisheries. In *Changing Tides: Gender, Fisheries and Globalization*, ed. Barbara Neis, Marian Binkley, Siri Gerrard, and Maria Cristina Maneschy, 18-28. Winnipeg: Fernwood.

MacDonald, Phil. 2005. Plant Biosafety Office, Canadian Food Inspection Agency. Telephone interview by Peter Andrée, 13 July.

MacEachern, Alan. 2001. *Natural Selections: National Parks in Atlantic Canada, 1935-1970*. Montreal and Kingston: McGill-Queen's University Press.

MacGregor, Sherilyn. 2006. *Beyond Mothering Earth: Ecological Citizenship and the Politics of Care*. Vancouver: UBC Press.

MacKenzie, Jody, and Naomi Krogman. 2005. Public Involvement Processes, Conflict, and Challenges for Rural Residents near Intensive Hog Farms. *Local Environment* 10: 513-24.

Mackie, Richard. 1999. Natives Oppose Plans for Northern Ontario. *Toronto Globe and Mail*, 30 March, A13.

Macnaghten, Phil, and John Urry. 1998. *Contested Natures*. London: Sage.

MacNair, Emily, and Shannon MacDonald. 2001. *A Path Less Taken: Planning for Smart Growth at the University of Victoria*. Victoria: POLIS Project on Ecological Governance.

MacRae, R., H. Penfound, and C. Margulis. 2002. *Against the Grain: The Threat of Genetically Engineered Wheat*. Toronto: Greenpeace Canada.

Magnusson, W., and K. Shaw. 2003. Introduction: The Puzzle of the Political. In *A Political Space: Reading the Global through Clayoquot Sound*, ed. W. Magnusson and K. Shaw, 1-20. Minneapolis: University of Minnesota Press.

Mallen, Caroline. 2004. Moraine Land Deal a Boost for Green Space. *Toronto Star*, 24 September, A1, A7.

Marquand, David. 2004. *Decline of the Public: The Hollowing-Out of Citizenship.* Cambridge: Polity.

Marshall, D., Sr., A. Denny, and S. Marshall. 1989. The Covenant Chain [of the Mi'kmaq]. In *Drumbeat: Anger and Renewal in Indian Country,* ed. Boyce Richardson. Toronto: Summerhill Press.

Marshall, T.H. 1963. *Sociology at the Crossroads.* London: Heinemann Educational Books.

–. 1968. *Class, Citizenship, and Social Development.* New York: Anchor Books.

Martin, Jenn. 2004. Media Frames and the Fight for the Oak Ridges Moraine: Framing, Tactics and Lessons to Be Learned. Paper prepared for ENVS 5073, New Social Movements, York University.

Marx, Leo. 1999. Environmental Degradation and the Ambiguous Social Role of Science and Technology. In *Earth, Air, Fire, Water,* ed. J.K. Conway, K. Keniston, and L. Marx, 320-38. Boston: University of Massachusetts Press.

Marzall, Katia. 2005. Environmental Extension: Promoting Ecological Citizenship. *Environments Journal* 33, 3: 65-77.

Mascarenhas, Michael. 2002. Material-Semiotic Practices of Water Quality Testing and Standards: The Constitution of Water Contamination in Walkerton, Ontario, Canada. *Interdisciplinary Environmental Review* 4, 2: 66-79.

–. 2005. Where the Waters Divide: Environmental Justice, Neoliberalism, and Aboriginal Voices. An Ethnography of the Changing Canadian Water Sector. PhD diss., Department of Sociology, Michigan State University, East Lansing.

Mason, Michael. 1999. *Environmental Democracy.* London: Earthscan.

Mauro, I., and S.M. McLachlan. 2003. Risk Analysis of Genetically Modified Crops on the Canadian Prairies. Technical Workshop on Management of Herbicide Tolerant Crops Report, Canadian Food Inspection Agency. http://www.inspection.gc.ca/english/plaveg/bio/consult/herbtolrepe.shtml. Accessed 22 February 2005; page now discontinued.

May, E. 1998. *At the Cutting Edge: The Crisis in Canada's Forests.* Toronto: Key Porter Books.

McCann, L.D. 1987. Heartland and Hinterland: A Framework for Regional Analysis. In *Heartland and Hinterland: A Geography of Canada,* 2nd ed., ed. L.D. McCann, 22-33. Scarborough: Prentice-Hall.

McCarthy, James. 1998. Environmentalism, Wise Use and the Nature of Accumulation in the Rural West. In *Remaking Reality: Nature at the Millennium,* ed. Bruce Braun and Noel Castree, 126-49. New York: Routledge.

McCay, B.J., and J.M. Acheson. 1987. Human Ecology of the Commons. In *The Question of the Commons: The Culture and Ecology of Communal Resources,* ed. B.J. McCay and J.M. Acheson, 1-34. Tucson: University of Arizona Press.

McFarlane, B.L., and P.C. Boxall. 2000. *Forest Values and Attitudes of the Public, Environmentalists, Professional Foresters, and Members of Public Advisory Groups in Alberta.* Information Report NOR-X-37. Edmonton: Canadian Forest Service, Northern Forestry Centre.

McGinnis, M.V., ed. 1999. *Bioregionalism.* London: Routledge.

McGrath, Darrin. 1998. *Economic Recovery and Social Conflict in Newfoundland.* St. John's: Institute of Social and Economic Research (ISER).

McGraw, R. 2003. Aboriginal Fisheries Policy in Atlantic Canada. *Marine Policy* 27: 417-24.

McIlwraith, T.F. 1992. *The Bella Coola Indians.* 2 vols. Toronto: University of Toronto Press. (Orig. pub. 1948.)

McInnis, John, and Ian Urquhart. 1995. Protecting Mother Earth or Business? Environmental Politics in Alberta. In *The Trojan Horse: Alberta and the Future of Canada,* ed. Trevor Harrison and Gordon Laxer, 239-44. Montreal: Black Rose Books.

McIntosh, P., and J. Kearney. 2002. Enhancing Natural Resources and Livelihoods Globally through Community-Based Resource Management. Proceedings from the Learning and Innovations Institute Conference, Antigonish, NS, 6-9 November. http://ibcperu.nuxit.net/doc/isis/3008.pdf.

McLaughlin, Judith Block. 2004. Leadership, Management, and Governance. *New Directions for Higher Education* 128: 5-13.

McMichael, P., ed. 1994. *The Global Restructuring of Agro-food Systems.* Ithaca, NY: Cornell University Press.

McMurtry, John. 1998. *Unequal Freedoms: The Global Market as an Ethical System.* Toronto: Garamond Press.

Meili, Dianne. 2007. Oilsands Boom Creates Uneasy Wealth in North. *Alberta Sweetgrass,* May. http://www.ammsa.com/sweetgrass/Sweet-May1-2007.html.

Merchant, Carolyn. 1980. *The Death of Nature: Women, Ecology, and the Scientific Revolution.* San Francisco: Harper and Row.

–. 1996. The Death of Nature: Women and Ecology in the Scientific Revolution. In *Earthcare: Women and the Environment,* ed. Carolyn Merchant, 75-90. New York: Routledge.

Meredith, T. 1997. Information Limitations in Participatory Impact Assessment. In *Canadian Environmental Assessment in Transition.* Monograph Series no. 49, ed. John Sinclair, 125-54. Waterloo, ON: Department of Geography, University of Waterloo.

M'Gonigle, R. Michael. 1997. Behind the Green Curtain. *Alternatives Journal* 23, 4: 16-21.

–. 2003. Somewhere between Center and Territory: Exploring a Nodal Site in the Struggle against Vertical Authority and Horizontal Flows. In *A Political Space: Reading the Global through Clayoquot Sound,* ed. W. Magnusson and K. Shaw, 121-38. Minneapolis: University of Minnesota Press.

M'Gonigle, R. Michael, and Jessica Dempsey. 2003. Ecological Innovation in an Age of Bureaucratic Closure: The Case of the Global Forest. *Studies in Political Economy* 70: 97-124.

M'Gonigle, R. Michael, and Justine Starke. 2006a. Minding Place: Towards a (Rational) Political Ecology of the Sustainable University. *Environment and Planning D: Society and Space* 24: 325-48.

–. 2006b. *Planet U: Sustaining the World, Reinventing the University.* Gabriola Island: New Society.

Milette, Gérard. 1997. L'entente MEF-UPA sur la pollution, c'est pas du sérieux. *Producteur Plus* (Farnham, QC), January-February.

Milley, Chris, and Anthony Charles. 2001. Mi'kmaq Fisheries in Atlantic Canada: Traditions, Legal Decisions and Community Management. Paper presented at "People and the Sea: Maritime Research in the Social Sciences: An Agenda for the 21st Century," University of Amsterdam and SISWO (Netherlands Institute for the Social Sciences), Amsterdam, 30 August-1 September. http://husky1.stmarys.ca/~charles/PDFS_2005/072.pdf.

Mittelstaedt, Martin. 1999. Group Flays Ontario's Record on Pollution. *Toronto Globe and Mail,* 12 April, A4.

Moffet, John, and François Bergha. 1999. Non-regulatory Environmental Measures: What Are They and What Makes Them Work? In *Voluntary Initiatives: The New Politics of Corporate Greening,* ed. Robert Gibson, 15-31. Peterborough, ON: Broadview Press.

Monsanto. 2001. 2001 Annual Report. St. Louis, MO. http://www.monsanto.com/monsanto/content/media/pubs/2001/2001-Monsanto_Annual_Report.pdf. Accessed 6 July 2005; page now discontinued.

–. 2004a. Dialogue Leads to Wheat Decision. Press release, St. Louis, MO, 10 May.

–. 2004b. Monsanto to Realign Research Portfolio, Development of Roundup Ready Wheat Deferred. Press release, St. Louis, MO, 10 May.

Montpetit, Éric, and William D. Coleman. 1999. Policy Communities and Policy Divergence in Canada: Agro-environmental Policy Development in Quebec and Ontario. *Canadian Journal of Political Science* 32, 4: 691-714.

More, Thomas A. 2002. "The Parks Are Being Loved to Death" and Other Frauds and Deceits in Recreation Management. *Journal of Leisure Research* 34, 1: 52-78.

Morris, Dave. 2005. A Brief History of Failure: You'd Think Being a City on a Lake Would Be a Good Thing. *Eye Magazine* (Toronto), 13 January: n.p.

Morrison, James. 1993. *Protected Areas and Aboriginal Interests in Canada.* Toronto: World Wildlife Fund.

Mouffe, Chantal. 1993. *The Return of the Political.* London: Verso.

–. 2000. *The Democratic Paradox.* London: Verso.

–. 2005. *On the Political*. Milton Park, Oxford: Routledge.

Mouffe, Chantal, and Ernesto Laclau. 1985. *Hegemony and Socialist Strategy*. London: Verso.

Mulkewich, Jane. 2003. The Red Hill Valley and Aboriginal Rights. *Hammer* (Hamilton Independent Media Centre newsletter) September, 5-7.

Murdie, Robert A. 2003. Housing Affordability and Toronto's Rental Market: Perspectives from the Housing Careers of Jamaican, Polish and Somali Newcomers. *Housing, Theory, and Society* 20, 4: 183-96.

Murdie, Robert A., and Carlos Teixeira. 2003. Towards a Comfortable Neighbourhood and Appropriate Housing: Immigrant Experiences in Toronto. In *The World in a City*, ed. P. Anisef and M. Lanphier, 132-91. Toronto: University of Toronto Press.

Murdoch, Jonathan, and Graham Day. 1998. Middle Class Mobility, Rural Communities and the Politics of Exclusion. In *Migration into Rural Areas: Theories and Issues*, ed. P. Boyle and K. Halfacree, 186-99. New York: John Wiley and Sons.

Murphy, David. 2002. Environmental Activists Should Stop Complaining. *St. John's Express*, 27 June-3 July, 18.

Murray, W.E. 2005. *Geographies of Globalization*. Contemporary Human Geography Series. London: Routledge.

Muszynski, Alicja. 1996. *Cheap Wage Labour: Race and Gender in the Fisheries of British Columbia*. Montreal and Kingston: McGill-Queen's University Press.

Nadasdy, Paul. 1999. The Politics of TEK: Power and the 'Integration' of Knowledge. *Arctic Anthropology* 36, 1-2: 1-18.

Naess, Arne. 1983. The Shallow and the Deep, Long-Range Ecology Movement: A Summary. *Inquiry* 16: 95-100.

Nash, Nicholas, and Alan Lewis. 2006. Overcoming Obstacles to Ecological Citizenship: The Dominant Social Paradigm and Local Environmentalism. In *Environmental Citizenship*, ed. Andrew Dobson and Derek Bell, 153-84. Cambridge, MA: MIT Press.

Natcher, David C. 2001. Land Use Research and the Duty to Consult: A Misrepresentation of the Aboriginal Landscape. *Land Use Policy* 18: 113-22.

Neis, B., M. Binkley, and S. Gerrard. 2005. *Changing Tides: Gender, Fisheries and Globalization*. Halifax: Fernwood.

Neumann, Roderick P. 1998. *Imposing Wilderness: Struggles over Livelihood and Nature in Africa*. Berkeley: University of California Press.

New Brunswick. Department of the Environment. 1993. *Site Selection Guidelines for Municipal and Industrial Sanitary Landfills*. Saint John: Department of the Environment.

Newfoundland. 1982. *Green Paper on Recreation*. St. John's: Department of Culture, Recreation and Youth.

Newfoundland and Labrador. 1992. *Change and Challenge: A Strategic Economic Plan for Newfoundland and Labrador*. St. John's: Queen's Printer.

–. 1994a. Newfoundland and Labrador Round Table on Environment: Submission of the Working Group on Resource Management. In *Newfoundland and Labrador Round Table on the Environment and the Economy: Working Group Report*. St. John's: Department of Environment and Lands.

–. 1994b. *Privatization and Commercialization of Certain Public Sector Operations*. St. John's: Economic Recovery Commission.

–. 1994c. *A Vision for Tourism in Newfoundland and Labrador*. St. John's: Department of Tourism, Culture and Recreation.

–. 1995. *A White Paper on Proposed Reforms to the Environmental Assessment Process*. St. John's: Department of Environment.

–. 1996. *Parks and Natural Areas Division Program Review*. St. John's: Parks and Natural Areas Division.

–. 1999. *Our Smiling Land: Government's Vision for the Protection and Use of Newfoundland's Outdoor Resources*. St. John's: Queen's Printer.

Newman, Julie. 2005. The Onset of Creating a Model Sustainable Institution: A Case Study Analysis of Yale University. In *Committing Universities to Sustainable Development*. Proceedings of the Conference on the International Launch in Higher Education: Education for Sustainable Development United Nations Decade 2005-2014, 22-29. Graz, Austria: Graz University and the United Nations Educational, Scientific and Cultural Organization.

NFU (National Farmers Union). 2003. Factsheet: Ten Reasons Why We Don't Want GM Wheat. February (on file with Peter Andrée and Lucy Sharratt).

–. 2004. Widespread Opposition to Roundup Ready Wheat Forces Monsanto to Back Off. Press release, Saskatoon, 10 May. http://www.healthcoalition.ca/gew-update2.pdf.

NFU (National Farmers Union), Manitoba Keystone Agricultural Producers, the Agricultural Producers Association of Saskatchewan, the Saskatchewan Organic Directorate, the Canadian Wheat Board, Greenpeace Canada, the Council of Canadians, and the Canadian Health Coalition. 2001. Groups Oppose Approval of Genetically Modified Wheat. Press release, 31 July (on file with Peter Andrée and Lucy Sharratt).

Norwegian, Herb, and Roland Wilson. 2006. Northern Water Rights Issue Heats Up. *Edmonton Journal,* 16 October, A18.

NRTEE (National Round Table on the Environment and the Economy). 2004. Lands for Life Process, Ontario (Case Study). Conservation Documents. http://www.nrtee-trnee.ca/eng/programs/Current_Programs/Nature/Natural-Heritage/Documents/Lands-for-Life-Case-Study-Brief_e.htm. Accessed June 2007; page now discontinued.

Nue, Dean, and Richard Therrien. 2003. *Accounting for Genocide: Canada's Bureaucratic Assault on Aboriginal People.* New York: Zed Books.

O'Connor, Dennis. 2002a. *The Report of the Walkerton Inquiry, Part One: The Events of May 2000 and Related Issues.* Ontario Ministry of the Attorney General. Toronto: Queen's Printer for Ontario.

–. 2002b. *The Report of the Walkerton Inquiry, Part Two: A Strategy for Safe Drinking Water.* Ontario Ministry of the Attorney General. Toronto: Queen's Printer for Ontario.

O'Connor, Julia S. 1996. Welfare State Analysis: From Women as an Issue to Gender as a Dimension of Analysis. *Current Sociology* 44, 2-3 (Summer): 101-8.

Oddie, Richard. 2003. Top Down Planning from the Bottom Up? The Challenges of Participatory Environmental Governance. In *A Dialogue on Development, Displacement and Democracy.* EDID Working Papers No. 3, ed. Pablo Bose, 18-34. Toronto: Centre for Refugee Studies, York University.

Offe, Claus. 1984. *Contradictions of the Welfare State.* Ed. John Keane. Cambridge, MA: MIT Press.

Oleniuk, Lucas. 2005. Former Industrial Hub Goes Chic: Developers Revitalizing Tired Neighbourhood. *Metro News* (*Toronto Star* News Service), 14 July. http://www.metronews.ca.html. Accessed 2 August 2005; website now discontinued.

Ontario. 2000. Ontario's Living Legacy: Ours to Enjoy ... Ours to Protect. Full-colour poster, distributed as a supplement in major Ontario newspapers.

Ontario. Ministry of Environment. 2005. Backgrounder: Redeveloping Brownfields. 22 June. http://www.ene.gov.on.ca/envision/news/2005/062201mb.htm. Website now discontinued.

Ontario. Ministry of Municipal Affairs and Housing. 2004. Brownfields Showcase website. http://www.mah.gov.on.ca/userfiles/HTML/nts_1_3095_1.html. Accessed 9 August 2004; page now replaced with http://www.mah.gov.on.ca/Page220.aspx.

–. 2001. *Share Your Vision.* Toronto: Queen's Printer.

Ontario. Ministry of Natural Resources. 1997. Lands for Life: A Commitment to the Future. Press release, 6 November.

–. 1998. *Consolidated Recommendations of the Boreal East, Boreal West and Great Lakes – St. Lawrence Round Tables* (October). Toronto: Queen's Printer for Ontario.

Ontario. Ministry of Public Infrastructure Renewal. 2004. *Places to Grow: Better Choices, Better Future: A Discussion Paper.* Toronto: Ontario Ministry of Public Infrastructure Renewal.

–. 2005. *Places to Grow. Better Choices, Brighter Future: A Growth Plan for the Greater Golden Horseshoe.* Toronto: Queen's Printer.

Ontario Coalition Against Poverty. 2002a. Pope Squat Update: What Happened and Why Squat More. *OCAP News.* Updates on the Pope Squat are available on the OCAP website. http://www.ocap.ca/ocapnews/pope_squat.html.

–. 2002b. Poverty of Planning: Tent City, City Hall and Toronto's New Official Plan. *OCAP News.* http://members.rogers.com/planningaction/popups/ocapsnewsletter.htm. Website now discontinued.

Orkin, Andrew, and Murray Klippenstein. 2003. Hamilton's Proposed Red Hill Valley Expressway Project Violates Important 1701 Crown-Iroquois Treaty Rights. Draft version of article in *Hamilton Spectator,* December. http://www.hwcn.org/link/forhv/expressway/law_specArticle.htm.

Orr, David. 1992. *Ecological Literacy: Education and the Transition to a Postmodern World.* Albany: State University of New York Press.

Overton, James. 1996. *Making a World of Difference: Essays on Tourism, Culture and Development in Newfoundland.* St. John's: Institute for Social and Economic Research.

–. 1999. Economic Recovery or Scorched Earth? The State of Newfoundland in the Late 1990s. In *Citizens or Consumers? Social Policy in a Market Society,* ed. Dave Broad and Wayne Anthony, 99-115. Halifax: Fernwood.

–. 2001. Official Act of Vandalism: Privatizing Newfoundland's Provincial Parks in the 1990s. In *From Red Ochre to Black Gold,* ed. D. McGrath, 76-121. St. John's: Flanker Press.

Owen, S. 1993. Overcoming Dysfunction in Public Policy. In *From Conflict to Consensus: Shared Decision-Making in British Columbia. Proceedings of a Symposium Held on March 5, 1993 at Simon Fraser Harbour Centre Campus,* ed. M. Roseland, 1-3. Vancouver: School of Resource and Environmental Management, Simon Fraser University.

–. 1998. Landuse Planning in the Nineties: CORE Lessons. *Environments: A Journal of Interdisciplinary Studies* 25, 2-3: 14-26.

PAA (Protected Areas Association of Newfoundland and Labrador). 1993. *Mechanisms for Protection: A Review of Legislative Options for Protecting Natural Areas.* St. John's: Protected Areas Association.

–. 1997. Privatization Alarms Parks Supporters. *Fresh Tracks* (Spring): 1.

Parker, Martin. 2002. *Against Management: Organization in the Age of Management.* Oxford: Blackwell.

Parkins, J. 2002. Forest Management and Advisory Groups in Alberta: An Empirical Critique of an Emergent Public Sphere. *Canadian Journal of Sociology* 27, 2: 163-84.

Parkins, J., and R. Mitchell. 2005. Public Participation as Public Debate: A Deliberative Turn in Natural Resource Management. *Society and Natural Resources* 18: 529-40.

Parkins, J.R., and D.J. Davidson. 2008. Constructing the Public Sphere in Compromised Settings: A Case Study of Environmental Decision-Making in the Alberta Forest Sector. *Canadian Review of Sociology* 45, 2: 177-96.

Partnership for Public Lands. 1998. *Planning for Prosperity.* Toronto: Partnership for Public Lands.

–. 2000. *A New Way in the Woods.* Toronto: Partnership for Public Lands.

Pateman, Carole. 1988. The Patriarchal Welfare State. In *Democracy and the Welfare State,* ed. Amy Gutmann, 231-60. Princeton: Princeton University Press.

Patton, P. 2000. *Deleuze and the Political.* London: Routledge.

PBO (Plant Biosafety Office). 2003. Letter to Monsanto Canada Inc. signed by Phil MacDonald and Monica Ficker, 11 September (on file with Peter Andrée and Lucy Sharratt).

Peace, Walter, ed. 1998. *From Mountain to Lake: The Red Hill Creek Valley.* Hamilton: W.L. Griffin.

Peck, Jamie, and Adam Tickell. 2002. Neoliberalizing Space. *Antipode* 34, 3: 380-404.

Peeples, Jennifer A., and Kevin M. DeLuca. 2006. The Truth of the Matter: Motherhood, Community, and Environmental Justice. *Women's Studies in Communication* 29, 1 (Spring): 59-87.

Peet, Richard, and Michael Watts. 1993. Development Theory and Environment in an Age of Market Triumphalism – Introduction. *Economic Geography* 69, 3: 227-53.

–, eds. 1996. *Liberation Ecologies.* London: Routledge.

Pellerin, Laurent. 1996. Cliche: ministre de l'Environnement ou fossoyeur de l'agriculture? *La Terre de chez nous,* 19-25 September.

Pembina Institute. 2007. Government Protects Oil Sands Industry, Fails to Protect Athabasca River. Press release, 2 March. http://www.pembina.org/media-release/1384.

Percy Schmeiser and Schmeiser Enterprises Ltd. v. Monsanto Canada Inc. and Monsanto Company, 2004 SCC 34.

Peters, Evelyn. 1996. Urban and Aboriginal: An Impossible Contradiction? In *City Lives and City Forms: Critical Research and Canadian Urbanism,* ed. J. Caulfield and L. Peake, 47-62. Toronto: University of Toronto Press.

Pinkerton, E., and M. Weinstein. 1995. *Fisheries That Work: Sustainability through Community-Based Management.* Vancouver: David Suzuki Foundation.

Plumwood, Val. 2005. Inequality, Ecojustice, and Ecological Rationality. In *Debating the Earth,* 2nd ed., ed. J. Dryzek and D. Schlosberg, 608-32. Oxford: Oxford University Press.

Polaris Institute. 2008. Boiling Point: Six Community Profiles of the Water Crisis Facing First Nations within Canada. A synopsis prepared by the Polaris Institute in collaboration with the Assembly of First Nations and supported by the Canadian Labour Congress. Ottawa, May. http://www.polarisinstitute.org/files/Boiling%20Point_0.pdf.

POLIS Project on Ecological Governance. 2003. *Planning for a Change: Innovation and Sustainability for the 21st Century at the University of Victoria.* Victoria: POLIS Project on Ecological Governance.

PollutionWatch. 2006. Reforming the *Canadian Environmental Protection Act:* Submission to the Parliamentary Review of CEPA, 1999. Toronto, June.

Porter, David, and Chester L. Mirsky. 2003. *Megamall on the Hudson: Planning, Wal-Mart, and Grassroots Resistance.* Victoria: Trafford.

Power, Michael. 1997. *The Audit Society: Rituals of Verification.* Oxford: Oxford University Press.

Pratt, Larry, and Ian Urquhart. 1994. *The Last Great Forest: Japanese Multi-nationals and Alberta's Northern Forests.* Edmonton: NeWest Press.

Prins, H. 1996. *The Mi'kmaq: Resistance, Accommodation, and Cultural Survival.* Fort Worth: Harcourt Brace College.

Procter, James D., and Stephanie Pincell. 1996. Nature and the Reproduction of Endangered Space: The Spotted Owl in the Pacific Northwest and Southern California. *Environment and Planning D: Society and Space* 14: 683-708.

Pronovost Commission. 2008. *Agriculture and Agrifood: Securing and Building the Future. Proposals for Sustainable and Healthy Agriculture.* Report by Commission sur l'avenir de l'agriculture et de l'agroalimentaire québécois, Gouvernement du Québec. http://www.mapaq.gouv.qc.ca/NR/rdonlyres/BBE52F54-B9EB-4BEC-A7C5-6C27049DF519/0/RapportCAAAQ_AN.pdf.

Prudham, Scott. 2004. Poisoning the Well: Neoliberalism and the Contamination of Municipal Water in Walkerton, Ontario. *Geoforum* 35: 343-59.

Public Spaces. 2001. Public Spaces Appreciation Association of Ontario. http://www.utoronto.ca/en/st/pspaces/publicspaces.htm.

Quesnel, Shannon. 2005. Windmill Bight Golf Idea Dies: Stakeholders Leave, Park Might Not Open for Summer Season. *Gander Beacon,* 16 May, A5.

R. v. Marshall, [1999] 3 S.C.R. 533. http://scc.lexum.umontreal.ca/en/1999/1999rcs3-533/1999rcs3-533.html.

R. v. Sparrow, [1990] 1 S.C.R. 1075. http://scc.lexum.umontreal.ca/en/1990/1990rcs1-1075/1990rcs1-1075.html.

Rabeau, Yves. 2006. *Échec aux projets créateurs de richesse au Québec.* Montreal: Fédération des chambres de commerce du Québec. http://www.fccq.ca/documents/pdf/lettres_economiques/Etude_du_professeur_Yves_Rabeau.pdf.

Rampton, R. 2004. Be Wary of GM Wheat: ADM. *Farmer's Independent Weekly,* 15 April, 3.

Rance, Laura. 2004. Monsanto Told to Expand Its Buffer Zone. *Farmer's Independent Weekly,* 4 March, 2.

Rawls, John. 1971. *A Theory of Justice.* Harvard: Belknap Press of Harvard University Press.

Renn, Ortwin, Thomas Webler, and Peter Wiedemann, eds. 1995. *Fairness and Competence in Citizen Participation: Evaluating Models for Environmental Discourse.* Boston: Kluwer Academic.

Richardson, Mary, Joan Sherman, and Michael Gismondi. 1993. *Winning Back the Words: Confronting Experts in an Environmental Public Hearing.* Toronto: Garamond Press.

Ricker, D. 1997. *L'sitkuk: The Story of the Bear River Mi'kmaw Community.* Lockeport, NS: Roseway.

Rising, David. 1995. Threat to Spawning Salmon Prompts Review of Dump Site. *Saint John Telegraph Journal,* 15 December, A1, A2.

Roberts, S.M. 2000. Realizing Critical Geographies of the University. *Antipode* 32, 3: 230-44.

Robinson, M.P., and M.M. Ross. 1997. Traditional Land Use and Occupancy Studies and Their Impact on Forest Planning and Management in Alberta. *Forestry Chronicle* 73, 5: 596-605.

Rodrigue, Jean-Paul, Claude Comtois, and Brian Slack. 2006. *The Geography of Transport Systems.* New York: Routledge. http://people.hofstra.edu/geotrans/.

Roosa, John. 1997. The Enclosure of Nature and the Nature of Enclosures. *Science as Culture* 6, 27 (Part 2): 284-300.

Rose, Damaris. 2004. *The Uneasy Cohabitation of Gentrification and "Social Mix": A Case Study of Infill Condominiums in Montreal, Working Paper.* Montreal: Institut National de la Recherche Scientifique (INRS), Urbanisation, Culture et Société.

Rosenberg, Harriet. 2004. From Trash to Treasure: Housewife Activists and the Environmental Justice Movement. In *Feminisms and Womanisms,* ed. A. Prince and S. Silva-Wayne, 447-60. Toronto: Women's Press.

Rosher, Patty. 2005. Senior program manager, Market Development, Canadian Wheat Board. Telephone interview by Peter Andrée, 25 May.

Ross, Michael Lee. 2005. *First Nations Sacred Sites in Canada's Courts.* Vancouver: UBC Press.

Ross, Monique M., and Peggy Smith. 2002. Accommodation of Aboriginal Rights: The Need for an Aboriginal Forest Tenure (Synthesis Report). Prepared for the Sustainable Forest Management Network, University of Alberta, Edmonton, April.

Rouffignat, Joël. 1995. Penser global, agir local ou les contradictions de l'aménagement rural au Québec. In *The Sustainbility of Rural Systems. Le développement durable de systèmes ruraux,* ed. C. Bryant and C. Marois, 229-47. Montreal: Département de géographie, Université de Montréal.

Rowell, Andrew. 1996. *Green Backlash: Global Subversion of the Environmental Movement.* London: Routledge.

Rowley, D.J., and H. Sherman. 2001. *From Strategy to Change: Implementing the Plan in Higher Education.* San Francisco: Jossey-Bass.

Rowsell, Wade. 2004. A Deadly Threat. *St. John's Telegram,* 3 November, A6.

RRCA (River Road Community Alliance). 2007. Presentation to Saint John Common Council, Saint John, 22 October (on file with Susan Lee).

RSG (Regional Steering Group for the NES Strategy). 2000. *Final Terms of Reference: Northern East Slopes Sustainable Resource and Environmental Planning Management Strategy.* Edmonton: Alberta Environment.

–. 2001. *The NES Strategy: Interim Report.* Edmonton: Alberta Environment.

–. 2003. *Draft Northern East Slopes Sustainable Resource Management Strategy.* Edmonton: Alberta Environment.

Rutkowsky, J. 1992. Letter to British Columbia Commissioner of Resources and Environment S. Owen, 12 February (used by permission).

Rynard, P. 2000. Welcome In, but Check Your Rights at the Door: The James Bay and Nisga'a Agreements in Canada. *Canadian Journal of Political Science* 32: 211-43.

Sainsbury, Diane, ed. 1994. *Gendering Welfare States.* London: Sage.

Saint John Board of Trade *Business Today.* 1996. [No title.] 1.

Saint John Times Globe. 1995. Council Passes Landfill Rezoning. 8 November, A1.

Saint-Pierre, Annie. 2007. Olymel: La direction va fermer l'usine. *Journal de Montréal,* 2 February.

Saint-Pierre, Marc. 1996. Tous ligués contre le péril brun. *Le Soleil* (Quebec City), 8 June.

Sáiz, Angel Valencia. 2005. Globalisation, Cosmopolitanism and Ecological Citizenship. *Environmental Politics* 14, 2 (April): 163-78.

Salazar, D., and D. Alper, eds. 2000. *Sustaining the Forests of the Pacific Coast: Forging Truces in the War in the Woods.* Vancouver: UBC Press.

Sandercock, Leonie. 1998. *Towards Cosmopolis: Planning for Multicultural Cities.* Chichester, UK; New York: J. Wiley.

Sandilands, Catriona. 2003. Between the Local and the Global: Clayoquot Sound and Simulacral Politics. In *A Political Space: Reading the Global through Clayoquot Sound,* ed. W. Magnusson and K. Shaw, 139-67. Minneapolis: University of Minnesota Press.

Sassen, Saskia. 1991. *The Global City.* New York: Princeton University Press.

–. 2000. Whose City Is It? Globalization and the Formation of New Claims. In *The Globalization Reader,* ed. F. Lechner and J. Boli, 70-76. London: Blackwell.

–. 2002. The Repositioning of Citizenship: Emergent Subjects and Spaces for Politics. *Berkeley Journal of Sociology* 46: 4-25.

Schlosberg, David, Stuart Shulman, and Stephen Zavestoski. 2006. Virtual Environmental Citizenship: Web-Based Public Participation in Rule Making in the United States. In *Environmental Citizenship,* ed. Andrew Dobson and Derek Bell, 207-36. Cambridge, MA: MIT Press.

Schneider, R.R. 2002. *Alternative Futures: Alberta's Boreal Forest at the Crossroads.* Edmonton: Alberta Centre for Boreal Research and Federation of Alberta Naturalists.

Schrecker, Ted. 2001. Using Science in Environmental Policy: Can Canada Do Better? In *Governing the Environment: Persistent Challenges, Uncertain Innovations,* ed. E. Parson, 31-72. Toronto: University of Toronto Press.

Schuster, Jack H., Daryl G. Smith, Kathleen Corak Sund, and Myrtle M. Yamada. 1994. *Strategic Governance: How to Make Big Decisions Better.* Phoenix, AZ: American Council on Education/Oryx Press Series on Higher Education.

Scott, Allan. 1990. *Ideology and the New Social Movements.* London: Unwin Hyman.

Scott, Joan W. 1992. Experience. In *Feminists Theorize the Political,* ed. J. Butler and J. Scott, 22-40. New York: Routledge.

Searle, Rick. 2000. *Phantom Parks: The Struggle to Save Canada's National Parks.* Toronto: Key Porter.

Semenchuk, Glen, and Cliff Wallis. 2002. The Alberta ENGOs Support the Alberta Forest Conservation Strategy as the Base for Land Use Planning of the Forested Regions of Our Province. Press release signed by eleven Alberta environmental associations, 5 February. http://news.albertawilderness.ca/2002NR/NR020205/NR020205.pdf.

Senge, P.M. 1990. *The Fifth Discipline: The Art and Practice of the Learning Organization.* New York: Doubleday/Currency.

Sewell, John. 1993. *The Shape of the City: Toronto's Struggle with Modern Planning.* Toronto: University of Toronto Press.

SFI (Sustainable Forestry Initiative). 2002. 2002-2004 Edition Sustainable Forestry Initiative (SFI) Program. http://www.sfiprogram.org/miscPDFs/H-sfistandard02.pdf.

Shapiro, Robert. 1999. How Genetic Engineering Will Save Our Planet. *Futurist* 33, 4 (April): 28-29.

Sharratt, Lucy. 2001. No to Bovine Growth Hormone: Ten Years of Resistance in Canada. In *Redesigning Life? The Worldwide Challenge to Genetic Engineering,* ed. Brian Tokar, 385-96. Montreal and Kingston: McGill-Queen's University Press.

Sheedy, Amanda, Mary Pat Mackinnon, Sonia Pitre, and Judy Watling. 2008. *Handbook on Citizen Engagement: Beyond Consultation.* Ottawa: Canadian Policy Research Networks. http://www.cprn.org/documents/49583_EN.pdf.

Shields, John, and B. Mitchell Evans. 1998. *Shrinking the State: Globalization and Public Administration "Reform."* Halifax: Fernwood.

Shiva, Vandana. 1988. *Staying Alive: Women, Ecology and Survival in India.* New York: Zed Books.

–. 2004. Water Democracy. In *Cochabamba! Water War in Bolivia,* ed. Oscar Olivera and Tom Lewis, iv-xi. Cambridge: South End Press.

Shoard, Marion. 1987. *This Land Is Our Land: The Struggle for Britain's Countryside.* London: Paladin.

Shragge, E. 2003. *Activism and Social Change: Lessons for Community and Local Organizations.* Peterborough, ON: Broadview Press.

Shriberg, M.P. 2002. Sustainability in US Higher Education: Organizational Factors Influencing Campus Environmental Performance and Leadership. PhD diss., University of Michigan.

Shultis, John. 2005. The Effects of Neo-conservatism on Park Science, Management, and Administration: Examples and a Discussion. *George Wright Forum* 22, 2: 51-58.

Sierra Club of Canada. 2005. Making Mountain Park a Reality. http://www.cpaws-edmonton.org/cheviot.

Siltanen, Janet. 2002. Paradise Paved? Reflections on the Fate of Social Citizenship in Canada. *Citizenship Studies* 6, 4: 395-414.

Silvey, Rachel. 2004. A Wrench in the Global Works: Anti-sweat Shop Activism on Campus. *Antipode* 36, 2: 191-97.

Simpson, S., and C. Wilson. 2002. Salmon Farmers Seek Review after Escapes from Net Pens: Thousands of Atlantic Salmon Are Missing from Clayoquot Sound Farms. *Vancouver Sun,* 9 January, B8.

Six Nations and New Credit Tekawennake News. 2003. Man Launches $100 million Suit to Red Hill Creek Valley, Claiming Aboriginal Rights. 3 December, 1.

Skogstad, Grace. 1990. The Farm Policy Community and Public Policy in Ontario and in Quebec. In *Policy Communities and Public Policy in Canada,* ed. W.D. Coleman and Grace Skogstad, 59-90. Mississauga: Copp Clark Pitman.

Slocan Valley Resource Society. 1974. *Slocan Valley Community Forest Management Project: Final Report.* Winlaw, BC: Slocan Valley Resource Society.

Smith, Mark J. 2005. Obligation and Ecological Citizenship. *Environments Journal* 33, 3: 9-23.

Smith, Mick. 2005. Ecological Citizenship and Ethical Responsibility: Arendt, Benjamin, and Political Activism. *Environments Journal* 33, 3: 51-63.

Smith, Neil. 1987. Of Yuppies and Housing: Gentrification, Social Structuring and the Urban Dream. *Environment and Planning D: Society and Space* 5: 151-72.

Smith, Peggy. 1995. *Aboriginal Participation in Forest Management: Not Just Another "Stakeholder."* Ottawa: National Aboriginal Forestry Association.

–. 1998. Aboriginal and Treaty Rights and Aboriginal Participation: Essential Elements of Sustainable Forest Management. *Forestry Chronicle* 74, 3: 327-33.

Social Planning and Research Council. 2003. *Progress Report on Homelessness in Hamilton 2003.* Hamilton: Social Planning and Research Council. http://www.sprc.hamilton.on.ca/Reports/pdf/ProgressReportOnHomlessness2003.pdf.

–. 2004. *Incomes and Poverty in Hamilton.* Hamilton: SPRC. http://www.sprc.hamilton.on.ca/Poverty/IncomesandPovertyinHamiltonReport.php.

Soper, Kate. 1995. *What Is Nature? Culture, Politics and the Non-human.* Oxford: Blackwell.

SourceWatch. 2006. Privatization of Parks, Vancouver Island, British Columbia, Canada. http://www.sourcewatch.org/index.php?title=Privatization_of_Parks%2CVancouver_Island%2C_British_Columbia%2C_Canada.

Spangle, Michael, and David Knapp. 1996. Ways We Talk about the Earth: An Exploration of Persuasive Tactics and Appeals in Environmental Discourse. In *Earthtalk: Empowerment for Environmental Action,* ed. S.A. Muir and T.L. Veenendall, 3-26. Westport, CT: Praeger.

Spanou, Calliope. 1991. *Fonctionnaires et militants, L'administration et les nouveaux mouvements sociaux.* Paris: L'Harmattan.

Stacey, Jean Edwards. 1995. Environment Approach Correct Despite WWF, Grimes Says. *St. John's Evening Telegram,* 20 April, 8.

Stackhouse, John. 2001. Trouble in Paradise. *Toronto Globe and Mail,* 19 November, A8.

Steel, Debora. 2007. The Story. *Ha-Shilth-Sa,* 7 June, 5.

Stefanick, Lorna, and Kathleen Wells. 2000. Alberta's Special Places 2000: Conservation, Conflict, and the Castle-Crown Wilderness. In *Biodiversity in Canada: Ecology, Ideas and Action,* ed. S. Bocking, 367-90. Peterborough, ON: Broadview Press.

Stern, P., T. Dietz, N. Dolsak, E. Ostrum, and S. Stonich. 2002. Knowledge and Questions after 15 Years of Research. In *The Drama of the Commons,* ed. Elinor Ostrom and the National Research Council (US) Committee on the Human Dimensions of Global Change, 445-90. Washington, DC: National Academy Press.

Stevenson, M. 1996. Indigenous Knowledge in Environmental Assessment. *Arctic* 49, 3: 278-91.

Stiegman, M. 2003. United We Fish. *Alternatives Journal* 29, 4: 38-41.

Stokstad, Erik. 2004. Monsanto Pulls the Plug on Genetically-Modified Wheat. *Science* 304, 21 May: 1088-89.

Stonehouse, Peter. 1997. Reducing Pollution from Canada's Farms. *Forum for Applied Research and Public Policy* 12, 4: 102-7.

Struck, Doug. 2006. Canada Pays Environmentally for U.S. Thirst for Oil. *Seattle Times,* 2 June. http://community.seattletimes.nwsource.com/archive/?date=20060602&slug=oilsands02.

Sullivan, Kathleen. 2006. (Re)landscaping Sovereignty in British Columbia, Canada. *PoLAR: Political and Legal Anthropology Review* 29, 1: 44-65.

Sumner, Jennifer. 2005. *Sustainability and the Civil Commons: Rural Communities in the Age of Globalization.* Toronto: University of Toronto Press.

Swainson, Gail. 2000. Step Closer to Development in Sensitive Ecological Area: Richmond Hill Backs Rezoning of Moraine. *Toronto Star,* 13 January, NE1.

–. 2001. Salamander in Middle of Road Extension Battle. *Toronto Star,* 24 October, 1.

–. 2002. Moraine Hearing Called a "Sham." *Toronto Star,* 6 November, A24.

Swann, Dennis. 1993. Privatisation, Deregulation and the New Right. In *Public Policy and the Impact of the New Right,* ed. Grant Jordan and Nigel Ashford, 120-43. London: Pinter.

Swyngedouw, Erik, Ben Page, and Maria Kaika. 2002. Sustainability Policy Innovation in a Multi-level Context: Crosscutting Issues in the Water Sector. In *Participatory Governance in Multi-level Context: Concepts and Experience,* ed. Hubert Heinelt, P. Getimis, G. Kafkalas, R. Smith, and E. Swyngedouw, 107-31. Leverkusen Opladen, Germany: Leske and Budrich.

Tarrow, Sidney. 1996. States and Opportunities: The Political Structuring of Social Movements. In *Comparative Perspectives on Social Movements: Political Opportunities, Mobilizing Structures and Cultural Framings,* ed. D. McAdam, J. D. McCarthy, and M. N. Zald, 41-61. Cambridge: Cambridge University Press.

–. 1998. *Power in Movement.* Cambridge: Cambridge University Press.

Tarter, Jim. 2002. Some Live More Downstream Than Others: Cancer, Gender and Environmental Justice. In *The Environmental Justice Reader,* ed. J. Adamson, M.M. Evans, and R. Stein, 213-28. Tucson: University of Arizona Press.

Taylor, Dorceta. 1993. Environmentalism and the Politics of Inclusion. In *Confronting Environmental Racism: Voices from the Grassroots,* ed. Robert Bullard, 53-62. Boston: South End Press.

Taylor, Gordon. 2005. R.A.P. An Exercise in Futility? International Joint Commission Web Dialogue. Intervention posted 1 December 2005. http://www.webdialogues.net/cs/ijc-greatlakes-discussions/view/dm/1043.

Teelucksingh, Cheryl. 2001. In Somebody's Backyard: Racialized Space and Environmental Justice in Toronto (Canada). PhD diss., York University, Toronto.

–. 2002. Spatiality and Environmental Justice in Parkdale (Toronto). *Ethnologies* 24, 1: 119-41.

Teeple, Gary. 1995. *Globalization and the Decline of Social Reform.* Toronto: Garamond Press.

Thomas, A.G., J.Y. Leeson, and R.C. Van Acker. 1999. Farm Management Practices in Manitoba: 1997 Weed Survey Questionnaire Results. *Agriculture and Agri-Food Canada Weed Survey Series 99-3.* Saskatoon.

Tierney, W.G., and J.T. Minor. 2004. A Cultural Perspective on Communication and Governance. *New Directions for Higher Education* 127: 85-94.

Timoney, Kevin P. 2007. *A Study of Water and Sediment Quality as Related to Public Health Issues, Fort Chipewyan, Alberta.* Prepared on behalf of the Nunee Health Board Society, Fort Chipewyan, Alberta, 11 November. http://www.borealbirds.org/resources/timoney-fortchipwater-111107.pdf.

Toner, Glen. 1996-97. Environment Canada's Continuing Roller Coaster Ride. In *How Ottawa Spends 1996-97: Life under the Knife,* ed. Gene Swimmer, 99-132. Ottawa: Carleton University Press.

Treseder, Leslie, and Naomi Krogman. 2001. Forest Co-management in Northern Alberta: Does It Challenge the Industrial Model? *International Journal of Environment and Sustainable Development* 1, 3: 210-23.

Treseder, Leslie, Naomi Krogman, and Frank Tough. 2005. Toward a More Precise Definition of Co-management in Canada. Unpublished paper, University of Alberta, Edmonton.

Tri-Council: Medical Research Council of Canada, Natural Sciences and Engineering Research Council of Canada, and Social Sciences and Humanities Research Council of Canada. 1998. Section 6: Research Involving Aboriginal Peoples. In *Tri-council Policy Statement: Ethical Conduct for Research Involving Humans,* 6.1-6.4. Ottawa: Public Works and Government Services Canada.

Trom, Danny. 1999. De la réfutation de l'effet *NIMBY* considérée comme une pratique militante. *Revue française de science politique* 49, 1: 31-50.

Turcotte, Claude. 2004a. De producteur à transformateur. *Le Devoir* (Montreal), 4-5 December.

–. 2004b. Modifications au *règlement sur les exploitations agricoles.* L'UPA est très déçue. *Le Devoir* (Montreal), 18 December.

–. 2007. Vallée-Jonction-Olymel: Québec et la Coop fédérée préparent l'après-fermeture. *Le Devoir* (Montreal), 3-4 February.

Turnbull, Lori, and Peter Aucoin. 2006. *Fostering Canadians' Role in Public Policy: A Strategy for Institutionalizing Public Involvement in Policy.* Ottawa: Canadian Policy Research Networks. http://www.cprn.org/doc.cfm?doc=1404&l=en.

Turner, Alicia. 2005. Intensification for Hanna Avenue: Local Residents Lament Lack of Green Space. *Liberty Gleaner* 5, 8 (April-May): 1, 4.

Turner, Bryan. 1992. Outline of a Theory of Citizenship. In *Dimensions of Radical Democracy: Pluralism, Citizenship and Community,* ed. Chantal Mouffe, 33-62. London: Verso Books.

UMQ (Union des municipalités du Québec). 1996. Projet de loi 23. Loi modifiant la Loi sur la protection du territoire agricole et d'autres dispositions législatives afin de favoriser la protection des activités agricoles. A brief presented to the commission de l'Agriculture, des Pêcheries et de l'Alimentation.

–. 2005. *Rapport d'activité 2004. S'approprier et assumer la nouvelle gouvernance.* Montreal: Union des municipalités du Québec.

UMRCQ (Union des municipalités régionales de comté du Québec). 1995. Mémoire de l'UMRCQ sur l'avant-projet de loi modifiant la Loi sur la protection du territoire agricole et d'autres dispositions législatives afin de favoriser la protection des activités agricoles. A brief presented to the Commission parlementaire de l'Agriculture, des Pêcheries et de l'Alimentation.

–. 1996. Droit de produire. L'UMRCQ dénonce le tordage de bras de l'UPA. Press release, Sainte-Foy, QC.

United Church of Christ Commission for Racial Justice. 1987. *Toxic Waste and Race in the United States: A National Report on the Racial and Socio-economic Characteristics of Communities Surrounding Hazardous Waste Sites.* New York: United Church of Christ.

United States. General Accounting Office. 1983. Siting of Hazardous Waste Landfills and Their Correlation with Racial and Economic Status of Surrounding Communities. Washington, DC: US General Accounting Office.

University of Victoria. 2003. Campus Plan 2003. http://web.uvic.ca/vpfin/campusplan/.

UP (Union paysanne). 2001. *Manifeste: pour une agriculture paysanne.* St. Germain de Kamouraska, QC: Union paysanne.

UPA (Union des producteurs agricoles). 1993. *L'UPA à l'heure des choix durables. Des orientations en matière d'environnement.* Longueuil: Union des producteurs agricoles.

–. 1995. Mémoire sur l'avant-projet de loi sur la protection du territoire agricole et le développement durable des activités agricoles en zone agricole. A brief presented to the Commission parlementaire de l'Agriculture, des Pêcheries et de l'Alimentation.

–. 1997. *Bilan des pratiques agroenvironnementales. L'environnement au premier rang.*

–. 1999. Nourrir le monde. A special forty-page section published in Quebec papers on the seventy-fifth anniversary of the UCC-UPA.

–. 2007. Sector profile. http://www.upa.qc.ca/eng/agriculture_quebec/sector_profile.asp.

UQCN. 1996. Manifeste des sept (AQWLPA, COSAPUE, CQDE, Greenpeace Québec, Mouvement Vert Mauricie, RNCRE, UQCN). Press release, 24 October.

Urban Development Institute/Ontario. 2004. Province Releases Growth Management Discussion Paper. Ontario Issue Update. Toronto, 13 July. http://www.udiontario.com/issupd/upd040713.htm.

Urquhart, Ian. 2001. Blind Spots in Rearview Mirrors. In *Writing Off the Rural West: Globalization, Governments, and the Transformation of Rural Communities*, ed. R. Epp and D. Whitson, 127-44. Edmonton: University of Alberta Press.

Van Acker, Rene. 2005. Weed scientist, University of Manitoba. Telephone interview by Peter Andrée, 24 May.

Van Acker, Rene, A.L. Brulé-Babel, and L.F. Friesen. 2003. *An Environmental Risk Assessment of Roundup Ready Wheat: Risks for Direct-Seeding Systems in Western Canada*. Report prepared for the Canadian Wheat Board for submission to the Plant Biosafety Office of the Canadian Food Inspection Agency, Winnipeg.

Van Acker, Rene, and M. Entz. 2001. Agronomic Benefits and Risks of Using Roundup Ready Wheat in Western Canada. In *Proceedings of the Manitoba Agronomist Conference*, 173-80. Winnipeg: Faculty of Agricultural and Food Sciences, University of Manitoba.

Van Sickle, Kerry, and Paul F.J. Eagles. 1998. Budgets, Pricing Policies and User Fees in Canadian Parks' Tourism. *Tourism Management* 19, 3: 225-35.

Van Tatenhove, J.P.M., and P. Leroy. 2003. Environment and Participation in a Context of Political Modernization. *Environmental Values* 12: 155-74.

Vancouver Province. 1998. Fishery Relief Urgent: Streifel. 18 October, A12.

Vancouver Sun. 1998. Fish Farming Tops Tourism on Island's West Coast. 26 March, B6.

Veltmeyer, H. 1990. The Restructuring of Capital and the Regional Problem. In *Restructuring and Resistance from Atlantic Canada*, ed. B. Fairley, C. Leys, and J. Sacouman, 79-103. Toronto: Garamond Press.

Vodden, Kelly, and Brenda Kuecks. 2003. Clayoquot Green Economic Opportunities Project. Report prepared for Friends of Clayoquot Sound, Ahousaht First Nation and Clayoquot Biosphere Trust. Simon Fraser University Community Economic Development Centre and Ecotrust Canada. http://www.focs.ca/reports/cgeocontents.html.

Voller, J., and S. Harrison. 1998. *Conservation Biology Principles for Forested Landscapes*. Vancouver: UBC Press.

von Mirbach, Martin. 2004. *The Canadian Standards Association (CSA) Sustainable Forest Management (SFM) Standard: Review and Analysis*. Prepared for Forests and the European Union Resource Network (FERN) by the Sierra Club of Canada. Toronto: Sierra Club of Canada. http://sierraclub.ca/national/programs/biodiversity/forests/csa-forest-standards.pdf. (This report was amalgamated into a cross-national study by FERN entitled *Footprints in the Forest: Current Practice and Future Challenges in Forest Certification*, ed. Saskia Ozinga. Moreton in Marsh, Gloucestershire, UK: FERN, 2004. http://www.fern.org/media/documents/document_1890_1900.pdf.)

Walker, Gerald. 2000. Urbanites Creating New Ruralities: Reflections on Social Action and Struggle in the Greater Toronto Area. *Great Lakes Geographer* 7, 2: 106-18.

Walker, Peter. 1998. Politics of Nature: An Overview of Political Ecology. *Capitalism, Nature, Socialism* 33, 1: 131-44.

–. 2003. Reconsidering Regional Political Ecologies: Toward a Political Ecology of the Rural American West. *Progress in Human Geography* 27, 1: 7-24.

Wals, A., and Bob Jickling. 2002. Sustainability in Higher Education: From Doublethink and Newspeak to Critical Thinking and Meaningful Learning. *International Journal of Sustainability in Higher Education* 3, 3: 221-32.

Walther, Pierre. 1987. Against Idealistic Beliefs in the Problem-Solving Capacities of Integrated Resource Management. *Environmental Management* 11, 4: 439-46.

Watts, M., and D. Goodman. 1994. Reconfiguring the Rural or Fording the Divide? Capitalist Restructuring and the Global Agro-food System. *Journal of Peasant Studies* 22, 1: 1-43.

Weaver, John. 1982. *Hamilton: An Illustrated History.* Toronto: James Lorimer.

Webb Yackee, Jason, and Susan Webb Yackee. 2006. A Bias towards Business? Assessing Interest Group Influence on the U.S. Bureaucracy. *Journal of Politics* 68, 1: 128-39.

Weber, Bob. 2006. Band Skeptical of Study Showing Normal Cancer Rate in Village. *Native Journal,* August. http://www.nativejournal.ca/pages/frameset.html.

Webler, T. 1995. 'Right' Discourse in Citizen Participation: An Evaluative Yardstick. In *Fairness and Competence in Citizen Participation: Evaluating Models for Environmental Discourse,* ed. O. Renn, T. Webler, and P. Wiedemann, 35-86. Boston: Kluwer Academic.

Weersink, Alfons, John Livernois, Jason F. Shogren, and James S. Shortle. 1998. Environmental Instruments and Environmental Policy in Agriculture. *Canadian Public Policy* 24, 3: 309-27.

Weis, Tony, and Anita Krajnc. 1999. Greenwashing Ontario's Lands for Life. *Canadian Dimension* (December): 34-38.

Wekerle, Gerda R., and Teresa V. Abbruzzese. Forthcoming. Producing Regionalism: Regional Movements, Ecosystems and Equity in a Fast and Slow Growth Region. *Geojournal.*

Wekerle, Gerda R., L. Anders Sandberg, and Liette Gilbert. Forthcoming. Regional Resistances in an Exurban Region: Intersections of the Politics of Place and the Politics of Scale. In *Leviathan Undone? A Political Economy of Scale,* ed. Rianne Mahon and Roger Keil, 247-67. Vancouver: UBC Press.

Wekerle, Gerda R., L. Anders Sandberg, Liette Gilbert, and Matthew Binstock. 2007. Nature as the Cornerstone of Growth: Regional and Ecosystems Planning in the Greater Golden Horseshoe. *Canadian Journal of Urban Research,* Canadian Planning and Policy 16, 1, supplement: 20-38.

Wells, Clyde. 1995. Closing Remarks. Proceedings of the Conference "Rethinking the Way We Do Business: A Conference about Public-Private Partnering," St. John's, NL, November 15-16. St. John's: Newfoundland and Labrador Chamber of Commerce.

Wells, Stewart. 2005. Letter to François Guimont, president of the Canadian Food Inspection Agency, 20 September (on file with Peter Andrée and Lucy Sharratt).

Wells, Stewart, and Holly Penfound. 2003. The Risks of Modified Wheat. Editorial in the *Toronto Star,* 25 February, A26.

Whatmore, Sarah, P. Lowe, and T. Marsden, eds. 1991. *Rural Enterprise: Shifting Perspectives on Small-Scale Production.* Critical Perspectives on Rural Change Series 3. London: David Fulton.

Wiber, M., F. Berkes, A. Charles, and J. Kearney. 2004. Participatory Research Supporting Community-Based Fishery Management. *Marine Policy* 28, 6: 459-68.

Wiber, M., and J. Kennedy. 2001. Impossible Dreams: Reforming Fisheries Management in the Canadian Maritimes after the Marshall Decision. *Law and Anthropology* 11: 282-97.

Wildlands League. 1998a. *New Directions in Ontario's Forests: Nurturing Diversity through Ecotourism.* Toronto: Our Times.

–. 1998b. *A Sense of Place: People and Communities on the Road to a New Northern Economy.* Toronto: Wildlands League.

Willems-Braun, B. 1997. Buried Epistemologies: The Politics of Nature in (Post)colonial British Columbia. *Annals of the Association of American Geographers* 87, 1: 3-31.

Williams, M.B. 1936. *Guardians of the Wild.* London: Nelson.

Williams, R., and G. Theriault. 1990. Crisis and Response: Underdevelopment in the Fishery and the Evolution of the Maritime Fishermen's Union. In *Restructuring and Resistance from Atlantic Canada,* ed. B. Fairley, C. Leys, and J. Sacouman, 104-29. Toronto: Garamond Press.

Williams, Raymond. 1989. Socialism and Ecology. In *Resources of Hope: Culture, Democracy, Socialism,* ed. Robin Gable, 210-26. London: Verso.

Wilson, Alexander. 1992. *The Culture of Nature: North American Landscape from Disney to the Exxon Valdez.* Cambridge, MA: Blackwell.

Wilson, Arlene. 1996. Letter to Nancy Colpitts, BCLRUP Chairperson, April 4, 1996. In *Bella Coola Local Resource Use Plan,* ed. LRUP Committee, n.p. Hagensborg, BC: BC Ministry of Forests.

Wilson, Jeremy. 1998. *Talk and Log: Wilderness Politics in British Columbia.* Vancouver: UBC Press.

Wilson, Jim. 1998. The Aboriginal Presence: The Red Hill Creek Valley, 9000 B.C. – 1615 A.D. In *From Mountain to Lake: The Red Hill Creek Valley,* ed. Walter Peace, 105-28. Hamilton: W.L. Griffin Printing.

Winfield, Mark S., and Greg Jenish. 1998. Ontario's Environment and the 'Common Sense Revolution.' *Studies in Political Economy* 57 (Autumn): 129-47.

Winseck, Dwayne. 2006. Canadian Newspaper Ownership in the Era of Convergence. *Canadian Journal of Political Science* 39, 3 (September): 703-8.

Wiwchar, David. 2005a. Ahousaht Asks for Help. *Ha-Shilth-Sa,* 16 June, 1.

–. 2005b. Fisheries Case Gains Momentum. *Ha-Shilth-Sa,* 6 October, 4.

Wood, Ellen Meiksins. 1999. *The Origin of Capitalism.* New York: Monthly Review Press.

Wood, Greg. 1997. Privatizing Parks Complete Reverse of Province's 'Green Space' Policies. *St. John's Express,* 12 March, 10.

Wynne, Brian. 1994. Scientific Knowledge and the Global Environment. In *Social Theory and the Global Environment,* ed. M. Redclift and T. Benton, 169-89. London: Routledge.

Yakabuski, Konrad. 2002. High on the Hog. *Report on Business Magazine (Toronto Globe and Mail),* September. http://www.konradyakabuski.com/articles/2002_03.html.

Young, David. 1997. Landfill Disaster Predicted. *Saint John Evening-Times Globe,* 4 February, A1.

Young, Iris M. 1990. *Justice and the Politics of Difference.* Princeton: Princeton University Press.

–. 1999. Residential Segregation and Differentiated Citizenship. *Citizenship Studies* 3, 2: 237-51.

Zuckerman, Seth. 2002. Can Salmon Farms Be Redeemed? BC Fish Farmers Address Some Environmental Problems, but Others Remain. *Tidepool,* 19 December. http://www.tidepool. org/dispatches/salmonfarmsolutions.cfm. Accessed 27 January 2005; website now discontinued.

–. 2003. If You Can't Beat Them, Join Them: Ahousaht Natives Drop Long Standing Battle against Salmon Farming. *Tidepool,* 22 January. http://www.tidepool.org/original_content. cfm?articleid=63947. Accessed 6 June 2005; website now discontinued.

Contributors

Laurie E. Adkin is an Associate Professor in the Department of Political Science at the University of Alberta, where she teaches in the fields of comparative politics and gender and politics. She is the author of *Politics of Sustainable Development: Citizens, Unions, and the Corporations* (1998), as well as essays on social movements and political ecology. Her current research projects concern the French Green Party, Canadian family policy, and an edited volume on the political ecology and governance of Alberta.

Peter Andrée is an Assistant Professor in the Department of Political Science at Carleton University in Ottawa. Peter's current research has two main components. The first focuses on the regulation of agricultural biotechnology and the global politics of genetically modified organisms. The second area of research is directed toward food quality assurance programs, eco-labelling strategies, and the marketing of sustainably produced foods. He is the author of *Genetically Modified Diplomacy* (2007).

Patricia Ballamingie is an Assistant Professor in the Department of Geography and Environmental Studies at Carleton University. A broad range of experience in the public, private, and non-profit sectors informs both her research and teaching. Her interests include political ecology, engaged scholarship, urban sustainability, human health and the environment, and sustainable consumption.

Darren R. Bardati is Associate Professor and Director of Environmental Studies at the University of Prince Edward Island. His research projects examine the creation of local-expert knowledge partnerships, source water protection at the watershed scale, and coastal community adaptation to climate change.

Nathalie Berny completed her doctorate in political science at Université Laval (Quebec) and at Université de Bordeaux (France) in 2005. She is an Assistant Professor at Université de Bordeaux (Sciences Po Bordeaux), where she lectures on public policy studies and European integration. As a researcher at the Centre SPIRIT (Science Politique Relations Internationales Territoire), she is interested in multi-level processes of decision making, green activism, and the capacity building of NGOs.

Colette Fluet completed an MS in rural sociology at the University of Alberta in 2003. She has worked as a Research Officer at the Alberta Alcohol and Drug Abuse Commission, and is currently looking after her two young boys.

Jason Found holds a master's in environmental policy from Oxford University and currently works as an independent consultant on sustainable housing and food systems. He has worked as a researcher for the POLIS Project on Ecological Governance at the University of Victoria and with FarmFolk/CityFolk Society on alternative models of land tenure for farms in BC. He recently co-produced *Island on the Edge,* a documentary film on Vancouver Island's food security (http://www.dvcuisine.com).

Liette Gilbert is Associate Professor and Associate Dean at the Faculty of Environmental Studies at York University, Toronto. Her research interests include urban environmental citizenship, immigration and multiculturalism politics, and urban and regional planning.

Donna Harrison is completing a doctorate in sociology from York University, Toronto. Her areas of specialization include the political economy of the West Coast salmon fishery, as well as gender and fisheries. She also has interests in feminist sociology and methodologies, race relations, and cultural racialization processes. Recently, she has been combining lecturing at the University of Victoria with parenting her two small children.

William T. Hipwell has a doctorate in geography from Carleton University (2001). He is an Honorary Research Associate with the Victoria University of Wellington, New Zealand. After holding full-time faculty positions in South Korea and New Zealand, he has returned to Ottawa with his family and is now working as an international development consultant. Specializing in eco-political geography, he has researched environmental conflict and the development of sustainability involving indigenous peoples, including the Nuxalk and Mi'kmaq in Canada, the Tsou, Taroko, and L'olu Tayal in Taiwan, the Ngarrindjeri in Australia, and the Māori Ngāti Hauiti iwi in New Zealand.

Raymond Hudon teaches political science at Université Laval, Quebec. He is the founding chair of a post-graduate program in public affairs and interests representation managed jointly by Université Laval and Institut d'études politiques de Bordeaux. His main areas of teaching are political forces, interest groups, and lobbying. He is presently working on research projects funded by the Canadian Institutes of Health Research (lobbies and modes of influence in Quebec health policies) and by the Social Sciences and Humanities Research Council of Canada (coalitions and lobbying).

Naomi Krogman is an Associate Professor in the Department of Rural Economy, Faculty of Agricultural, Life, and Environmental Sciences, at the University of Alberta. Her specializations include environmental and resource sociology, and international development and gender. She is presently working on longitudinal research of social impacts of industrial development and land-use change, and she has a keen interest in how rural people engage in resource conflicts and in changing allocations to resources, such as water, minerals, and land.

Susan W. Lee is an Associate Director of Academic Planning at the Centre for Initiatives of Education (CIE), Carleton University, Ottawa. This centre provides equity and access to higher education for "non-traditional" students. She holds a master's degree in comparative literature from Carleton University and is currently doing graduate studies in social work.

Michael Mascarenhas completed his doctorate in sociology at Michigan State University, East Lansing (2005), and is an Assistant Professor in the Science and Technology Studies Department at Rensselaer Polytechnic Institute in Troy, New York. His areas of specialization include environmental and rural sociology, political economy, science studies, and race and ethnicity. His current research examines the relationship between neo-liberal water reform and social reproduction and environmental justice, and the effects of these reforms on First Nations and minority communities.

R. Michael M'Gonigle is the Eco-Research Professor in Environmental Law and Policy in the Faculty of Law at the University of Victoria. A lawyer and political ecologist, he is currently researching the development of a new field of "green legal theory." Dr. M'Gonigle has co-founded numerous organizations, including Greenpeace International, the urban-planning NGO SmartGrowth BC, and Forest Futures (Dogwood Initiative). In 2000 he founded the POLIS Project on Ecological Governance at UVic.

Jane Mulkewich recently completed a law degree at the University of Western Ontario. Among her legal interests are environmental and Aboriginal law. Jane received her BA in geography at McMaster University, did post-graduate research as an MS candidate in geography, and worked for twenty years in the areas of community development, human rights, and social justice, including political work and police-related work, before going back to school to study law.

Richard Oddie recently completed a doctorate in the Faculty of Environmental Studies at York University, Toronto. His research interests include urban political ecology, social movements, environmental governance and democracy, colonialism, and environmental justice.

Maxime Ouellet has a master's degree in political science from Université Laval (2006). His research has focused on the influence of religious interest groups on American foreign policy, on media and democracy, and on parliamentary procedure. He is currently working as an analyst of performance in governmental health organizations and their networks.

James Overton is a Professor in the Department of Sociology, Memorial University of Newfoundland. His fields of specialization are economic and social development, and culture. He is currently completing a book on the resettlement programs in Newfoundland and is involved in research on rural preservation movements.

John R. Parkins is an Associate Professor in the Department of Rural Economy, Faculty of Agricultural, Life, and Environmental Sciences, at the University of Alberta. Until recently, he was Senior Sociologist at Natural Resource Canada (Canadian Forest Service) in Edmonton. His research interests include rural community and natural-resource industry interactions, deliberative democracy, environmental politics, and social impact assessment.

L. Anders Sandberg is Professor and Associate Dean of the Faculty of Environmental Studies at York University. He is the author (with Peter Clancy) of *Against the Grain: Foresters and Politics in Nova Scotia* (2000) and editor (with Sverker Sörlin) of *Sustainability: The Challenge* (1998). His fields of specialization include environmental politics, forest policy, and environmental history.

Lucy Sharratt has an MA in political economy from Carleton University (2000). She has worked as a researcher for the BioJustice Project of the Polaris Institute in Ottawa, as Project Manager for Voices from the South, a project of the Working Group on Canadian Science and Technology Policy, and as the Coordinator for the International Ban Terminator Campaign. Presently, she is the Coordinator for the Canadian Biotechnology Action Network in Ottawa.

Martha Stiegman is a Maritime-bred, Montreal-based academic and filmmaker. She is currently a PhD candidate at Concordia University. Her documentary film *In the Same Boat?* (2007) (http://inthesameboat.net), produced as part of her doctoral research, examines the grounds for solidarity between Mi'kmaq and non-Native communities in the fight against the privatization of the Nova Scotia fisheries. A number of her short documentary films are featured on the National Film Board of Canada's CitizenShift website at http://citizenshift.nfb.ca.

Cheryl Teelucksingh is an Associate Professor in the Department of Sociology, Ryerson University, Toronto. Her work focuses on environmental justice and racism in Canada. She has recently co-authored *Environmental Justice and Racism in Canada: An Introduction* (2008), together with Andil Gosine.

Gerda R. Wekerle is a Professor in the Faculty of Environmental Studies at York University and Coordinator of the Planning Program. She is co-author of *Safe Cities: Guidelines for Planning, Design and Management,* and co-editor of *New Space for Women, Women and the Canadian Welfare State,* and *Local Places in the Age of the Global City.* Ongoing research includes projects related to exurban movements and urban and regional planning, growth management policies, gender and cities, and urban agriculture and food justice movements.

Index

a = appendix; f = figure; m = map; n = note